HARMONIC AND MINIMAL MAPS:
With Applications in Geometry and Physics

ELLIS HORWOOD SERIES IN MATHEMATICS AND ITS APPLICATIONS

Series Editor: G. M. BELL, Chelsea College, University of London

Statistics and Operational Research Editor:
B. W. CONOLLY, Professor of Mathematics (Operational Research), Chelsea College, University of London

Baldock, G. R. & Bridgeman, T.	**Mathematical Theory of Wave Motion**
de Barra, G.	**Measure Theory and Integration**
Berry, J. S., Burghes, D. N., Huntley, I. D., James, D. J. G. & Moscardini, A. O.	**Teaching and Applying Mathematical Modelling**
Burghes, D. N. & Borrie, M.	**Modelling with Differential Equations**
Burghes, D. N. & Downs, A. M.	**Modern Introduction to Classical Mechanics and Control**
Burghes, D. N. & Graham, A.	**Introduction to Control Theory, including Optimal Control**
Burghes, D. N., Huntley, I. & McDonald, J.	**Applying Mathematics**
Burghes, D. N. & Wood, A. D.	**Mathematical Models in the Social, Management and Life Sciences**
Butkovskiy, A. G.	**Green's Functions and Transfer Functions Handbook**
Butkovskiy, A. G.	**Structural Theory of Distributed Systems**
Chorlton, F.	**Textbook of Dynamics, 2nd Edition**
Chorlton, F.	**Vector and Tensor Methods**
Dunning-Davies, J.	**Mathematical Methods for Mathematicians, Physical Scientists and Engineers**
Eason, G., Coles, C. W. & Gettinby, G.	**Mathematics and Statistics for the Bio-sciences**
Exton, H.	**Handbook of Hypergeometric Integrals**
Exton, H.	**Multiple Hypergeometric Functions and Applications**
Exton, H.	***q*-Hypergeometric Functions and Applications**
Faux, I. D. & Pratt, M. J.	**Computational Geometry for Design and Manufacture**
Firby, P. A. & Gardiner, C. F.	**Surface Topology**
Gardiner, C. F.	**Modern Algebra**
Gasson, P. C.	**Geometry of Spatial Forms**
Goodbody, A. M.	**Cartesian Tensors**
Goult, R. J.	**Applied Linear Algebra**
Graham, A.	**Kronecker Products and Matrix Calculus: with Applications**
Graham, A.	**Matrix Theory and Applications for Engineers and Mathematicians**
Griffel, D. H.	**Applied Functional Analysis**
Hanyga, A.	**Mathematical Theory of Non-linear Elasticity**
Hoskins, R. F.	**Generalised Functions**
Hunter, S. C.	**Mechanics of Continuous Media, 2nd (Revised) Edition**
Huntley, I. & Johnson, R. M.	**Linear and Nonlinear Differential Equations**
Jaswon, M. A. & Rose, M. A.	**Crystal Symmetry: The Theory of Colour Crystallography**
Johnson, R. M.	**Linear Differential Equations and Difference Equations: A Systems Approach**
Kim, K. H. & Roush, F. W.	**Applied Abstract Algebra**
Kosinski, W.	**Field Singularities and Wave Analysis in Continuum Mechanics**
Marichev, O. I.	**Integral Transforms of Higher Transcendental Functions**
Meek, B. L. & Fairthorne, S.	**Using Computers**
Muller-Pfeiffer, E.	**Spectral Theory of Ordinary Differential Operators**
Nonweiler, T. R. F.	**Computational Mathematics: An Introduction to Numerical Analysis**
Oldknow, A. & Smith, D.	**Learning Mathematics with Micros**
Ogden, R. W.	**Non-linear Elastic Deformations**
Rankin, R.	**Modular Forms**
Ratschek, H. & Rokne, Jon	**Computer Methods for the Range of Functions**
Scorer, R. S.	**Environmental Aerodynamics**
Schendel, U.	**Introduction to Numerical Methods for Parallel Computers**
Smith, D. K.	**Network Optimisation Practice: A Computational Guide**
Srivastava, H. M. & Karlsson, P. W.	**Multiple Gaussian Hypergeometric Series**
Srivastava, H. M. & Manocha, H. L.	**A Treatise on Generating Functions**
Stoodley, K. D. C.	**Applied and Computational Statsitics: A First Course**
Sweet, M. V.	**Algebra, Geometry and Triginometry Science, Engineering and Mathematics**
Temperley, H. N. V. & Trevena, D. H.	**Liquids and Their Properties**
Temperley, H. N. V.	**Graph Theory and Applications**
Thom, R.	**Mathematical Models of Morphogenesis**
Thomas, L. C.	**Games Theory and Applications**
Townend, M. Stewart	**Mathematics in Sport**
Twizell, E. H.	**Computational Methods for Partial Differential Equations**
Wheeler, R. F.	**Rethinking Mathematical Concepts**
Willmore, T. J.	**Total Curvature in Riemannian Geometry**
Willmore, T. J. & Hitchin, N.	**Global Riemannian Geometry**

HARMONIC AND MINIMAL MAPS: With Applications in Geometry and Physics

GÁBOR TÓTH Ph.D.

Mathematical Institute of the
Hungarian Academy of Sciences
Budapest, Hungary.

Translation Editor:
T. J. WILLMORE, M.Sc.,Ph.D.,D.Sc.,FIMA
Professor of Pure Mathematics
University of Durham

ELLIS HORWOOD LIMITED
Publishers · Chichester

Halsted Press: a division of
JOHN WILEY & SONS
New York · Brisbane · Chichester · Toronto

First published in 1984 by

ELLIS HORWOOD LIMITED
Market Cross House, Cooper Street, Chichester, West Sussex,
PO19 1EB, England

The publisher's colophon is reproduced from James Gillison's drawing of the ancient Market Cross, Chichester.

Distributors:
Australia, New Zealand, South-east Asia:
Jacaranda-Wiley Ltd., Jacaranda Press,
JOHN WILEY & SONS INC.,
G.P.O. Box 859, Brisbane, Queensland 4001, Australia.

Canada:
JOHN WILEY & SONS CANADA LIMITED
22 Worcester Road, Rexdale, Ontario, Canada.

Europe, Africa:
JOHN WILEY & SONS LIMITED
Baffins Lane, Chichester, West Sussex, England.

North and South America and the rest of the world:
Halsted Press: a division of
JOHN WILEY & SONS
605 Third Avenue, New York, N.Y. 10158, U.S.A.

British Library Cataloguing in Publication Data

British Library Cataloguing in Publication Data
Toth, Gabor
Harmonic and Minimal Maps:
With Applications in Geometry and Physics.—
Ellis Horwood Series in mathematics and its applications
1. Physics 2. Geometry 3. Harmonic maps
4. Minimal maps
I. Title
530'.01'51474 QC20.7.H3

ISBN 0-85312-738-7 (Ellis Horwood Limited)
ISBN 0-470-20127-4(Halsted Press)

Printed in Great Britain by Unwin Brothers of Woking.

Table of Contents

This book is dedicated with
affection and love
to my wife Erika

PREFACE

Ever since the year 1964, when the fundamental work of Eells-Sampson, "*Harmonic mappings of Riemannian manifolds*" appeared, the theory of harmonic maps has far outgrown its original scope to develop the first general properties of harmonic maps and to study the existence of harmonic maps in given homotopy classes of maps between Riemannian manifolds (existence theory). To mention only a few related branches of mathematics in which harmonic maps occur not only on the level of illustrating examples but on the level of rethinking the basic ideas, I refer to the recent developments of complex analysis, Morse theory with calculus of variations and the theory of minimal maps.
The rapid expansion of the theory of harmonic maps, as mentioned above, is to be thanked for the advantage due to its universal features but this fact, as far as writing a book on the subject is concerned, plays the role of certain difficulty as well.
The present book is therefore, by no means, intended to be a comprehensive introduction to harmonic maps but, rather, an expository which studies the geometry of harmonic and minimal maps into spaces of constant curvature. Though I have tried to keep the prerequisites to a minimum, and to make the book accessible for students of mathematics, mathematicians and physicists, the presentation can only be considered to be almost self-contained as basic Riemannian geometry and representation theory are applied rather than developed here; nevertheless, only those proofs are omitted which would be incompatible with the general scope. The background materials are given in their right places without bogging the reader down with a long introductory chapter. Each chapter begins with a concise introduction and ends with a set of problems to help the reader's comprehension of the material. The exposition has a gradually increasing speed which, I hope, would keep pace with the increasing interest.

The first (introductory) chapter displays a basic material on the general theory of differential operators on vector bundles. The stress being on applicability, I give here a detailed account on first and second order classical differential operators of Riemannian geometry. For the convenience of the reader, two specific sections summarize the basic facts of the spectrum and the (full) isometry group of Riemannian manifolds.

In Chapter II particular emphasis is placed on the generalized Hodge-de Rham theory by developing a differential calculus on twisted tensor bundles which is then specified leading to the concept of harmonicity of maps.The great majority of this chapter is devoted to the works of Eells-Sampson and Lichnerowicz on harmonic maps into flat codomains; a theory which has also proved to be fruitful in describing the geometric structure of the nonnegatively Ricci curved manifolds. Basic fibre bundle theory is used here.

The principal part of the book is essentially contained in ChaptersIII-IV, where the variational theories of harmonic and minimal maps are developed in a rather similar way offered by the calculus of variations. The basic existence theory of harmonic maps between spheres, due to R.T. Smith, is presented in Chapter III in full details. Based on different ideas, the counterpart of this theory for minimal maps, the Do Carmo-Wallach classification and rigidity of minimal immersions between spheres, is treated in Chapter IV with the necessary representation theoretical background material summarized in a specific section.

In view of Chapter IV, a particularly interesting problem in the theory of harmonic maps is their application to the study of their rigidity properties. In Chapter V, introducing the concept of infinitesimal rigidity, I deal with infinitesimal deformations and classification of harmonic maps of constant energy density into spheres. A variety of examples is also given here illustrating the fundamental concepts.

The differential geometry developed in Chapters I-II can also be served to give global formulation of various physical theories. In Chapter VI, as a theory of specific interest, I describe some basic concepts of Yang-Mills fields. As a detailed exposition on the subject could fill a whole monograph, I restrict myself to point out the close relationship between energy, volume and Yang-Mills functionals by giving a calculus of variations for Yang-Mills fields and proving Simons' instability theorem.

I owe a great deal of help and encouragement to Professor James Eells who, with a deep insight in the theory combined with his kind personality, has guided me towards the problems that determined my research activity over the last 5 years. I would like to express my sincere gratitude to him. I am especially grateful to Professor T.J. Willmore for his hard and dedicated work who has acted as a referee and translation editor. Their comments on the preliminary draft of the manuscript were invaluable aids for preparing the final version.
I had many long and informative discussions with Á. Elbert on differential equations used in the book; I am also indebted to him. Thanks are due to numerous mathematicians and physicists who, by their lectures, private talks and papers, taught me a lot of aspects of differential geometry; I am particularly grateful to J.C. Wood and L. Lemaire for their useful suggestions to improve the original text and to G. D'Ambra, P. Forgács, A. Lichnerowicz, S. Rallis, H.J.C. Sealey, J.H. Sampson, J. Szenthe and N. Wallach. Last (but not least) I would like to express my gratitude to my wife for her boundless patience and excellent work of typing the whole manuscript.

Gábor Tóth
The Ohio State University
231 West 18th Avenue
Columbus, Ohio 43210
U.S.A.
January 1984

CHAPTER I

Differential Operators on Vector Bundles

In this chapter we present some preliminary material on differential operators acting on (real) vector bundles. In Section 1 we review, in some detail, the elementary theory of differential operators. The proof of the finiteness theorem and the Hodge-de Rham decomposition for elliptic self-adjoint differential operators is omitted, since proving it in details would take us too far afield. The general framework is specified in Sections 2-6 to present a rapid elementary course on first and second order classical differential operators occurring in Riemannian geometry whose inevitable importance is reflected throughout this book as well. For later reference, to close the chapter, we attach two additional sections dealing with the spectrum and the (full) isometry group of Riemannian manifolds. Some of the well-known proofs will only be sketched or even omitted since all these are readily available in standard books on differential geometry, such as Berger, Gauduchon and Mazet (1971), Helgason (1978), Kobayashi and Nomizu (1963) and (1969), Lichnerowicz (1955), de Rham (1955), Wells (1973).

1. DIFFERENTIAL OPERATORS AND SYMBOLS

Let V and W be (real, finite dimensional) vector spaces and denote by ε_U^V and ε_U^W the trivial vector bundles over an open set $U \subset \mathbf{R}^m$ with typical fibres V and W, respectively. *A differential operator of order* r is a linear map

$$D_0 : C^\infty(\varepsilon_U^V) \to C^\infty(\varepsilon_U^W)$$

(sending a section of ε_U^V to a section of ε_U^W) of the form

$$D_o = \sum_{|\rho|\leq r} A_\rho D^\rho , \tag{1.1}$$

where the summation runs over all multi-indices $\rho=(\rho^1,\ldots,\rho^m)\in \in Z_+^m$ with $|\rho| = \sum_{i=1}^m \rho^i \leq r$, $D^\rho = (\frac{\partial}{\partial x^1})^{\rho_1} \ldots (\frac{\partial}{\partial x^m})^{\rho_m}$ and

$$A_\rho : U \to \mathrm{Hom}(V,W)$$

is a (smooth) map.† The *symbol* $\sigma_r(D_o)$ of the differential operator D_o is the map

$$\sigma_r(D_o) : U\times\mathbf{R}^m \to \mathrm{Hom}(V,W)$$

defined by

$$\sigma_r(D_o)(x,y) = \sum_{|\rho|=r} y^\rho A_\rho(x),\ x\in U ,$$
$$y = (y^1,\ldots,y^m)\in\mathbf{R}^m , \tag{1.2}$$

where y^ρ stands for $(y^1)^{\rho_1}\ldots(y^m)^{\rho_m}\in\mathbf{R}$. For fixed $x\in U$, the symbol $\sigma_r(D_o)(x,.)$ is then a homogeneous polynomial of degree r with coefficients in $\mathrm{Hom}(V,W)$. The differential operator D_o is said to be *elliptic* if $\sigma_r(D_o)(x,y)\in\mathrm{Hom}(V,W)$ is an isomorphism for all $x\in U$ and $0\neq y\in\mathbf{R}^m$.

Example 1.3. Set $V=W=\mathbf{R}$ and $r=2$. Then (1.1) can be rewritten as

$$D_o = \sum_{i,j=1}^m C_{ij}\frac{\partial^2}{\partial x^i \partial x^j} + \text{linear part,}$$

with $C_{ij} = C_{ji}:U\to\mathbf{R}$, $i,j=1,\ldots,m$, and its symbol, for fixed $x\in U$, is the quadratic form

$$\sigma_2(D_o)(x,y) = \sum_{i,j=1}^m C_{ij}(x)y^i y^j ,\ y = (y^1,\ldots,y^m)\in\mathbf{R}^m .$$

†Throughout this book, unless stated otherwise, all manifolds, maps, bundles, etc. will be smooth, i.e. of class C^∞ . For the notations explained briefly in the text, see Summary of Basic Notations.

Thus, D_o is elliptic if and only if the matrix $C(x) = (C_{ij}(x))_{i,j=1}^{m} \in M(m,m)$ is nonsingular for all $x \in U$.
Let ξ and η be (real) vector bundles over an m-dimensional manifold M † with typical fibres V and W, respectively. A linear map

$$D : C^{\infty}(\xi) \to C^{\infty}(\eta)$$

is said to be a *local operator* if for any section $v \in C^{\infty}(\xi)$ vanishing on an open set $U \subset M$ we have $(Dv)|U=0$. Then, for each open set $U \subset M$, there is induced a local operator

$$D : C^{\infty}(\xi|U) \to C^{\infty}(\eta|U) .$$

In fact, given $x_o \in U$, choose a scalar $\mu \in C^{\infty}(M)$ on M with compact support in U such that μ equals 1 in a neighbourhood of x_o. Then, for $v \in C^{\infty}(\xi|U)$, we set

$$(Dv)(x_o) = (D(\mu v))(x_o) ,$$

where

$$(\mu v)(x) = \begin{cases} \mu(x)v(x) & \text{if} \quad x \in U \\ 0 & \text{if} \quad x \notin U . \end{cases}$$

Further, if (U,ϕ) is a local coordinate neighbourhood on M such that the restrictions $\xi|U$ and $\eta|U$ are trivial then, choosing trivializations

$$h_U : \xi|U \to \varepsilon_U^V \quad \text{and} \quad k_U : \eta|U \to \varepsilon_U^W , \tag{1.4}$$

the local operator D gives rise to a linear map

$$D_o : C^{\infty}(\varepsilon_{\phi(U)}^V) \to C^{\infty}(\varepsilon_{\phi(U)}^W)$$

by setting

† Unless stated otherwise, all manifolds considered in this book are assumed to be connected.

$$D_0(v) = \{k_U(D(h_U^{-1}(v\circ\phi)))\}\circ\phi^{-1}\ ,\ v\in C^\infty(\varepsilon^V_{\phi(U)})\ . \qquad (1.5)$$

The local operator $D:C^\infty(\xi)\to C^\infty(\eta)$ is said to be a *differential operator of order* r if for every local coordinate neighbourhood (U,ϕ) and trivializations h_U, k_U (in (1.4)) the corresponding local operator D_0 is a differential operator of order r, i.e. it has the form given in (1.1). D_0 is then called a *local coordinate representation of* D. The set of all differential operators $D:C^\infty(\xi)\to C^\infty(\eta)$ of (a given) order $r\in\mathbf{Z}_+$ is denoted by $\mathrm{Diff}_r(\xi,\eta)$; it is, in fact, a module over the algebra $C^\infty(M)$ of scalars on M. Clearly, $\mathrm{Diff}_0(\xi,\eta) = \mathrm{Hom}(\xi,\eta)$ and, for each $r\in\mathbf{Z}_+$, we have the canonical inclusion $\mathrm{Diff}_r(\xi,\eta)\subset\mathrm{Diff}_{r+1}(\xi,\eta)$.

To define the symbol of a differential operator $D\in\mathrm{Diff}_r(\xi,\eta)$ of order $r\in\mathbf{Z}_+$, we consider the fibre bundle $T'(M)$ over M with typical fibre $\mathbf{R}^m\setminus\{0\}$ obtained from the *cotangent bundle* $T^*(M)$ of M by removing the zero section. The *symbol* $\sigma_r(D)$ of D will then associate to each cotangent vector $\alpha_x\in T'_x(M)$ $(=T^*_x(M)\setminus\{0\})$, $x\in M$, a linear map

$$\sigma_r(D)(\alpha_x) : \xi_x \to \eta_x$$

between the fibres of ξ and η at x, respectively, in the following way. First, choose a scalar $\mu\in C^\infty(M)$ with $\mu(x)=0$ such that α_x equals the total differential $d\mu$ at $x\in M$. Then, for each $v_x\in\xi_x$, we set

$$\sigma_r(D)(\alpha_x)v_x = (D(\frac{1}{r!}\mu^r\cdot v))(x)\in\eta_x\ , \qquad (1.6)$$

where $v\in C^\infty(\xi)$ is an extension of v_x. We claim that the symbol $\sigma_r(D)$ is well-defined (i.e. the value $\sigma_r(D)(\alpha_x)v_x$ does not depend on the choice of μ and v) and that, for the trivial bundles $\xi=\varepsilon^V_U$ and $\eta=\varepsilon^W_U$, with $U\subset\mathbf{R}^m$ open, we recover our earlier agreement (1.2). Assume that $D\in\mathrm{Diff}_r(\xi,\eta)$ has local coordinate representation (1.1) with respect to a local coordinate neighbourhood (U,ϕ) and trivializations h_U, k_U in (1.4). Then, setting $w = \frac{1}{r!}\mu^r\cdot v\in C^\infty(\xi)$, by (1.5), we have

$$k_U\{(D(w))(x)\} = (k_U D\{h_U^{-1}(h_U w)\})(x) = D_o((h_U w)\circ\phi^{-1}))(\phi(x)) =$$

$$= \sum_{|\rho|\leq r} A_\rho(\phi(x))(D^\rho((h_U w)\circ\phi^{-1}))(\phi(x)) .$$

As the scalar $\mu^r\circ\phi^{-1}$ vanishes in r-th order at $\phi(x)$,

$$D^\rho((h_U w)\circ\phi^{-1}) = \{D^\rho(\frac{1}{r!}\mu^r\circ\phi^{-1}\cdot(h_U v)\circ\phi^{\,1})\}(\phi(x)) -$$

$$= D^\rho(\frac{1}{r!}\mu^r\circ\phi^{-1})(\phi(x))\cdot h_U(v_x),$$

for every multi-index $\rho\in\mathbf{Z}_+^m$ with $|\rho|\leq r$. Further, by $d(\mu\circ\phi^{-1})_{\phi(x)} = (\phi^{-1})^* d\mu_x = (\phi^{-1})^*\alpha_x \in T^*_{\phi(x)}(\mathbf{R}^m)$, an elementary calculation yields

$$D^\rho(\frac{1}{r!}\mu^r\circ\phi^{-1})(\phi(x)) = \begin{cases} (((\phi^{-1})^*\alpha_x)^\vee)^\rho & \text{if } |\rho|=r \\ 0 & \text{if } |\rho|<r \end{cases}$$

where $^\vee : T^*(\mathbf{R}^m) \to \mathbf{R}^m$ stands for the canonical identification defined by parallel transport of cotangent vectors in $T^*(\mathbf{R}^m)$ to the origin. Summarizing, we obtain

$$k_U\{(D(\frac{1}{r!}\mu^r v))(x)\} = \sum_{|\rho|=r} y^\rho A_\rho(\phi(x))h_U(v_x) =$$

$$= \sigma_r(D_o)(\phi(x),\ y)h_U(v_x) ,$$

where $y = ((\phi^{-1})^*\alpha_x)^\vee$. In other words, the diagram

$$\begin{array}{ccc} \xi_x & \xrightarrow{\ \sigma_r(D)(\alpha_x)\ } & \eta_x \\ {\scriptstyle h_U}\big\downarrow & & \big\downarrow{\scriptstyle k_U} \\ V & \xrightarrow{\ \sigma_r(D_o)(\phi(x),y)\ } & W \end{array}$$

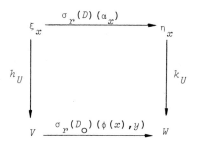

commutes and the claim follows.

Extending the earlier definition, a differential operator $D \in \mathrm{Diff}_r(\xi,\eta)$ is said to be *elliptic* if the linear map $\sigma_r(D)(\alpha_x):\xi_x \to \eta_x$ is an isomorphism for all $\alpha_x \in T'(M)$.

Remark. A result of Peetre asserts that any local operator $D \in C^\infty(\xi) \to C^\infty(\eta)$ has local coordinate representation (1.1) over suitably chosen coordinate neighbourhoods (U,ϕ) of M . (For a proof, see Narasimhan (1968) pages 175-176.)
Differential operators compose well as stated in the following.

Proposition 1.7. *Let* ξ,η *and* ζ *be vector bundles over a manifold* M . *If* $D \in \mathrm{Diff}_r(\xi,\eta)$ *and* $D' \in \mathrm{Diff}_s(\eta,\zeta)$ *then the linear map*

$$D' \circ D : C^\infty(\xi) \to C^\infty(\zeta)$$

is a differential operator of order $r+s$ *and its symbol is given by*

$$\sigma_{r+s}(D' \circ D) = \sigma_s(D') \circ \sigma_r(D) \ .$$

Proof. The composition $D' \circ D$ being a local operator, we work out a local coordinate representation of $D' \circ D$ over a coordinate neighbourhood (U,ϕ) . Assume that D and D' have local coordinate representations

$$D = \sum_{|\rho| \le r} A_\rho D^\rho$$

and

$$D' = \sum_{|\sigma| \le s} B_\sigma D^\sigma$$

over U , respectively. By making use of the Leibniz rule, we have

$$D' \circ D = \sum_{|\sigma| \le s} B_\sigma D^\sigma \{ \sum_{|\rho| \le r} A_\rho D^\rho \} =$$

$$= \sum_{|\rho| \le r} \sum_{|\sigma| \le s} B_\sigma \sum_{\gamma \le \sigma} \binom{\sigma}{\gamma} (D^{\sigma-\gamma} A_\rho) D^{\rho+\gamma} =$$

$$= \sum_{|\rho|\leq r} \sum_{|\sigma|\leq s} B_\sigma \sum_{\tau\leq\rho+\sigma} \binom{\sigma}{\tau-\rho} (D^{\rho+\sigma-\tau} A_\rho) D^\tau = \sum_{|\tau|\leq r+s} C_\tau D^\tau ,$$

where

$$C_\tau = \sum_{\substack{|\rho|\leq r, |\sigma|\leq s \\ \rho\leq\tau\leq\rho+\sigma}} \binom{\sigma}{\tau-\rho} B_\sigma D^{\rho+\sigma-\tau} A_\rho . \qquad (1.8)$$

Hence, $D'\circ D$ is a differential operator of order $r+s$. To compute its symbol, we first note that, for $|\tau|=r+s$, (1.8) reduces to the form

$$C_\tau(x) = \sum_{\substack{|\rho|=r \\ \rho+\sigma=\tau}} B_\sigma(x) A_\rho(x) , \quad x\in\phi(U) .$$

Thus, by the identity $y^{\rho+\sigma} = y^\rho y^\sigma$, $y\in\mathbf{R}^m$, (1.2) entails

$$\sigma_{r+s}(D'\circ D)(x,y) = \sum_{|\tau|=r+s} y^\tau C_\tau(x) =$$

$$= \sum_{|\tau|=r+s} y^\tau \sum_{\substack{|\rho|=r \\ \rho+\sigma=\tau}} B_\sigma(x) A_\rho(x) = \sum_{|\rho|=r} \sum_{|\sigma|=s} y^\rho y^\sigma B_\sigma(x) A_\rho(x) =$$

$$= \{\sum_{|\sigma|=s} y^\sigma B_\sigma(x)\}\{\sum_{|\rho|=r} y^\rho A_\rho(x)\} = \sigma_s(D')(x,y)\circ\sigma_r(D)(x,y) . \qquad \checkmark$$

To introduce the notion of adjoint for differential operators we have to endow the linear subspace $C_0^\infty(\xi)\subset C^\infty(\xi)$ of sections of compact support with a suitable prehilbert structure. For this, we first denote by $\mathcal{D}^r(M) (=C^\infty(\Lambda^r(T^*(M))))$, $r\in\mathbf{Z}_+$, the *vector space of forms of degree* r *on* M. Clearly, $\mathcal{D}^r(M)=\{0\}$ for $r>m$. The direct sum $\mathcal{D}(M) = \bigoplus_{r=0}^m \mathcal{D}^r(M)$ with the *exterior multiplication* Λ becomes an algebra over $\mathbf{R}$ (usually called the *exterior algebra of* M). Taking M to be a Riemannian manifold, the metric tensor $g = (,)\in C^\infty(S^2(T^*(M)))$, being nonsingular (in fact, positive definite) on each of the fibres of

$T(M)$, induces a canonical isomorphism $\gamma : T(M) \to T^*(M)$ by setting

$$(\gamma(X_x))(Y_x) = g(X_x, Y_x) = (X_x, Y_x) \, , \qquad X_x, Y_x \in T_x(M) \, , \quad x \in M \, . \tag{1.9}$$

(The induced isomorphism between the $C^\infty(M)$-modules of vector fields $V(M)(=C^\infty(T(M)))$ and 1-forms $\mathcal{D}^1(M)$ on M is also denoted by γ .) Furthermore, assuming that M is oriented, the *volume form* $\mathrm{vol}(M,g) \in \mathcal{D}^m(M)$ can be locally given by

$$\mathrm{vol}(M,g) = \gamma(E^1) \wedge \ldots \wedge \gamma(E^m) \, , \tag{1.10}$$

where $\{E^i\}_{i=1}^m$ is an oriented orthonormal moving frame of $T(M)$ over a local coordinate neighbourhood (U,ϕ) . (More generally, if $\{E^i\}_{i=1}^m \subset V(U)$ is *any* local oriented moving frame with

$$g = \sum_{i,j=1}^m g_{ij} \gamma(E^i) \otimes \gamma(E^j) \, , \quad g_{ij} \in C^\infty(U) \, , \quad i,j=1,\ldots,m \, ,$$

then

$$\mathrm{vol}(M,g) = \sqrt{\det(g_{ij})_{i,j=1}^m} \; \gamma(E^1) \wedge \ldots \wedge \gamma(E^m) \, .$$

For example, classically, if (U,ϕ) is oriented with associated local base $\{dx^i\}_{i=1}^m \subset \mathcal{D}^1(U)$ then the metric tensor and volume form can be written as

$$g = \sum_{i,j=1}^m g_{ij} dx^i \otimes dx^j \, , \quad g_{ij} \in C^\infty(U) \, , \qquad i,j=1,\ldots,m \, , \tag{1.11}$$

and

$$\mathrm{vol}(M,g) = \sqrt{g} \cdot dx^1 \wedge \ldots \wedge dx^m \, , \qquad g = \det(g_{ij})_{i,j=1}^m \, , \tag{1.12}$$

respectively.) By standard Riemannian geometry, the existence of the volume form enables to integrate scalars of compact support on M and, as usual,

$$\int_M \mu \operatorname{vol}(M,g) \ , \quad \mu \in C_0^\infty(M) \ ,$$

stands for the integral of μ over M.

A fibre metric on a (real) vector bundle ξ over M is a section $(,)_\xi \in C^\infty(S^2(\xi^*))$ whose restriction to each of the fibres of ξ is positive definite. The vector bundle ξ with $(,)_\xi$ (or rather, the pair $(\xi, (,)_\xi)$) is said to be a *Euclidean vector bundle* over M. Then, for $v, v' \in C^\infty(\xi)$, the *scalar product* $(v,v')_\xi$ is a scalar on M and, assuming supp $v \cap \cap$ supp $v' \subset M$ to be compact, the *global scalar product of* v and v' is defined by

$$((v,v'))_\xi = \int_M (v,v')_\xi \operatorname{vol}(M,g) \ . \tag{1.13}$$

The linear space $C_0^\infty(\xi)$ endowed with the global scalar product $((,))_\xi$ becomes a prehilbert space.

Proposition 1.14. *Let ξ and η be Euclidean vector bundles with fibre metrics $(,)_\xi$ and $(,)_\eta$, respectively, over an m-dimensional oriented Riemannian manifold M. Then, for any differential operator $D \in \mathrm{Diff}_r(\xi,\eta)$ there exists a unique differential operator $D^* \in \mathrm{Diff}_r(\eta,\xi)$ such that*

$$((Dv,w))_\eta = ((v,D^*w))_\xi \tag{1.15}$$

for all $v \in C_0^\infty(\xi)$ and $w \in C_0^\infty(\eta)$. Moreover, the symbols are related by the formula

$$\sigma_r(D^*) = (-1)^r \cdot {}^t\sigma_r(D) \ , \tag{1.16}$$

where ^{t} denotes the transpose.

Proof. As $C_0^\infty(\xi)$ is a prehilbert space, uniqueness of the adjoint *on* $C_0^\infty(\xi)$ is obvious. Also, by standard argument, D^* is unique on the whole of $C^\infty(\xi)$. The question of existence, by uniqueness, can easily be reduced to local existence of D^*, i.e. we may set $M=\mathbf{R}^m$, $\xi=\varepsilon^V_{\mathbf{R}^m}$, $\eta=\varepsilon^W_{\mathbf{R}^m}$ and the integration on $\mathbf{R}^m$

is taken by a positive scalar multiple $p \in C^{\infty}(\mathbf{R}^m)$ of the Lebesque measure $d\lambda$. Fixing inner products $(,)_V$ and $(,)_W$ on V and W, respectively, the Euclidean structures of ξ and η translate into maps

$$P:\mathbf{R}^m \to \mathrm{Hom}(V,V) \quad \text{and} \quad Q:\mathbf{R}^m \to \mathrm{Hom}(W,W)$$

(with values in the subspace of symmetric, positive definite linear endomorphisms) satisfying

$$(v,v')_{\xi} = (Pv,v')_V , \quad v,v' \in C_0^{\infty}(\xi) ,$$

and

$$(w,w')_{\eta} = (Qw,w')_W , \quad w,w' \in C_0^{\infty}(\eta) .$$

Assume now that $D \in \mathrm{Diff}_r(\xi,\eta)$ has the local coordinate representation

$$D = \sum_{|\rho| \leq r} A_{\rho} D^{\rho} ,$$

where $A_{\rho}:\mathbf{R}^m \to \mathrm{Hom}(V,W)$, for $|\rho| \leq r$. If $v,v' \in C_0^{\infty}(\xi)$ then the Leibniz rule gives

$$\frac{\partial}{\partial x^i}(v,v')_V = \left(\frac{\partial v}{\partial x^i}, v'\right)_V + \left(v, \frac{\partial v'}{\partial x^i}\right)_V ,$$

$$i=1,\ldots,m ,$$

and so, by an easy induction, we obtain

$$\int_M (D^{\rho}v,v')_V \, d\lambda = \int_M (v, (-1)^{|\rho|}D^{\rho}v')_V \, d\lambda .$$

Using this, for $v \in C_0^{\infty}(\xi)$ and $w \in C_0^{\infty}(\eta)$, we have

$$(Dv,w)_{\eta} = \sum_{|\rho| \leq r} \int_M (A_{\rho}D^{\rho}v,w)_{\eta} \, p d\lambda =$$

$$= \sum_{|\rho| \leq r} \int_M (A_{\rho}D^{\rho}v,Qw)_W \, p d\lambda =$$

$$= \sum_{|\rho|\leq r} \int_M (D^\rho v \ , \ p\cdot {}^tA_\rho\cdot Qw)_V \ d\lambda =$$

$$= \sum_{|\rho|\leq r} \int_M (v \ , \ (-1)^{|\rho|} D^\rho\{p\cdot {}^tA_\rho\cdot Qw\})_V d\lambda =$$

$$= \int_M (v \ , \ P \cdot \sum_{|\rho|\leq r} (-1)^{|\rho|} \frac{1}{p} P^{-1} D^\rho\{p\cdot {}^tA_\rho\cdot Qw\})_V \ pd\lambda =$$

$$= \int_M (v \ , \ D^*w)_\xi \ pd\lambda = ((v \ , \ D^*w))_\xi \ ,$$

where

$$D^*w = \sum_{|\rho|\leq r} (-1)^{|\rho|} \frac{1}{p} P^{-1} D^\rho\{p\cdot {}^tA_\rho\cdot Qw\} =$$

$$= \sum_{|\rho|\leq r} \{ \sum_{\sigma\leq\rho} (-1)^{|\rho|} \binom{\rho}{\sigma} \frac{1}{p} P^{-1} D^{\rho-\sigma}(p\cdot {}^tA_\rho\cdot Q)\} D^\sigma w \ .$$

Changing the indices, we get

$$D^*w = \sum_{|\rho|\leq r} B_\rho D^\rho \ ,$$

where

$$B_\rho = \sum_{\substack{\rho\leq\sigma \\ |\sigma|\leq r}} (-1)^{|\sigma|} \binom{\sigma}{\rho} \frac{1}{p} P^{-1} D^{\sigma-\rho}(p\cdot {}^tA_\rho\cdot Q) \ . \qquad (1.17)$$

Hence, D^* is a differential operator of order r satisfying (1.15). To compute its symbol, we first note that, for $|\rho|=r$, (1.17) reduces to the form

$$B_\rho(x) = (-1)^r P(x)^{-1}\cdot {}^tA_\rho(x)\cdot Q(x) \ , \ x\in\mathbf{R}^m \ .$$

Thus, by (1.2),

$$\sigma_r(D^*)(x,y) = (-1)^r \sum_{|\rho|=r} y^\rho \ P(x)^{-1}\cdot {}^tA_\rho(x)\cdot Q(x), \ x,y\in\mathbf{R}^m.$$

Finally, to check the validity of (1.16), we take $v_x \in \xi_x$ and $w_x \in \eta_x$ and compute

$$(\sigma_r(D)(x,y)v_x, w_x)_\eta = \sum_{|\rho|=r} y^\rho (A_\rho(x)v_x, w_x)_\eta =$$

$$= \sum_{|\rho|=r} y^\rho (A_\rho(x)v_x, Q(x)w_x)_W =$$

$$= \sum_{|\rho|=r} y^\rho (v_x, {}^tA_\rho(x)\cdot Q(x)w_x)_V =$$

$$= \sum_{|\rho|=r} y^\rho (v_x, P(x)\cdot P(x)^{-1}\cdot {}^tA_\rho(x)Q(x)w_x)_V =$$

$$= (v_x, (-1)^r P(x)\cdot\sigma_r(D^*)(x,y))_V =$$

$$= (v_x, (-1)^r \sigma_r(D^*)(x,y))_\xi \ . \ \checkmark$$

The differential operator $D^* \in \mathrm{Diff}_r(\eta,\xi)$ occurring in 1.14. is said to be the *(formal) adjoint* of $D \in \mathrm{Diff}_r(\xi,\eta)$.

Remark. The global scalar product (and hence the existence of the adjoint) requires only a Radon measure with support M on the base manifold and so M need not be oriented (cf. Palais (1965)). Nevertheless, concerning our forthcoming applications, such as Hodge-de Rham theory, it is convenient to restrict ourselves to oriented Riemannian base and take the volume form instead of a Radon measure on M .

A differential operator $D \in \mathrm{Diff}_r(\xi,\xi)$ is said to be *self-adjoint* if $D^*=D$. The *spectrum* $\mathrm{Spec}(D,M)$ of a self-adjoint differential operator D is defined to be the set of reals $\lambda \in \mathbf{R}$ such that there exists $0 \neq v \in C^\infty(\xi)$ with $Dv=\lambda v$. The elements of $\mathrm{Spec}(D,M)$ are said to be the *eigenvalues* of D and, for $\lambda \in \mathrm{Spec}(D,M)$, the vector space $V_\lambda = \{v \in C^\infty(\xi) \mid Dv=\lambda v\}$ is the *eigenspace* of D corresponding to the eigenvalue λ . We then have the orthogonality relation

$$((V_\lambda \cap C_0^\infty(\xi),\ V_{\lambda'} \cap C_0^\infty(\xi))) = 0\ ,\quad \lambda,\lambda' \in \mathrm{Spec}(D,M),\quad \lambda \neq \lambda'\ .$$

The following theorem is of fundamental importance and summarizes the basic properties of the elliptic self-adjoint differential operators.

Theorem 1.18. *(Finiteness and Hodge-de Rham decomposition). Let* $D\in\mathrm{Diff}_r(\xi,\xi)$ *be an elliptic self-adjoint differential operator acting on the Euclidean vector bundle* ξ *over a compact oriented Riemannian manifold* M *. Then,* $\mathrm{Spec}(D,M)\subset\mathbf{R}$ *has no finite accumulation point,* $\ker D\subset C^\infty(\xi)$ *is finite dimensional (in particular, each eigenspace* V_λ *,* $\lambda\in\mathrm{Spec}(D,M)$ *, is finite dimensional) and*

$$C^\infty(\xi) = \mathrm{im}\, D\oplus\ker D \tag{1.19}$$

is an orthogonal direct sum.

For the proof, see Agmon (1965) (discreteness of the spectrum), Boutot (1974), Wells (1973) (finiteness and Hodge-de Rham decomposition) and Palais (1965) (discrete Sobolev chains).

2. LIE DIFFERENTIATION

Let M be a manifold and denote by $T^s_r(M)$, $r,s\in\mathbf{Z}_+$, the tensor bundle of type (s,r) on M . If $\phi:M\to M$ is a diffeomorphism, its differential ϕ_* extends to an isomorphism $\tilde{\phi}:T^s_r(M)_{\phi^{-1}(x)}\to T^s_r(M)_x$, $x\in M$, between the corresponding fibres of $T^s_r(M)$. Given a tensor field $\tau\in T^s_r(M)\,(=C^\infty(T^s_r(M)))$ of type (s,r) on M , we define $\tilde{\phi}(\tau)\in T^s_r(M)$ (pointwise) by $\tilde{\phi}(\tau)_x = \tilde{\phi}(\tau_{\phi^{-1}(x)})$, $x\in M$. Let $X\in V(M)$ be a vector field on M and consider its induced local 1-parameter group of local transformations $(\phi_t)_{t\in\mathbf{R}}$. Then, for $\tau\in T^s_r(M)$, we define the *Lie derivative* $L_X\tau\in T^s_r(M)$ *of* τ *with respect to the vector field* X by

$$(L_X\tau)_x = \lim_{t\to 0}\frac{1}{t}\{\tau_x-(\tilde{\phi}_t(\tau))_x\}\ ,\quad x\in M\ .$$

(Note that $(\tilde{\phi}_t(\tau))_x\in T^s_r(M)_x$ makes sense, in general, only for small values of t .) The linear map $L_X:T^s_r(M)\to T^s_r(M)$ (called the *Lie differentiation with respect to* X) is, in fact, a differential operator of order 1 (cf. e.g. Problem 1 for this chapter). Furthermore, L_X is a *derivation of the full tensor algebra* $\bigoplus_{r,s=0}^{\infty} T_r^s(M)$, i.e. we have

$$L_X(\tau\otimes\tau') = (L_X\tau)\otimes\tau' + \tau\otimes(L_X\tau') ,$$
$$\tau\in T_r^s(M) , \quad \tau'\in T_{r'}^{s'}(M) . \tag{2.1}$$

For a vector field $Y\in V(M)(=T_0^1(M))$ the Lie derivative L_XY is nothing but the Lie bracket $[X,Y]$ of X and Y . (For proofs, see Kobayashi and Nomizu (1963) page 15 or Boothby (1975) page 153.) Hence, by (2.1), for $\alpha\in T_r^0(M)$, we obtain

$$(L_X\alpha)(Y^1,\ldots,Y^r) = X(\alpha(Y^1,\ldots,Y^r)) -$$
$$- \sum_{i=1}^{r} \alpha(Y^1,\ldots,[X,Y^i],\ldots,Y^r), \quad Y^1,\ldots,Y^r\in V(M) , \tag{2.2}$$

in particular, for $\mu\in C^\infty(M)(=T_0^0(M))$, we have $L_X\mu=X(\mu)=d\mu(X)$.

Proposition 2.3. *For* $X,Y\in V(M)$ *, the identity*

$$L_{[X,Y]} = [L_X,L_Y] \tag{2.4}$$

holds on $T_r^s(M)$, $r,s\in\mathbf{Z}_+$.

Proof. As both $L_{[X,Y]}$ and $[L_X,L_Y]$ are local operators, we may restrict ourselves to decomposable tensors of $T_r^s(M)$, $r,s\in\mathbf{Z}_+$. Also, $L_{[X,Y]}$ and $[L_X,L_Y]$ being derivations of the full tensor algebra, we have to check the validity of (2.4) on scalars and vector fields. Finally, on $C^\infty(M)$, (2.4) translates to the definition of the Lie bracket and, on $V(M)$, reduces to the Jacobi identity. √

To illustrate the general concepts introduced in Section 1, we compute the symbol $\sigma_1(L_X)$ of the Lie differentiation L_X with respect to $X\in V(M)$. Take $\alpha_x\in T'_x(M)$, $x\in M$, and choose $\mu\in C^\infty(M)$ with $\mu(x)=0$ and $(d\mu)_x=\alpha_x$. Then, by (2.1), for

$\tau \in T^s_r(M)$, we have

$$\sigma_1(L_X)(\alpha_x)\tau_x = (L_X(\mu\cdot\tau))_x = (L_X\mu)(x)\cdot\tau_x +$$

$$+ \mu(x)(L_X\tau)_x = (d\mu)(X_x)\cdot\tau_x = \alpha_x(X_x)\cdot\tau_x \ .$$

Thus, $\sigma_1(L_X)(\alpha_x): T^s_r(M)_x \to T^s_r(M)_x$ is multiplication with $\alpha_x(X_x)\in\mathbf{R}$. In particular, L_X is elliptic if and only if M is 1-dimensional and X nowhere vanishes on M .

3. EXTERIOR DIFFERENTIATION

Let M be a manifold of dimension m and consider the exterior algebra $\mathcal{D}(M)$ of M . A linear map $D:\mathcal{D}(M)\to\mathcal{D}(M)$ is said to be a *derivation* (resp. a *skew-derivation*) *of* $\mathcal{D}(M)$ if

$$D(\alpha\wedge\beta) = (D\alpha)\wedge\beta + \alpha\wedge(D\beta)$$

(resp. $D(\alpha\wedge\beta)=(D\alpha)\wedge\beta+(-1)^r\cdot\alpha\wedge(D\beta)$) holds for all $\alpha\in\mathcal{D}^r(M)$, $\beta\in\mathcal{D}^s(M)$, $r,s=0,\dots,m$. A derivation or a skew-derivation D of $\mathcal{D}(M)$ has *degree* k provided that $D(\mathcal{D}^r(M))\subset\mathcal{D}^{r+k}(M)$ is valid for $r=0,\dots,m$.

Example 3.1. The Lie differentiation L_X with respect to a vector field $X\in\mathcal{V}(M)$, by (2.1), is (or rather, restricts to) a derivation of $\mathcal{D}(M)$ of degree 0 .

For $\alpha\in\mathcal{D}^r(M)$, $r=0,\dots,m$, we define the *exterior differential* $d\alpha\in\mathcal{D}^{r+1}(M)$ of α by

$$(d\alpha)(X^1,\dots,X^{r+1}) = \sum_{i=1}^{r+1}(-1)^{i+1}X^i(\alpha(X^1,\dots,\hat{X}^i,\dots,X^{r+1})) +$$

$$+ \sum_{i<j}(-1)^{i+j}\alpha([X^i,X^j],X^1,\dots,\hat{X}^i,\dots,\hat{X}^j,\dots,X^{r+1}), \qquad (3.2)$$

$$X^1,\dots,X^{r+1}\in\mathcal{V}(M)$$

where $\hat{}$ means that the term is omitted. In particular, for a scalar $\mu\in C^\infty(M)$, formula (3.2) reduces to $d\mu(X)=X(\mu)$, $X\in\mathcal{V}(M)$, i.e. the 1-form $d\mu$ is nothing but the total differential of μ. The linear map $d:\mathcal{D}(M) \to \mathcal{D}(M)$ which associates to an r-form α its exterior differential $d\alpha$ is said to be the *exterior differentiation* on M. Clearly d is a local operator. Further, we claim that d is a skew-derivation of $\mathcal{D}(M)$ of degree 1. Indeed, the question being local, we set

$$\alpha = \mu^o \cdot d\mu^1 \wedge \dots \wedge d\mu^r \in \mathcal{D}^r(M) \quad , \quad \mu^o, \dots, \mu^r \in C^\infty(M) \quad ,$$

and compute $d\alpha$. By virtue of (3.2), for $X^1, \dots, X^{r+1} \in V(M)$, we have

$$(d\alpha)(X^1, \dots, X^{r+1}) =$$

$$= \sum_{i=1}^{r+1} (-1)^{i+1} X^i \{\mu^o (d\mu^1 \wedge \dots \wedge d\mu^r)(X^1, \dots, \hat{X}^i, \dots, X^{r+1})\} +$$

$$+ \sum_{i<j} (-1)^{i+j} \mu^o \cdot (d\mu^1 \wedge \dots \wedge d\mu^r)([X^i, X^j], X^1, \dots, \hat{X}^i, \dots, \hat{X}^j, \dots, X^{r+1}) =$$

$$= \sum_{i=1}^{r+1} (-1)^{i+1} (d\mu^o)(X^i)(d\mu^1 \wedge \dots \wedge d\mu^r)(X^1, \dots, \hat{X}^i, \dots, X^{r+1}) +$$

$$+ \sum_{i=1}^{r+1} (-1)^{i+1} \mu^o \cdot X^i \det \begin{bmatrix} X^1(\mu^1) \dots X^i(\mu^1)\hat{} \dots X^{r+1}(\mu^1) \\ \\ X^1(\mu^r) \dots X^i(\mu^r)\hat{} \dots X^{r+1}(\mu^r) \end{bmatrix} +$$

$$+ \sum_{i<j} (-1)^{i+j} \mu^o \det \begin{bmatrix} [X^i, X^j](\mu^1) X^1(\mu^1) \dots X^i(\mu^1)\hat{} \dots X^j(\mu^1)\hat{} \dots X^{r+1}(\mu^1) \\ \\ [X^i, X^j](\mu^r) X^1(\mu^r) \dots X^i(\mu^r)\hat{} \dots X^j(\mu^r)\hat{} \dots X^{r+1}(\mu^r) \end{bmatrix}$$

$$= (d\mu^o \wedge d\mu^1 \wedge \dots \wedge d\mu^r)(X^1, \dots, X^{r+1}) \quad ,$$

since the sums of determinants are easily seen to cancel. Thus,

$$d\alpha = d\mu^o \wedge d\mu^1 \wedge \dots \wedge d\mu^r \quad , \tag{3.3}$$

in particular, $d^2 = 0$. Since d is a local operator, to check the skew-derivative property, we may restrict ourselves to decomposable forms. Let $\alpha = \mu^o d\mu^1 \wedge \dots \wedge d\mu^r$ and $\beta = \nu^o d\nu^1 \wedge \dots \wedge d\nu^s$, where $\mu^0,\ldots,\mu^r\ ;\ \nu^0,\ldots,\nu^s \in C^\infty(M)$. Then, by (3.3),

$$d(\alpha\wedge\beta) = d(\mu^0\cdot\nu^0\cdot d\mu^1\wedge\ldots\wedge d\mu^r\wedge d\nu^1\wedge\ldots\wedge d\nu^s) =$$

$$= d(\mu^0\nu^0)\wedge d\mu^1\wedge\ldots\wedge d\mu^r\wedge d\nu^1\wedge\ldots\wedge d\nu^s =$$

$$= (d\mu^0\wedge d\mu^1\wedge\ldots\wedge d\mu^r)\wedge(\nu^0 d\nu^1\wedge\ldots\wedge d\nu^s) +$$

$$+ \mu^0\cdot d\nu^0\wedge d\mu^1\wedge\ldots\wedge d\mu^r\wedge d\nu^1\wedge\ldots\wedge d\nu^s =$$

$$= (d\alpha)\wedge\beta + (-1)^r\ \alpha\wedge(d\beta)$$

and the claim follows. (For different treatments, see Boothby (1975) pages 218-220 or Lichnerowicz (1955) pages 31-36.) Again by (3.3), over a local coordinate neighbourhood (U,ϕ) with associated base $\{\partial_i\}_{i=1}^m \subset V(U)$ and dual $\{dx^i\}_{i=1}^m \subset D^1(U)$, the exterior differential of an r-form $\alpha\in D^r(M)$ represented locally by

$$\alpha = \frac{1}{r!}\sum_{i_1,\ldots,i_r=1}^{m} \alpha_{i_1\ldots i_r}\ dx^{i_1}\wedge\ldots\wedge dx^{i_r}\ ,$$

$$\alpha_{i_1\ldots i_r}\in C^\infty(U)\ ,\quad i_1,\ldots,i_r = 1,\ldots,m\ ,$$

takes the form

$$d\alpha = \frac{1}{r!}\sum_{i_0,\ldots,i_r=1}^{m} \partial_{i_0}\alpha_{i_1\ldots i_r}\cdot dx^{i_0}\wedge dx^{i_1}\wedge\ldots\wedge dx^{i_r}\ .$$

It follows that $d\in \mathrm{Diff}_1(\Lambda^r T^*(M),\ \Lambda^{r+1}T^*(M))$, $r=0,\ldots,m$ (cf. also Problem 1 for this chapter). To compute its sybol $\sigma_1(d)(\alpha_x)$, $\alpha_x\in T'_x(M)$, choose $\mu\in C^\infty(M)$ with $\mu(x)=0$ and $(d\mu)_x=\alpha_x$. Then, for $\beta\in D^r(M)$, we have

$$\sigma_1(d)(\alpha_x)\beta_x = (d(\mu\cdot\beta))_x =$$

$$= (d\mu)_x\wedge\beta_x + \mu(x)(d\beta)_x = \alpha_x\wedge\beta_x$$

and we obtain that the linear map $\sigma_1(d)(\alpha_x):\Lambda^r(T^*_x(M)) \to \Lambda^{r+1}(T^*_x(M))$ is nothing but exterior multiplication with the covector $\alpha_x \in T'_x(M)$.

For each $X \in V(M)$ we define a skew-derivation ι_X (called the *interior product with respect to* X) of degree -1 of $\mathcal{D}(M)$ by setting

$$(\iota_X \alpha)(X^1,\dots,X^r) = \alpha(X,X^1,\dots,X^r) \ ,$$
$$\alpha \in \mathcal{D}^{r+1}(M) \ , \quad X,X^1,\dots,X^r \in V(M) \ . \tag{3.4}$$

(Note that ι_X is defined to be zero on scalars.) As $\mathcal{D}(M)$ consist of skew-symmetric tensors, we deduce $\iota_X^2=0$.

Proposition 3.5. *On* $\mathcal{D}(M)$,

$$L_X = d \circ \iota_X + \iota_X \circ d \ , \tag{3.6}$$

for each $X \in V(M)$.

Proof. Both sides of (3.6) are local operators and hence it is enough to evaluate them on decomposable r-forms, $r=0,\dots,m$. Also, L_X and $d \circ \iota_X + \iota_X \circ d$ being derivations of $\mathcal{D}(M)$ of degree 0 , we have only to check the validity of (3.6) on scalars and 1-forms. For $\mu \in C^\infty(M)$, $\iota_X \mu$ being zero, we get $L_X \mu = (d\mu)X = (\iota_X \circ d)\mu$. For $\alpha \in \mathcal{D}^1(M)$, by (2.2), (3.2) and (3.4), we have

$$((d \circ \iota_X + \iota_X \circ d)\alpha)Y = (d(\alpha(X)))Y + d\alpha(X,Y) =$$
$$= Y(\alpha(X)) + X(\alpha(Y)) - Y(\alpha(X)) - \alpha([X,Y]) =$$
$$= X(\alpha(Y)) - \alpha([X,Y]) = (L_X \alpha)Y \ , \quad Y \in V(M) \ . \ \checkmark$$

Corollary 3.7. *The exterior differentiation* d *commutes with the Lie differentiation* L_X *for all* $X \in V(M)$.

Proof. Obvious, since $d^2=0$. ✓

Let M and N be manifolds of dimensions m and n , respectively. A map $f: M \to N$ pulls back r-forms on N to r-forms on M , $r \in \mathbf{Z}_+$, and hence induces a linear map $f^*: \mathcal{D}(N) \to \mathcal{D}(M)$ defined by

$$(f^*\alpha)(X_x^1,\ldots,X_x^r) = \alpha(f_*(X_x^1),\ldots,f_*(X_x^r)) , \quad \alpha\in\mathcal{D}^r(N) ,$$

$$X_x^1,\ldots,X_x^r\in T_x(M) , \quad x\in M ,$$

where f_* stands for the differential of f .

Proposition 3.8. *The pull-back* f^* *commutes with the exterior differentiations on* M *and* N *, i.e.*

$$d(f^*\alpha) = f^*(d\alpha) \tag{3.9}$$

for every $\alpha\in\mathcal{D}(N)$.

Proof. As d is a local operator, for $\alpha\in\mathcal{D}^r(N)$, $r\in\mathbf{Z}_+$, we may set

$$\alpha = \mu^0\cdot d\mu^1 \wedge\ldots\wedge d\mu^r ,$$

where $\mu^0,\mu^1,\ldots,\mu^r\in C^\infty(N)$. For $r=0$ and $X\in V(M)$, we obtain

$$(d(f^*\mu))X = (d(\mu\circ f))X = X(\mu\circ f) = (f_*(X))\mu =$$

$$= (d\mu)f_*(X) = (f^*(d\mu))X , \quad \mu\in C^\infty(N) ,$$

and (3.9) is satisfied for scalars. Hence, as d is a skew-derivation of $\mathcal{D}(M)$ with $d^2=0$, by making use of (3.3), we have

$$d(f^*\alpha) = d(f^*(\mu^0 d\mu^1\wedge\ldots\wedge d\mu^r)) = d\{(f^*\mu^0)(f^*d\mu^1)\wedge\ldots\wedge(f^*d\mu^r)\} =$$

$$= d\{(f^*\mu^0)d(f^*\mu^1)\wedge\ldots\wedge d(f^*\mu^r)\} = d(f^*\mu^0)\wedge\ldots\wedge d(f^*\mu^r) =$$

$$= f^*(d\mu^0)\wedge f^*(d\mu^1)\wedge\ldots\wedge f^*(d\mu^r) = f^*\{d\mu^0\wedge d\mu^1\wedge\ldots\wedge d\mu^r\} =$$

$$= f^*\{d(\mu^0\cdot d\mu^1\wedge\ldots\wedge d\mu^r)\} = f^*(d\alpha) . \checkmark$$

Finally, we discuss some concepts of the de Rham cohomology theory. An r-form α on M , $r\in\mathbf{Z}_+$, is said to be *closed* if $d\alpha=0$. It is called *exact* if there exists an $(r-1)$-form β

such that $\alpha=d\beta$. Since $d^2=0$, every exact form is closed. The quotient of the vector space of closed r-forms modulo the subspace of exact r-forms is said to be the *r-th de Rham cohomology group* $H^r_{de\ R}(M;\mathbf{R})$ of M . If it is finite dimensional then $b_r(M) = \dim H^r_{de\ R}(M;\mathbf{R})$ is called the *r-th Betti number* of M . Clearly, $b_r(M)=0$ for $r>m=\dim M$. If $f:M\to N$ is a map then, by virtue of 3.8, it pulls back closed (resp. exact) r-forms on N to closed (resp. exact) r-forms on M , i.e. passing to the quotients, we obtain a linear map

$$f^*:H^r_{de\ R}(N;\mathbf{R}) \to H^r_{de\ R}(M;\mathbf{R}) \ , \qquad r\in\mathbf{Z}_+ \ .$$

One of the cornerstones of the de Rham cohomology theory is that the map f^* induced on cohomology depends only on the homotopy class of f .

Theorem 3.10. *If $f,f':M\to N$ are homotopic then the induced linear maps $f^*,f'^*:H^r_{de\ R}(N;\mathbf{R})\to H^r_{de\ R}(M;\mathbf{R})$, $r\in\mathbf{Z}_+$, coincide.*

Proof. The tangent space of the manifold (with boundary) $M\times[0,1]$ at (x,t) , $x\in M$, $t\in[0,1]$, splits into the sum

$$T_{(x,t)}(M\times[0,1]) = \ker\pi_* \oplus \ker\bar{\pi}_* \ ,$$

where $\pi:M\times[0,1]\to[0,1]$ and $\bar{\pi}:M\times[0,1]\to M$ denote the corresponding projections. Hence, by multilinear algebra, an r-form α on $M\times[0,1]$ decomposes as

$$\begin{aligned} &\alpha = \alpha' + dt\wedge\alpha'' \ , \ \alpha'\in\mathcal{D}^r(M\times[0,1]) \ , \\ &\alpha''\in\mathcal{D}^{r-1}(M\times[0,1]) \ , \end{aligned} \tag{3.11}$$

where $dt\in\mathcal{D}^1(M\times[0,1])$ stands for the 1-form $\pi^*(dt|[0,1])$ (with dt = dual of the canonical base field $\frac{\partial}{\partial t}\in V(\mathbf{R})$) and $\alpha'(X^1,\ldots,X^r)=0$ (resp. $\alpha''(X^1,\ldots,X^{r-1})=0$) if whenever $X^i\in$ $\in C^\infty(\ker\bar{\pi}_*)$ for at least one index $i=1,\ldots,r$ (resp. $i=1,\ldots$ $\ldots,r-1$) . Setting $i_t:M\to M\times[0,1]$, $t\in[0,1]$, with $i_t(x) =$ $=(x,t)$, $x\in M$, we define the $(r-1)$-form I_α on M by

$$I_\alpha(X^1_x,\dots,X^{r-1}_x) = \int_0^1 \alpha''((i_t)_* X^1_x,\dots,(i_t)_* X^{r-1}_x)\,dt ,$$

$$X^1_x,\dots,X^{r-1}_x \in T_x(M) ,$$

where $\alpha\in\mathcal{D}^r(M\times[0,1])$ has the decomposition (3.11) above. We claim that

$$i_1^*\alpha - i_0^*\alpha = dI_\alpha - I_{d\alpha} , \quad \alpha\in\mathcal{D}^r(M\times[0,1]) . \tag{3.12}$$

Let (U,ϕ) be a local coordinate neighbourhood on M and denote by $(U\times[0,1] , \bar\phi^1,\dots,\bar\phi^m,\pi)$ the local coordinate neighbourhood on $M\times[0,1]$ defined by $\bar\phi^i=\phi^i\circ\bar\pi$, $i=1,\dots,m$, with associated local frame and coframe fields $\{\bar\partial_1,\dots,\bar\partial_m , \frac{\partial}{\partial t}\}$ and $\{d\bar x^1,\dots,d\bar x^m,dt\}$, respectively. To prove (3.12), we distinguish two cases as follows:

(i) $\alpha = \mu d\bar x^1\wedge\dots\wedge d\bar x^{i_r}$, $1\le i_1<\dots<i_r\le m$, $\mu\in C^\infty(U\times[0,1])$.

By definition, $I_\alpha=0$ on U , and

$$d\alpha = \frac{\partial\mu}{\partial t}\, dt\wedge d\bar x^1\wedge\dots\wedge d\bar x^m + \text{terms not containing } 'dt' .$$

Thus, at $x\in U$, we have

$$I_{d\alpha}(x) = \{\int_0^1 \frac{\partial\mu}{\partial t}(x,t)\,dt\}d\bar x^{i_1}\wedge\dots\wedge d\bar x^{i_r} =$$

$$= \{\mu(x,1)-\mu(x,0)\}d\bar x^{i_1}\wedge\dots\wedge d\bar x^{i_r} = i_1^*\alpha_x - i_0^*\alpha_x .$$

(ii) $\alpha = \mu dt\wedge d\bar x^{i_1}\wedge\dots\wedge d\bar x^{i_{r-1}}$, $1\le i_1<\dots<i_{r-1}\le m$, $\mu\in C^\infty(U\times[0,1])$.

Clearly, $i_1^*\alpha = i_0^*\alpha = 0$ holds on U . On the other hand,

$$I_{d\alpha}(x) = -\sum_{j=1}^m \{\int_0^1(\bar\partial_j\mu)(x,t)\,dt\}d\bar x^j\wedge d\bar x^{i_1}\wedge\dots\wedge d\bar x^{i_{r-1}}$$

and

$$(dI_\alpha)(x) = \{d(\int_0^1\mu(\cdot,t)\,dt)\,d\bar x^{i_1}\wedge\dots\wedge d\bar x^{i_{r-1}}\}(x) =$$

$$= \sum_{j=1}^{m} \{\bar\partial_j (\int_0^1 \mu(\cdot,t)\,dt)\}_x d\bar x^j \wedge d\bar x^{i_1} \wedge \ldots \wedge d\bar x^{i_{r-1}} .$$

Thus, $dI_\alpha - I_{d\alpha} = 0$ holds on U and (3.12) follows. Now, let $H: M\times[0,1] \to N$ be a homotopy connecting f and f', i.e. by the notations above, $f = H\circ i_0$ and $f' = H\circ i_1$. If α is a closed r-form on N representing a cohomology class of $H^r_{de\ R}(N)$, then by virtue of (3.12), we have

$$f'^*(\alpha) - f^*(\alpha) = i_1^*(H^*\alpha) - i_0^*(H^*\alpha) =$$

$$= dI_{H^*\alpha} - I_{d(H^*\alpha)} = dI_{H^*\alpha}$$

since, by (3.9), $d(H^*\alpha) = H^*(d\alpha) = 0$. Thus, $f^*(\alpha)$ and $f'^*(\alpha)$ represent the same cohomology class in $H^r_{de\ R}(M;\mathbf{R})$. √

As a direct consequence of Theorem 3.10 we obtain the well-known Poincaré lemma.

Corollary 3.13. *If M is a manifold diffeomorphic with $\mathbf{R}^m$ then $H^r_{de\ R}(M;\mathbf{R}) = 0$, for $r\in\mathbf{N}$, or equivalently, every closed r-form on M is exact.*

Proof. The identity $id_M: M \to M$ is null-homotopic and thus, $(id_M)^*$ being the identity on cohomology, by 3.10, we obtain $H^r_{de\ R}(M;\mathbf{R}) = H^r_{de\ R}(\{\text{point}\},\mathbf{R}) = 0$, $r\in\mathbf{N}$. √

We close this section with a sharpening of 3.13 for 1-forms.

Corollary 3.14. *On a simply connected manifold M every closed 1-form α is exact.*

Proof. Note first that a straightforward proof can be given in the following way. By virtue of the Poincaré lemma, the equation $du=\alpha$ can be solved (uniquely up to an additive constant) on each of the members of a sufficiently fine open covering of M. Then, using simply connectedness of M, the Lichnerowicz factorization lemma (cf. Lichnerowicz (1955) pages 51-53 or Kobayashi and Nomizu (1963) pages 284-285) makes it possible to glue together the various solutions yielding a globally defined scalar u on M satisfying $du=\alpha$.

Following É. Cartan, we give an other proof based on the Frobenius theorem. Denote by $\pi:M\times\mathbf{R}\to\mathbf{R}$ and $\bar{\pi}:M\times\mathbf{R}\to M$ the projections and set $\eta=\bar{\pi}^*\alpha-dt\in\mathcal{D}^1(M\times\mathbf{R})$, where $dt\in\mathcal{D}^1(M\times\mathbf{R})$ stands for $\pi^*(dt)$ (with dt = dual of $\frac{\partial}{\partial t}\in\mathcal{V}(\mathbf{R})$). By definition,

$$\ker\eta_{(x,t)}\cap T_{(x,t)}(\{x\}\times\mathbf{R}) = \{0\} , \quad (x,t)\in M\times\mathbf{R} , \qquad (3.15)$$

and hence $\ker\eta$ is a codimension 1 distribution on $M\times\mathbf{R}$ which is invariant by translations $\tau_s:M\times\mathbf{R}\to M\times\mathbf{R}$, $s\in\mathbf{R}$, $\tau_s(x,t) = (x,t+s)$, $(x,t)\in M\times\mathbf{R}$. (For the concepts and results of the elementary theory of distributions used here we refer to Chevalley (1946) pages 92-94.) Furthermore, by virtue of 3.8, the 1-form η is closed on $M\times\mathbf{R}$ and hence, for $X,Y\in C^\infty(\ker\eta)$, (3.2) entails

$$0 = (d\eta)(X,Y) = X(\eta(Y)) - Y(\eta(X)) - \eta([X,Y]) = -\eta([X,Y]),$$

i.e. $[X,Y]\in C^\infty(\ker\eta)$. It follows that the distribution $\ker\eta$ is involutive and hence integrable. Fixing a (maximal) integral leaf $\hat{M}\subset M\times\mathbf{R}$, a straightforward argument in the use of the transversality condition (3.15) and the translation invariance shows that the restriction $\hat{\pi}=\bar{\pi}|\hat{M}:\hat{M}\to M$ is a covering projection onto M . (Indeed, by (3.15), $\hat{\pi}$ is a local diffeomorphism with im $\hat{\pi}\subset M$ open. If $x\in M$ is in the closure of im $\hat{\pi}$ then maximality of $\hat{M}$ and translation invariance of $\ker\eta$ imply that $x\in$ im $\hat{\pi}$ and so $\hat{\pi}$ is onto. Again by (3.15), $\hat{M}\cap\{x\}\times\mathbf{R}$ is discrete for all $x\in M$ and, $\hat{\pi}$ being a local diffeomorphism, translation invariance yields that $\hat{\pi}$ is a covering.) Now, as M is simply connected, $\hat{\pi}:\hat{M}\to M$ is a diffeomorphism, or equivalently, $\hat{M}\subset M\times\mathbf{R}$ is the graph of a scalar $u:M\to\mathbf{R}$. To accomplish the proof, we show that $du=\alpha$. Indeed, for $X_x\in T_x(M)$, the vector $(id_M,u)_* X_x\in T_{(x,u(x))}(M\times\mathbf{R})$ is tangent to $\hat{M}$ and hence belongs to $\ker\eta$. We compute

$$\alpha(X_x) = \alpha((\bar{\pi}\circ(id_M,u))_* X_x) = (\bar{\pi}^*\alpha)((id_M,u)_* X_x) =$$

$$= (\pi^*dt)((id_M,u)_* X_x) = dt((\pi\circ(id_M,u))_* X_x) =$$

$$= (dt)(u_*(X_x)) = du(X_x) \ . \ \surd$$

4. COVARIANT DIFFERENTIATION

Let ξ be a vector bundle over a manifold M. A *covariant differentiation on* ξ is a linear map

$$\nabla : C^\infty(\xi) \to C^\infty(\mathrm{Hom}(T(M),\xi))$$

satisfying the Leibniz rule

$$\nabla(\mu\cdot v) = (d\mu) \otimes \nu + \mu\cdot\nabla v \ , \ \mu\in C^\infty(M) \ , \ v\in C^\infty(\xi) \ . \qquad (4.1)$$

(For a detailed account on connection theory some results of which are stated here without proof we refer to Kobayashi and Nomizu (1963) or Lichnerowicz (1955).) A covariant differentiation ∇ is always a local operator. Indeed, given a section $v\in C^\infty(\xi)$ vanishing on an open set U of M, for $x\in U$, choose a scalar $\mu\in C^\infty(M)$ such that μ is zero in a neighbourhood of x and equals 1 outside U. Then, $v=\mu v$ and, applying (4.1), we have

$$(\nabla v)(x) = (\nabla(\mu v))(x) = (d\mu)_x \otimes v_x + \mu(x)(\nabla v)(x) = 0 \ ,$$

and the claim follows. Furthermore, $\nabla\in\mathrm{Diff}_1(\xi \ , \ \mathrm{Hom}(T(M), \ \xi))$ (as can be shown by expressing ∇ in local coordinates; cf. also Problem 1 for this chapter). To compute its symbol

$$\sigma_1(\nabla)(\alpha_x) : \xi_x \to \mathrm{Hom}(T_x(M),\xi_x) \ , \ \alpha_x\in T'_x(M) \ , \ x\in M \ ,$$

choose $\mu\in C^\infty(M)$ with $\mu(x)=0$ and $(d\mu)_x=\alpha_x$. Applying again (4.1), for $v\in C^\infty(\xi)$, we have

$$\begin{aligned}\sigma_1(\nabla)(\alpha_x)v_x &= (\nabla(\mu v))(x) = (d\mu)_x \otimes v_x + \\ &+ \mu(x)(\nabla v)(x) = \alpha_x \otimes v_x \ ,\end{aligned} \qquad (4.2)$$

i.e. the symbol $\sigma_1(\nabla)(\alpha_x)$ is nothing but tensor multiplication with the covector $\alpha_x\in T'_x(M)$.

Given a covariant differentiation $\nabla \in \mathrm{Diff}_1(\xi, \mathrm{Hom}(T(M),\xi))$ on the vector bundle ξ over M, for fixed $X \in V(M)$, we define

$$\nabla_X : C^\infty(\xi) \to C^\infty(\xi)$$

by

$$\nabla_X v = (\nabla v)(X) \ , \quad v \in C^\infty(\xi) \ . \tag{4.3}$$

Then, by 1.7, $\nabla_X \in \mathrm{Diff}_1(\xi,\xi)$ (called the *covariant differentiation in the direction* $X \in V(M)$) and its symbol $\sigma_1(\nabla_X)(\alpha_x)$, $\alpha_x \in T'_x(M)$, $x \in M$, is multiplication with $\alpha_x(X) \in \mathbf{R}$. Clearly, (4.1) can be rewritten as

$$\nabla_X(\mu v) = X(\mu)\cdot v + \mu\cdot\nabla_X v \ , \quad \mu \in C^\infty(M) \ , \quad v \in C^\infty(\xi) \ . \tag{4.4}$$

Furthermore, for fixed $X_x \in T_x(M)$, $x \in M$, we set

$$\nabla_{X_x} v = (\nabla v)(X_x) = (\nabla_X v)(x) \in \xi_x \ . \tag{4.5}$$

Then, $\nabla_{X_x} : C^\infty(\xi) \to \xi_x$ is a linear map satisfying (4.4) at $x \in M$ (with X replaced by X_x).

If $f: M \to N$ is a map between manifolds M and N and ξ is a vector bundle over N with covariant differentiation ∇^ξ then a covariant differentiation ∇^{ξ^f} is induced on the pull-back bundle $\xi^f = f^*(\xi)$ satisfying

$$\nabla^{\xi^f}_{X_x} w = \nabla^{\xi}_{f_*(X_x)} v \ , \quad X_x \in T_x(M) \ , \quad x \in M \ , \tag{4.6}$$

where $w = v\circ f \in C^\infty(\xi^f)$. As every section of ξ^f can locally be expressed as a sum of terms of type

$$\mu\cdot v\circ f \ , \quad \mu \in C^\infty(M) \ , \quad v \in C^\infty(\xi) \ ,$$

the required additivity and the Leibniz rule (4.4) extend $\nabla^{\xi^f}_{X_x}$ to the whole of $C^\infty(\xi^f)$. Now, formula (4.6) defines the pull-back covariant differentiation ∇^{ξ^f}.) In particular, let $c: I \to M$ be a curve (defined on an (open or closed) interval $I \subset \mathbf{R}$) and ξ a vector bundle over M with covariant differentiation ∇.

A section $w \in C^\infty(\xi^c)$ is said to be a *section of* ξ *along* c. Then, w associates to each $t \in I$ a vector $w_t \in \xi_{c(t)}$ and is therefore denoted as $t \to w_t$, $t \in I$. The *covariant differential* of w is a section of ξ along c defined by

$$t \to \nabla_{\frac{\partial}{\partial t}} w \ , \quad t \in I \ ,$$

(whose value at $t \in I$ will sometimes be denoted by $\nabla_{\frac{\partial}{\partial t}} w_t \in \xi_{c(t)}$) where $\nabla_{\frac{\partial}{\partial t}}$ stands for the pull-back covariant differentiation in the direction of the canonical base field $\frac{\partial}{\partial t} \in V(I)$. (Especially, for $v \in C^\infty(\xi)$, (4.6) implies that $\nabla_{\dot{c}(t)} v = \nabla_{c_* \frac{\partial}{\partial t}} v = \nabla_{\frac{\partial}{\partial t}} (v \circ c)$, i.e. the value $\nabla_{X_x} v$ is determined by X_x and v restricted to *any* curve through x with tangent vector X_x at $x \in M$.) If $[t', t''] \subset I$ and $w_{t'} \in \xi_{c(t')}$, then there is a *unique* vector $w_t \in \xi_{c(t)}$, $t' \leq t \leq t''$, such that, at $t'=t$, it reduces to $w_{t'}$ and $\nabla_{\frac{\partial}{\partial t}} w = 0$ (cf. Problem 3). We call $t \to w_t$, $t \in [t', t'']$, a *parallel section along* $c|[t', t'']$ and $w_{t''} \in \xi_{c(t'')}$ the *parallel translate* of $w_{t'} \in \xi_{c(t')}$. The map

$$\tau_{t''}^{t'} : \xi_{c(t')} \to \xi_{c(t'')} \ , \quad [t', t''] \subset I \ , \tag{4.7}$$

(called the *parallel translation along* $c|[t', t'']$) which associates to $w_{t'}$ its parallel translate $w_{t''}$, is a linear isomorphism. Conversely, the covariant differentiation $\nabla_{\dot{c}(t)} v$, $v \in C^\infty(\xi)$, in the direction $\dot{c}(t)$, $t \in I$, can be recovered from the knowledge of the parallel translation $\tau_{t''}^{t'}$, $t', t'' \in I$, via the formula

$$\nabla_{\dot{c}(t)} v = \lim_{s \to 0} \frac{1}{s} \{\tau_t^{t+s}(w_{t+s}) - w_t\} \ , \quad t \in I \ , \tag{4.8}$$

where $w = v \circ c \in C^\infty(\xi^c)$. (Note that the limit makes sense for boundary points of I as well.) Indeed, choose parallel sections $t \to E_t^j$, $t \in I$, $j = 1, \ldots, n$, of the vector bundle ξ along c

such that $\{E_t^j\}_{j=1}^n \subset \xi_{c(t)}$, $t \in I$, is a base. Then, w decomposes as

$$w = \sum_{j=1}^{n} w_j E^j ,$$

where $w_j : I \to \mathbf{R}$, $j=1,\ldots,n$. Using the Leibniz rule, we compute

$$\nabla_{\dot{c}(t)} v = \nabla_{\frac{\partial}{\partial t}} w = \sum_{j=1}^{n} \frac{\partial w_j(t)}{\partial t} E_t^j =$$

$$= \sum_{j=1}^{n} \lim_{s \to 0} \frac{1}{s} \{w_j(t+s) - w_j(t)\} E_t^j =$$

$$= \sum_{j=1}^{n} \lim_{s \to 0} \frac{1}{s} \{w_j(t+s) \tau_t^{t+s}(E_{t+s}^j) - w_j(t) E_t^j\} =$$

$$= \lim_{s \to 0} \frac{1}{s} \{\tau_t^{t+s}(w_{t+s}) - w_t\}$$

and (4.8) follows.

The *curvature* of a vector bundle ξ over M with covariant differentiation ∇ is a section $R \in C^\infty(\Lambda^2(T^*(M)) \otimes \mathrm{Hom}(\xi,\xi))$ defined by

$$R(X,Y)v = \nabla_X \nabla_Y v - \nabla_Y \nabla_X v - \nabla_{[X,Y]} v ,$$
$$X, Y \in V(M) , \quad v \in C^\infty(\xi) . \tag{4.9}$$

(In fact, a straightforward computation shows that the right hand side of (4.9) is $C^\infty(M)$-linear in each variable.)

If $f : M \to N$ is a map between manifolds M and N and ξ is a vector bundle over N with covariant differentiation ∇^ξ then the curvature tensors R^ξ and R^{ξ^f} of ξ and the pull-back bundle ξ^f, respectively, are related as

$$R^{\xi^f}(X_x, Y_x) w_x = R^\xi(f_*(X_x), f_*(Y_x)) v_{f(x)} ,$$
$$X_x, Y_x \in T_x(M) , \quad x \in M , \tag{4.10}$$

where $w = v \circ f \in C^\infty(\xi^f)$, $v \in C^\infty(\xi)$. The geometric significance of the curvature tensor lies in the fact that it measures infinitesimally the dependence of the parallel transport on curves of M. More precisely, specializing (4.10) to curves yields the following.

Proposition 4.11. *Let* $C:(-\varepsilon,\varepsilon)\times(-\varepsilon,\varepsilon) \to M$ *be a map,* $\varepsilon > 0$, *with*

$$C_*\left(\frac{\partial}{\partial s}\Big|_{(0,0)}\right) = X_x \quad \text{and} \quad C_*\left(\frac{\partial}{\partial t}\Big|_{(0,0)}\right) = Y_x ,$$

$$C(0,0) = x \in M ,$$

where $\frac{\partial}{\partial s}, \frac{\partial}{\partial t} \in V((-\varepsilon,\varepsilon)\times(-\varepsilon,\varepsilon))$ *denote the canonical coordinate fields, respectively. Given a vector bundle* ξ *over* M *with covariant differentiation* ∇, *for each section* w *of* ξ *along the curve* $s \to C(s,0)$, $s \in (-\varepsilon,\varepsilon)$, *we have*

$$\lim_{t\to 0} \frac{1}{t}\{(\tau_0^t(0)\circ\nabla_{\frac{\partial}{\partial s}\big|_{(0,t)}} \circ \tau_t^0(s))w_s - \nabla_{\frac{\partial}{\partial s}\big|_{(0,0)}} w_s\} = \tag{4.12}$$

$$= -R(X_x,Y_x)w_0 ,$$

where, for fixed $s \in (-\varepsilon,\varepsilon)$ *and* $[t',t''] \subset (-\varepsilon,\varepsilon)$, $\tau_{t''}^{t'}(s)$ *denotes the parallel translation along the curve* $t \to C(s,t)$, $t \in [t',t'']$.

Proof. Extend w to a section of $C^*(\xi)$ by setting

$$w_{(s,t)} = \tau_t^0(s)w_s , \quad t \in (-\varepsilon,\varepsilon) .$$

Then, $t \to w_{(s,t)}$, $t \in (-\varepsilon,\varepsilon)$, is parallel for fixed $s \in (-\varepsilon,\varepsilon)$, i.e. we have

$$\nabla_{\frac{\partial}{\partial t}\big|_{(s,0)}} w_{(s,t)} = 0 .$$

Now, by making use of (4.8), (4.10) and $[\frac{\partial}{\partial s}, \frac{\partial}{\partial t}] = 0$, we compute

$$R(X_x, Y_x)w_o = R(C_*(\frac{\partial}{\partial s}\Big|_{(o,o)}) , C_*(\frac{\partial}{\partial t}\Big|_{(o,o)}))w_o =$$

$$= R(\frac{\partial}{\partial s}\Big|_{(o,o)}, \frac{\partial}{\partial t}\Big|_{(o,o)})w_o = \nabla_{\frac{\partial}{\partial s}\big|_{(o,o)}} \nabla_{\frac{\partial}{\partial t}\big|_{(s,o)}} w_{(s,t)} -$$

$$- \nabla_{\frac{\partial}{\partial t}\big|_{(o,o)}} \nabla_{\frac{\partial}{\partial s}\big|_{(o,t)}} w_{(s,t)} - \nabla_{[\frac{\partial}{\partial s}, \frac{\partial}{\partial t}]_{(o,o)}} w_{(s,t)} =$$

$$= -\nabla_{\frac{\partial}{\partial t}\big|_{(o,o)}} \nabla_{\frac{\partial}{\partial s}\big|_{(o,t)}} w_{(s,t)} =$$

$$= -\lim_{t\to o} \frac{1}{t}\{\tau_o^t(O)(\nabla_{\frac{\partial}{\partial s}\big|_{(o,t)}} w_{(s,t)}) - \nabla_{\frac{\partial}{\partial s}\big|_{(o,o)}} w_{(s,o)}\} =$$

$$= -\lim_{t\to o} \frac{1}{t} \{(\tau_o^t(O)\circ\nabla_{\frac{\partial}{\partial s}\big|_{(o,t)}} \circ\tau_t^o(s))w_s - \nabla_{\frac{\partial}{\partial s}\big|_{(o,o)}} w_s\}. \ \checkmark$$

Assume now that ξ is a Euclidean vector bundle over a manifold M with fibre metric $(,)$. A covariant differentiation ∇ on ξ is said to be *orthogonal* if

$$X(v,v') = (\nabla_X v, v') + (v, \nabla_X v') \tag{4.13}$$

for all $X \in V(M)$ and $v, v' \in C^\infty(\xi)$. In this case, ξ (or rather, the triple $(\xi, (,), \nabla))$ is said to be a *Riemannian-connected* vector bundle over M .

Example 4.14. The trivial line bundle ε_M^1 over a manifold M, with $C^\infty(\varepsilon_M^1) = C^\infty(M)$, inherits a fibre metric from the (pointwise) multiplication of scalars on M . Further, the total differential d on $C^\infty(M)$ can be considered as a differential operator $d \in \mathrm{Diff}_1(\varepsilon_M^1 , \mathrm{Hom}(T(M), \varepsilon_M^1))$ with $\iota_X \circ d = X$, $X \in V(M)$. Then, the Leibniz rule

$$X(\mu\mu') = (X\mu)\cdot\mu + \mu\cdot(X\mu') , \quad \mu,\mu' \in C^\infty(M) , \quad X \in V(M) ,$$

implies that d is an orthogonal covariant differentiation on

ε_M^1 . Finally, for the curvature R of the Riemannian-connected vector bundle ε_M^1 , we have

$$R(X,Y)\mu = X(Y\mu) - Y(X\mu) - [X,Y]\mu = 0 ,$$

for any $X,Y\in V(M)$ and $\mu\in C^\infty(M)$.

Orthogonality of a covariant differentiation is preserved under pull-back as the next result asserts.

Theorem 4.15. *Let $f:M \to N$ be a map and ξ a Riemannian-connected vector bundle over N with fibre metric (,) and covariant differentiation ∇^ξ . Then the pull-back bundle ξ^f is also Riemannian-connected with respect to the induced fibre metric (,) and pull-back covariant differentiation ∇^{ξ^f} .*

Proof. We have to check that the orthogonality relation

$$X_x(w,w') = (\nabla^{\xi^f}_{X_x} w, w'_x) + (w_x , \nabla^{\xi^f}_{X_x} w') \tag{4.16}$$

holds for all $X_x \in T_x(M)$, $x\in M$, and $w,w'\in C^\infty(\xi^f)$. As any section in $C^\infty(\xi^f)$ can locally be expressed as a sum of sections of type $\mu\cdot v\circ f$, with $\mu\in C^\infty(M)$ and $v\in C^\infty(\xi)$, the bilinear nature of (4.16) allows to take

$$w = \mu\cdot v\circ f \quad\text{and}\quad w' = \mu'\cdot v'\circ f , \quad \mu,\mu'\in C^\infty(M) ,$$

$$v,v'\in C^\infty(\xi) .$$

Further, as (4.16) is derivative in both sides, without loss of generality we may assume that $\mu=\mu'=1$. Then, by making use of (4.6) and (4.13), we have

$$X_x(w,w') = X_x(v\circ f, v'\circ f) = X_x((v,v')\circ f) =$$

$$= f_*(X_x)(v,v') = (\nabla^\xi_{f_*(X_x)} v, v'_{f(x)}) + (v_{f(x)}, \nabla^\xi_{f_*(X_x)} v') =$$

$$= (\nabla^{\xi^f}_{X_x} w, w_x) + (w_x , \nabla^{\xi^f}_{X_x} w) . \checkmark$$

Specializing to curves, 4.15 reformulates the condition of orthogonality (4.13) in terms of the parallel translation.

Corollary 4.17. *A covariant differentiation* ∇ *on a Euclidean vector bundle* ξ *over* M *is orthogonal if and only if for every curve* $c:I \to M$ *the parallel translation*

$$\tau_{t''}^{t'} : \xi_{c(t')} \to \xi_{c(t'')} \quad , \quad [t',t''] \subset I \ ,$$

is an isometry (with respect to the fibre metric $(,)$ *of* ξ*).*

Proof. Assume first that ∇ is orthogonal on ξ . By 4.15, ∇^{ξ^c} is orthogonal on ξ^c and thus, for $w_{t'}$, $w'_{t'} \in \xi_{c(t')}$, we obtain

$$\frac{\partial}{\partial t}(\tau_t^{t'}(w_{t'}), \tau_t^{t'}(w'_{t'})) = (\nabla^{\xi^c}_{\frac{\partial}{\partial t}} \tau_t^{t'}(w_{t'}), \tau_t^{t'}(w'_{t'})) +$$

$$+ (\tau_t^{t'}(w_{t'}), \nabla^{\xi^c}_{\frac{\partial}{\partial t}} \tau_t^{t'}(w'_{t'})) = 0 \ , \quad t \in [t',t''] \subset I \ .$$

It follows that the function

$$t \to (\tau_t^{t'}(w_{t'}), \tau_t^{t'}(w'_{t'})), \quad t \in [t',t''] \ ,$$

is constant, in particular

$$(w_{t'} , w'_{t'}) = (\tau_{t''}^{t'}(w_{t'}), \tau_{t''}^{t'}(w'_{t'})) \ ,$$

i.e. $\tau_{t''}^{t'}$ is orthogonal. Conversely, let $X_x \in T_x(M)$, $x \in M$, and choose a curve $c:(-\varepsilon,\varepsilon) \to M$, $\varepsilon > 0$, with $\dot{c}(0) = X_x$. Then, for $v,v' \in C^\infty(\xi)$, we have

$$X_x(v,v') = \frac{\partial}{\partial t}(v \circ c , v' \circ c)\Big|_{t=0} = \frac{\partial}{\partial t}(\tau_0^t(v_{c(t)}), \tau_0^t(v'_{c(t)}))\Big|_{t=0} =$$

$$= (\frac{\partial}{\partial t}\{\tau_0^t(v_{c(t)})\}_{t=0} , v'_x) + (v_x , \frac{\partial}{\partial t}\{\tau_0^t(v'_{c(t)})\}_{t=0}) =$$

$$= (\nabla_{X_x} v , v'_x) + (v_x , \nabla_{X_x} v') \ ,$$

where the last equality holds because of (4.8). ✓

We close the general setting by the following.

Proposition 4.18. *The curvature tensor* $R \in C^\infty(\Lambda^2(T^*(M)) \otimes \otimes \mathrm{Hom}(\xi,\xi))$ *of a Riemannian-connected vector bundle* ξ *is skew-adjoint, i.e.*

$$ {}^tR(X,Y) = -R(X,Y) \ , \quad X,Y \in V(M) \ , \tag{4.19}$$

where ξ^* *is identified with* ξ *by the fibre metric of* ξ .

Proof. For $u,v \in C^\infty(\xi)$, orthogonality of ∇ yields

$$(\nabla_X \nabla_Y u, v) = X(\nabla_Y u, v) - (\nabla_Y u \ , \nabla_X v) =$$

$$= XY(u,v) - X(u \ , \nabla_Y v) - Y(u, \nabla_X v) + (u, \nabla_Y \nabla_X v) \ .$$

Changing the role of X and Y , we have

$$((\nabla_X \nabla_Y - \nabla_Y \nabla_X)u,v) = [X,Y](u,v) + (u,(\nabla_Y \nabla_X - \nabla_X \nabla_Y)v) =$$

$$= (\nabla_{[X,Y]}u,v) + (u,R(Y,X)v) \ ,$$

or equivalently,

$$(R(X,Y)u,v) = -(u,R(X,Y)v) \ . \ ✓$$

The considerations above will now be specified to the case $\xi = T(M)$, where M is a Riemannian manifold with metric tensor $g = (,)$. The existence of a covariant differentiation on the Euclidean vector bundle $T(M)$ is provided by the Fundamental Theorem of Riemannian geometry.

Theorem 4.20. *Let* M *be a Riemannian manifold. Then there exists a unique orthogonal covariant differentiation*

$$\nabla \in \mathrm{Diff}_1(T(M), \ \mathrm{Hom}(T(M), \ T(M)))$$

satisfying

$$\nabla_X Y - \nabla_Y X = [X,Y] \ , \quad X,Y \in V(M) \ . \tag{4.21}$$

Proof. For uniqueness, we prove that, for $X,Y,Z \in V(M)$, the scalar

$$(\nabla_X Y, Z) \in C^\infty(M)$$

is determined by orthogonality and (4.21). Indeed, we have

$$X(Y,Z) = (\nabla_X Y, Z) + (Y, \nabla_X Z) \ ,$$

$$Y(X,Z) = (\nabla_Y X, Z) + (X, \nabla_Y Z) \ ,$$

$$Z(X,Y) = (\nabla_Z X, Y) + (X, \nabla_Z Y) \ .$$

Adding the first two of these equations and substracting the third, by making use of (4.21), we obtain

$$2(\nabla_X Y, Z) = X(Y,Z) + Y(X,Z) - Z(X,Y) +$$

$$+ ([X,Y],Z) - ([X,Z],Y) - ([Y,Z],X) \ .$$

Hence if ∇ exists it is unique. Conversely, we define $\nabla_X Y$ by this formula. Then, it is a straightforward computation to check that ∇ is an orthogonal covariant differentiation satisfying (4.21). ✓

Remark 1. In general, a covariant differentiation ∇ on the tangent bundle of a manifold is said to be *torsionfree* if it satisfies (4.21). The covariant differentiation given in Theorem 4.20 is said to be the *Levi-Civita covariant differentiation* of the Riemannian manifold M .

Remark 2. In terms of a local coordinate neighbourhood (U,ϕ) with associated base fields $\{\partial_i\}_{i=1}^m \subset V(U)$ the Levi-Civita covariant differentiation ∇ is uniquely determined by the *Christoffel symbols* $\Gamma_{ij}^k \in C^\infty(U)$, $i,j,k=1,\ldots,m$, defined by

$$\nabla_{\partial_i}\partial_j = \sum_{k=1}^{m} \Gamma_{ji}^{k}\partial_k , \qquad i,j=1,\dots,m . \tag{4.22}$$

Since $[\partial_i,\partial_j]=0$, $i,j=1,\dots,m$, condition (4.21) translates into

$$\Gamma_{ij}^{k} = \Gamma_{ji}^{k} , \quad i,j,k=1,\dots,m . \tag{4.23}$$

A curve $c:I \to M$ on a Riemannian manifold M is said to be a *geodesic* if the velocity vector field $t \to \dot{c}(t)$, $t\in I$, is parallel along c . Given $X_x\in T_x(M)$, $x\in M$, there exists a unique maximal geodesic $c:I \to M$ satisfying the initial condition $\dot{c}(0)=X_x$. (For the results used here without proofs we refer to Kobayashi and Nomizu (1963) pages 138-140.) Further, for fixed $x\in M$, there exists $\varepsilon>0$ such that, for $X_x\in T_x(M)$ with $\| X_x\| < \varepsilon$, the geodesic $c:[0,1] \to M$ with initial condition $\dot{c}(0)=X_x$ is defined. Setting $\exp(X_x)=c(1)$, we obtain the *exponential map*

$$\exp:B(0,\varepsilon) \to M$$

defined on the open ball $B(0,\varepsilon)\subset T_x(M)$ around 0 with radius $\varepsilon>0$. (Note that is case when M is *complete*, i.e. every maximal geodesic is defined on $\mathbf{R}$, the domain of exp is the whole tangent space $T_x(M)$. In particular, this is the case for M compact.) As the differential of exp at $0\in T_x(M)$ is the identity, by choosing $\varepsilon>0$ small enough, $\exp:B(0,\varepsilon) \to M$ becomes a diffeomorphism of $B(0,\varepsilon)$ onto an open neighbourhood U of x in M . Fixing an orthonormal base $\{e^i\}_{i=1}^m\subset T_x(M)$, and thus preassigning an isometry $\Phi:T_x(M) \to \mathbf{R}^m$, we obtain a local coordinate neighbourhood $(U,\Phi\circ\exp^{-1})$ called *normal coordinate neighbourhood centered at* $x\in M$. For $i=1,\dots,m$, and $y\in U$, put $E_y^i=\tau(y)e^i$, where $\tau(y):T_x(M) \to T_y(M)$ is the parallel translation along the unique geodesic segment in U joining x and y . By 4.17, $\{E^i\}_{i=1}^m\subset V(U)$ is a pointwise orthonormal system, i.e.

$$(E^i,E^j) = \delta_{ij} , \quad i,j=1,\dots,m ,$$

and, furthermore,

$$\nabla_{e^i} E^j = (\nabla_{E^i} E^j)(x) = 0 \,, \qquad i,j=1,\ldots,m \,.$$

Then, $\{E^i\}_{i=1}^m \subset V(U)$ is said to be the *orthonormal moving frame on* U *adapted to the base* $\{e^i\}_{i=1}^m$.

Example 4.24. Let $M=V$ be an m-dimensional vector space. Denoting by $\check{}:T(V) \to V$ the canonical identification given by translating tangent vectors on V to the origin of V , any vector field $Y \in V(V)$ induces a vector function $\check{Y}:V \to V$ by setting $\check{Y}(x)=\check{Y}_x$, $x \in V$. Now, for $X_x \in T_x(V)$, $x \in V$, and $Y \in V(V)$, define $\nabla_{X_x} Y$ to be the tangent vector at x such that

$$(\nabla_{X_x} Y)\check{} = X_x(\check{Y}) \,. \qquad (4.25)$$

Clearly, $\nabla \in \mathrm{Diff}_1(T(V),\mathrm{Hom}(T(V),T(V)))$ is a covariant differentiation on $T(V)$. We claim that ∇ is torsionfree. Indeed, let $\{\partial_i\}_{i=1}^m \subset V(V)$ denote the base fields associated to a linear isomorphism $V \cong \mathbf{R}^m$. Then, for $Y,Z \in V(V)$, with $Y = \sum_{i=1}^m Y^i \partial_i$ and $Z = \sum_{i=1}^m Z^i \partial_i$, by (4.25), we have

$$[Y,Z]\check{} = \sum_{i=1}^m Y(Z^i)\check{\partial}_i - \sum_{i=1}^m Z(Y^i)\check{\partial}_i =$$

$$= Y(\check{Z}) - Z(\check{Y}) = (\nabla_Y Z - \nabla_Z Y)\check{}$$

since $[\partial_i,\partial_j]=0$, $i,j=1,\ldots,m$, and $\check{\partial}_i:V \to V$, $i=1,\ldots,m$, is constant. Furthermore, the curvature R of $T(V)$ vanishes identically, since, by (4.25),

$$(R(X,Y)Z)\check{} = (\nabla_X \nabla_Y Z)\check{} - (\nabla_Y \nabla_X Z)\check{} - (\nabla_{[X,Y]} Z)\check{} =$$

$$= X(Y(\check{Z})) - Y(X(\check{Z})) - [X,Y](\check{Z}) = 0 \,, \quad X,Y,Z \in V(V) \,.$$

Finally, a scalar product $(,)$ of V induces a Riemannian metric $(,)$ on V by setting $(Y_x,Z_x) = (\check{Y}_x,\check{Z}_x)$, $Y_x,Z_x \in T_x(V)$, $x \in V$.

Then, for $X,Y,Z\in V(V)$, by (4.25), we have

$$X_x(Y,Z) = X_x(\check{Y},\check{Z}) = (X_x(\check{Y}),\check{Z}_x) + (\check{Y}_x,X_x(\check{Z})) =$$

$$= ((\nabla_{X_x}Y)\check{},\check{Z}_x) + (\check{Y}_x,(\nabla_{X_x}Z)\check{}) = (\nabla_{X_x}Y,Z_x) + (Y_x,\nabla_{X_x}Z) ,$$

or equivalently, ∇ is orthogonal. Thus, ∇ is the Levi-Civita covariant differentiation on $T(V)$ with respect to the Riemannian metric (,) (with vanishing curvature and Christoffel symbols). V endowed with ∇ and (,) is said to be *flat*. (Clearly, different choices of the scalar product on V give rise to isometric Riemannian manifolds.)

The Levi-Civita covariant differentiation ∇ on the Riemannian manifold M extends to the full tensor bundle $\bigoplus_{r,s=0}^{\infty} T_r^s(M)$ as a derivation (and still called the Levi-Civita covariant differentiation on $\bigoplus_{r,s=0}^{\infty} T_r^s(M)$). (On $T_0^0(M)=\varepsilon_M^1$ we take $\nabla=d$ (cf. Example 4.14). Note that the condition of orthogonality (4.13) can then be written in the more concise form $\nabla g=0$, where $g==(,)\in C^\infty(S^2T^*(M))$ denotes the metric tensor of M .) In particular, for each $X\in V(M)$, the covariant differentiation ∇_X in the direction X acts on $\Lambda^r(T^*(M))$, $r=0,\ldots,m$, via the formula

$$(\nabla_X\alpha)(X^1,\ldots,X^r) = X(\alpha(X^1,\ldots,X^r)) - \sum_{i=1}^{r} \alpha(X^1,\ldots,\nabla_X X^i,\ldots,X^r) , \tag{4.26}$$

where $\alpha\in D^r(M)$, and $X^1,\ldots,X^r\in V(M)$. In particular, using the canonical isomorphism $\gamma:T(M)\to T^*(M)$, (4.13) reformulates as

$$\nabla_X\gamma(Y) = \gamma(\nabla_X Y) , \quad X,Y\in V(M) .$$

The metric tensor $g=(,)\in C^\infty(S^2T^*(M))$ extends to a fibre metric (,) of the vector bundle $\Lambda(T^*(M)) = \bigoplus_{r=0}^{m}\Lambda^r(T^*(M))$. Namely, given an orthonormal base $\{e^i\}_{i=1}^m\subset T_x(M)$, $x\in M$, the system

$$\{\gamma(e^{i_1})\wedge\ldots\wedge\gamma(e^{i_r})\,|\,1\le i_1<\ldots<i_r\le m\ ;\ r=1,\ldots,m\}\subset\Lambda(T_x^*(M))$$

is declared to be orthonormal and this determines the fibre metric $(,)$ on $\Lambda(T_x^*(M))$, uniquely. Thus $\Lambda(T^*(M))$ becomes a Euclidean vector bundle. To illustrate the use of local orthonormal moving frames, we claim that $\Lambda(T^*(M))$ is Riemannian-connected with respect to the Levi-Civita covariant differentiation ∇. Indeed, let (U,ϕ) be a normal coordinate neighbourhood centered at $x\in M$. Given an orthonormal base $\{e^i\}_{i=1}^m\subset T_x(M)$, denote by $\{E^i\}_{i=1}^m\subset V(U)$ the adapted orthonormal moving frame. Then, for $\alpha,\beta\in\mathcal{D}^r(M)$ and $i=1,\ldots,m$, we have

$$e^i(\alpha,\beta) = e^i\{\sum_{i_1<\ldots<i_r}\alpha(E^{i_1},\ldots,E^{i_r})\beta(E^{i_1},\ldots,E^{i_r})\} =$$

$$= \sum_{i_1<\ldots<i_r}\{e^i(\alpha(E^{i_1},\ldots,E^{i_r}))\}\beta(e^{i_1},\ldots,e^{i_r}) +$$

$$+ \sum_{i_1<\ldots<i_r}\alpha(e^{i_1},\ldots,e^{i_r})\{e^i(\beta(E^{i_1},\ldots,E^{i_r}))\} =$$

$$= \sum_{i_1<\ldots<i_r}(\nabla_{e^i}\alpha)(e^{i_1},\ldots,e^{i_r})\beta(e^{i_1},\ldots,e^{i_r}) +$$

$$+ \sum_{i_1<\ldots<i_r}\alpha(e^{i_1},\ldots,e^{i_r})(\nabla_{e^i}\beta)(e^{i_1},\ldots,e^{i_r}) =$$

$$= (\nabla_{e^i}\alpha,\beta_x) + (\alpha_x,\nabla_{e^i}\beta)\ ,$$

where in the last but one equality we used (4.26). Thus the claim follows, i.e. the vector bundle $\Lambda(T^*(M))$ is Riemannian-connected. In particular, by restriction, $T^*(M)$ and hence, by duality, $T(M)$ are Riemannian-connected vector bundles.

The Levi-Civita covariant differentiation ∇ on a Riemannian manifold M with metric tensor $g=(,)$ defines, by (4.9), the curvature $R\in C^\infty(\Lambda^2(T^*(M))\otimes\mathrm{Hom}(T(M),T(M)))$ of the bundle $T(M)$ which, in the present situation, is viewed as a $C^\infty(M)$-trilinear map

$$R: V(M)\times V(M)\times V(M)\to V(M)$$

(called the *Riemannian curvature tensor* of M). By definition, R is skew-symmetric in the first two arguments, i.e.

$$R(X,Y) = -R(Y,X) , \quad X,Y\in V(M) . \tag{4.27}$$

Furthermore, by (4.21), expressing the Lie brackets in the Jacobi identity in terms of the covariant differentiation, for $X,Y,Z\in V(M)$, we have

$$0 = [X,[Y,Z]] + [Z,[X,Y]] + [Y,[Z,X]] =$$

$$= \nabla_X[Y,Z] - \nabla_{[Y,Z]}X + \nabla_Z[X,Y] - \nabla_{[X,Y]}Z +$$

$$+ \nabla_Y[Z,X] - \nabla_{[Z,X]}Y = \{\nabla_X\nabla_Y Z-\nabla_Y\nabla_X Z-\nabla_{[X,Y]}Z\} +$$

$$+ \{\nabla_Z\nabla_X Y-\nabla_X\nabla_Z Y-\nabla_{[Z,X]}Y\} + \{\nabla_Y\nabla_Z X-\nabla_Z\nabla_Y X-\nabla_{[Y,Z]}X\}$$

and thus the Bianchi's identity

$$R(X,Y)Z + R(Z,X)Y + R(Y,Z)X = 0 , \quad X,Y,Z\in V(M), \tag{4.28}$$

follows. Summarizing (4.19), (4.27) and (4.28), for $X,Y,Z,W\in$ $\in V(M)$, we have the identities

$$(R(X,Y)Z,W) = -(R(X,Y)W,Z) = -(R(Y,X)Z,W)$$

and

$$(R(X,Y)Z,W) + (R(Z,X)Y,W) + (R(Y,Z)X,W) = 0 .$$

It follows, by standard multilinear algebra, that the identity

$$(R(X,Y)Z,W) = (R(Z,W)X,Y), \quad X,Y,Z,W\in V(M) , \tag{4.29}$$

is also valid (cf. Kobayashi and Nomizu (1963) page 198, or Helgason (1978) page 68).

For linearly independent tangent vectors $X_x,Y_x\in T_x(M)$, $x\in M$, the value

$$\kappa(X_x,Y_x) = \frac{(R(X_x,Y_x)Y_x,X_x)}{\|X_x\|^2\cdot\|Y_x\|^2-(X_x,Y_x)^2}$$

is easily seen to depend only on the 2-plane $\pi \subset T_x(M)$ spanned by the vectors X_x and Y_x. Then, $\kappa(\pi)=\kappa(X_x,Y_x)$ is said to be the *sectional curvature* of M for the 2-plane π.

The *Ricci tensor field* of a Riemannian manifold M with metric tensor $g=(,)$ is a tensor field $\mathrm{Ric} \in T_2^0(M)$ given as

$$\mathrm{Ric}(X_x,Y_x) = \{\text{trace of the linear endomorphism}$$

$$Z_x \to R(Z_x,X_x)Y_x \quad \text{of} \quad T_x(M)\} ,$$

where $X_x,Y_x \in T_x(M)$, $x \in M$.

In terms of an orthonormal base $\{e^i\}_{i=1}^m \subset T_x(M)$, we have

$$\mathrm{Ric}(X_x,Y_x) = \sum_{i=1}^m (R(e^i,X_x)Y_x,e^i), \quad X_x,Y_x \in T_x(M) .$$

Hence, using the curvature identities above,

$$\mathrm{Ric}(X_x,Y_x) = \sum_{i=1}^m (R(e^i,X_x)Y_x,e^i) =$$

$$= \sum_{i=1}^m (R(X_x,e^i)e^i,Y_x) = \sum_{i=1}^m (R(e^i,Y_x)X_x,e^i) =$$

$$= \mathrm{Ric}(Y_x,X_x) , \quad X_x,Y_x \in T_x(M) ,$$

i.e. the Ricci tensor field of M is symmetric and can then be considered as a section $\mathrm{Ric} \in C^\infty(S^2(T^*(M)))$. Equivalently, at any point $x \in M$, it is a symmetric linear endomorphism

$$\mathrm{Ric}: T_x(M) \to T_x(M), \quad ({}^t\mathrm{Ric}=\mathrm{Ric}) ,$$

where, as usual, $T^*(M)$ is identified with $T(M)$ via γ. For $X_x \in T_x(M)$, the tangent vector $\mathrm{Ric}(X_x) \in T_x(M)$ is said to be the *Ricci curvature of* M *in the direction* X_x. With respect to an orthonormal base $\{e^i\}_{i=1}^m \subset T_x(M)$,

$$\mathrm{Ric}(X_x) = \sum_{i=1}^m R(X_x,e^i)e^i , \quad X_x \in T_x(M) , \qquad (4.30)$$

since, for $Y_x \in T_x(M)$,

$$(\mathrm{Ric}(X_x),Y_x) = \sum_{i=1}^{m} (R(X_x,e^i)e^i,Y_x) =$$

$$= \sum_{i=1}^{m} (R(e^i,X_x)Y_x,e^i) = \mathrm{Ric}(X_x,Y_x) \ .$$

5. THE EXTERIOR CODIFFERENTIATION

Let M be an m-dimensional Riemannian manifold and consider the Riemannian-connected vector bundle $\Lambda(T^*(M))$ over M with Levi-Civita covariant differentiation ∇. We introduce the *exterior codifferentiation* ∂ *on* $\Lambda(T^*(M))$ as a differential operator

$$\partial \in \mathrm{Diff}_1(\Lambda^{r+1}(T^*(M)),\ \Lambda^r(T^*(M))) \ , \quad r=0,\ldots,m-1 \ ,$$

defined by

$$(\partial\alpha)_x = -\sum_{i=1}^{m} (\iota_{e^i}\circ\nabla_{e^i})\alpha \ , \quad \alpha\in\mathcal{D}^{r+1}(M) \ , \quad x\in M \ , \qquad (5.1)$$

where $\{e^i\}_{i=1}^m \subset T_x(M)$ is an orthonormal base. (On $\mathcal{D}^0(M)=C^\infty(M)$, we set $\partial=0$.) Equivalently,

$$(\partial\alpha)(X_x^1,\ldots,X_x^r) = -\sum_{i=1}^{m} (\nabla_{e^i}\alpha)(e^i,X_x^1,\ldots,X_x^r) \ ,$$

$$X_x^1,\ldots,X_x^r \in T_x(M) \ .$$

(Clearly, the definition of ∂ does not depend on the particular choice of the base.) Its symbol

$$\sigma_1(\partial)(\alpha_x):\Lambda^{r+1}(T_x^*(M)) \to \Lambda^r(T_x^*(M)) \ , \quad \alpha_x\in T_x'(M) \ , \quad x\in M \ ,$$

is the interior product $-\iota_{\gamma^{-1}(\alpha_x)}$. Indeed, choosing a scalar $\mu\in C^\infty(M)$ with $\mu(x)=0$ and $(d\mu)_x=\alpha_x$, for $\beta\in\mathcal{D}(M)$, with respect to an orthonormal base $\{e^i\}_{i=1}^m \subset T_x(M)$, we have

$$\sigma_1(\partial)(\alpha_x)\beta_x = (\partial(\mu\beta))_x = -\sum_{i=1}^{m} (\iota_{e^i}\circ\nabla_{e^i})(\mu\beta) =$$

$$= -\sum_{i=1}^{m} e^i(\mu)\iota_{e^i}\beta_x = -\sum_{i=1}^{m} \alpha_x(e^i)\iota_{e^i}\beta_x = -\iota_{\gamma^{-1}(\alpha_x)}\beta_x ,$$

since $\sum_{i=1}^{m} \alpha_x(e^i)e^i = \sum_{i=1}^{m} (\gamma^{-1}(\alpha_x),e^i)e^i = \gamma^{-1}(\alpha_x)$.

The main result of this section is to show that, for oriented M , the exterior codifferentiation ∂ is nothing but the adjoint of $d\in\mathrm{Diff}_1(\Lambda(T^*(M)),\Lambda(T^*(M)))$. We begin with the following.

Lemma 5.2. *For each vector field X on M ,*

$$^t(\gamma(X)\wedge) = \iota_X ,$$

or equivalently,

$$(\gamma(X)\wedge\alpha,\beta) = (\alpha,\iota_X\beta) , \quad \alpha,\beta\in\mathcal{D}(M) . \tag{5.3}$$

Proof. We first evaluate the exterior product $\gamma(X)\wedge\alpha$, $\alpha\in\mathcal{D}^r(M)$, on an $(r+1)$-tuple of vector fields $(X^1,\ldots,X^{r+1})$, $X^j\in V(M)$, $j=1,\ldots,r+1$. By multilinear algebra, we have

$$(\gamma(X)\wedge\alpha)(X^1,\ldots,X^{r+1}) =$$
$$= \sum_{j=1}^{r+1} (-1)^{j+1}\gamma(X)(X^j)\cdot\alpha(X^1,\ldots,\hat{X}^j,\ldots,X^{r+1}) , \tag{5.4}$$

where the factor $(-1)^{j+1}$ comes from the sign of the permutation $(j,1,\ldots,\hat{j},\ldots,r+1)$. By orthogonality, we may set $\beta\in\mathcal{D}^{r+1}(M)$ since otherwise both sides of (5.3) reduce to zero. Now, for an orthonormal base $\{e^i\}_{i=1}^m\subset T_x(M)$ at $x\in M$, by (5.4), we have

$$(\gamma(X)\wedge\alpha,\beta)_x = \sum_{i_1<\ldots<i_{r+1}} (\gamma(X)\wedge\alpha)(e^{i_1},\ldots,e^{i_{r+1}}) .$$

$$\cdot\ \beta(e^{i_1},\ldots,e^{i_{r+1}}) = \sum_{i_1<\ldots<i_{r+1}} \sum_{j=1}^{r+1} (-1)^{j+1} .$$

$$\cdot\ (X_x, e^{i_j})\alpha(e^{i_1},\ldots,(e^{i_j})\hat{},\ldots,e^{i_{r+1}})\beta(e^{i_1},\ldots,e^{i_{r+1}}) =$$

$$= \sum_{i_1<\ldots<i_{r+1}} \sum_{j=1}^{r+1} \alpha(e^{i_1},\ldots,(e^{i_j})\hat{},\ldots,e^{i_{r+1}})\cdot$$

$$\cdot\beta((X_x, e^{i_j})e^{i_j}, e^{i_1},\ldots,(e^{i_j})\hat{},\ldots,e^{i_{r+1}}) =$$

$$= \sum_{i_1<\ldots<i_{r+1}} \sum_{j=1}^{r+1} \alpha(e^{i_1},\ldots,(e^{i_j})\hat{},\ldots,e^{i_{r+1}})\cdot$$

$$\cdot(\iota_{(X_x, e^{i_j})e^{i_j}}\beta)(e^{i_1},\ldots,(e^{i_j})\hat{},\ldots,e^{i_{r+1}}) =$$

$$= \sum_{i_1<\ldots<\hat{i}_j<\ldots<i_{r+1}} \alpha(e^{i_1},\ldots,(e^{i_j})\hat{},\ldots,e^{i_{r+1}})\cdot$$

$$\cdot(\iota_X\beta)(e^{i_1},\ldots,(e^{i_j})\hat{},\ldots,e^{i_{r+1}}) = (\alpha,\iota_X\beta)\ .\ \surd$$

Theorem 5.5. *For an oriented Riemannian manifold* M ,

$$\partial = d^*\ .$$

Proof. We first show that, with respect to a local orthonormal moving frame $\{E^i\}_{i=1}^m$, the exterior differential of an r-form α on M can be expressed as

$$d\alpha = \sum_{i=1}^{m} \gamma(E^i)\wedge(\nabla_{E^i}\alpha)\ , \tag{5.6}$$

where $\gamma:T(M)\to T^*(M)$ is the canonical isomorphism induced by the metric tensor $g=(,)\in C^\infty(S^2(T^*(M)))$. Indeed, by making use of (4.26) and (5.4), for $X^1,\ldots,X^{r+1}\in V(M)$, we have

$$\sum_{i=1}^{m} (\gamma(E^i)\wedge(\nabla_{E^i}\alpha))(X^1,\ldots,X^{r+1}) =$$

$$= \sum_{i=1}^{m} \sum_{j=1}^{r+1} (-1)^{j+1} (\gamma(E^i)X^j)(\nabla_{E^i}\alpha)(X^1,\dots,\hat{X}^j,\dots,X^{r+1}) =$$

$$= \sum_{j=1}^{r+1} (-1)^{j+1} (\nabla_{X^j}\alpha)(X^1,\dots,\hat{X}^j,\dots,X^{r+1}) =$$

$$= \sum_{j=1}^{r+1} (-1)^{j+1} X_j(\alpha(X^1,\dots,\hat{X}^j,\dots,X^{r+1})) +$$

$$+ \sum_{j=1}^{r+1} (-1)^j \{ \sum_{i=1}^{j-1} \alpha(X^1,\dots,\nabla_{X^j}X^i,\dots,\hat{X}^j,\dots,X^{r+1}) +$$

$$+ \sum_{i=j}^{r} \alpha(X^1,\dots,\hat{X}^j,\dots,\nabla_{X^j}X^{i+1},\dots,X^{r+1})\} .$$

The last term can be rewritten as

$$\sum_{i<j} (-1)^j \alpha(X^1,\dots,\nabla_{X^j}X^i,\dots,\hat{X}^j,\dots,X^{r+1}) +$$

$$+ \sum_{j<i} (-1)^j \alpha(X^1,\dots,\hat{X}^j,\dots,\nabla_{X^j}X^i,\dots,X^{r+1}) =$$

$$= \sum_{i<j} (-1)^{i+j+1} \alpha(\nabla_{X^j}X^i, X^1,\dots,\hat{X}^i,\dots,\hat{X}^j,\dots,X^{r+1}) +$$

$$+ \sum_{j<i} (-1)^{i+j} \alpha(\nabla_{X^j}X^i, X^1,\dots,\hat{X}^j,\dots,\hat{X}^i,\dots X^{r+1}) =$$

$$= \sum_{i<j} (-1)^{i+j} \alpha(\nabla_{X^i}X^j - \nabla_{X^j}X^i, X^1,\dots,\hat{X}^i,\dots,\hat{X}^j,\dots X^{r+1}) .$$

Since the Levi-Civita covariant differentiation ∇ on $T(M)$ is torsionfree, by (3.2), formula (5.6) follows. Turning to the proof of the equality $\partial = d^*$, let $\alpha \in \mathcal{D}^r(M)$ and $\beta \in \mathcal{D}^{r+1}(M)$ such that $\operatorname{supp}\alpha \cap \operatorname{supp}\beta \subset M$ is compact. For $x \in M$, choose a local orthonormal moving frame $\{E^i\}_{i=1}^m$ (adapted to an orthonormal base $\{e^i\}_{i=1}^m \subset T_x(M)$) on a normal coordinate neighbourhood centered at x . Then, (5.1), (5.3) and orthogonality of ∇ imply

$$(\partial\beta,\alpha) = -\sum_{i=1}^{m}(\iota_{E^i}(\nabla_{E^i}\beta),\alpha) = -\sum_{i=1}^{m}(\nabla_{E^i}\beta,\gamma(E^i)\wedge\alpha) =$$

$$= -\sum_{i=1}^{m} E^i(\beta,\gamma(E^i)\wedge\alpha) + (\beta, \sum_{i=1}^{m}\nabla_{E^i}(\gamma(E^i)\wedge\alpha)) .$$

To rewrite the first sum on the right hand side, define a vector field X on M by

$$(X,Y) = (\beta,\gamma(Y)\wedge\alpha) , \quad Y\in V(M) .$$

Then, the *divergence* of X at $x\in M$, i.e. the trace of the linear endomorphism $Z_x \to \nabla_{Z_x} X$ of $T_x(M)$ (cf. Kobayashi and Nomizu (1963) page 282) is computed as

$$(\text{div } X)(x) = \sum_{i=1}^{m}(\nabla_{e^i}X,e^i) = \sum_{i=1}^{m} e^i(X,E^i) =$$

$$= \sum_{i=1}^{m} e^i(\beta,\gamma(E^i)\wedge\alpha)$$

and hence $-(\text{div } X)(x)$ equals to the first sum above at $x\in M$. Furthermore, by (5.6),

$$\sum_{i=1}^{m}\nabla_{e^i}(\gamma(E^i)\wedge\alpha) = \sum_{i=1}^{m}\gamma(e^i)\wedge\nabla_{e^i}\alpha = (d\alpha)_x$$

since, by orthogonality, $\nabla_{e^i}(\gamma(E^i)) = \gamma(\nabla_{e^i}E^i) = 0$, $i=1,\ldots,m$. Summarizing, we obtain

$$(\partial\beta,\alpha) = -\text{div } X + (\beta,d\alpha) .$$

Integrating and using Green's theorem (cf. Kobayashi and Nomizu (1963) pages 281-283), it follows that

$$((\partial\beta,\alpha)) = ((\beta,d\alpha)) . \quad \checkmark$$

Remark. For another proof of 5.5 based on classical local

calculus we refer to de Rham (1955) page 129, or Lichnerowicz (1955) page 183.

6. THE LAPLACIAN

Let M be an m-dimensional Riemannian manifold with metric tensor $g=(,)\in C^\infty(S^2(T^*(M)))$. We introduce the *Laplacian* $\Delta\in \in \mathrm{Diff}_2(\Lambda(T^*(M)),\Lambda(T^*(M)))$ as

$$\Delta = d\circ\partial + \partial\circ d\,. \tag{6.1}$$

From 1.7, the value of the symbol $\sigma_2(\Delta)(\alpha_x):\Lambda(T^*_x(M))\to\Lambda(T^*_x(M))$, $\alpha_x\in T'_x(M)$, $x\in M$, on a covector $\beta_x\in\Lambda(T^*_x(M))$ is computed as

$$\sigma_2(\Delta)(\alpha_x)\beta_x = \sigma_1(d)(\alpha_x)\{\sigma_1(\partial)(\alpha_x)\beta_x\} +$$

$$+ \sigma_1(\partial)(\alpha_x)\{\sigma_1(d)(\alpha_x)\beta_x\} = -\sigma_1(d)(\alpha_x)\{\iota_{\gamma^{-1}(\alpha_x)}\beta_x\} +$$

$$+ \sigma_1(\partial)(\alpha_x)\{\alpha_x\wedge\beta_x\} = -\alpha_x\wedge\{\iota_{\gamma^{-1}(\alpha_x)}\beta_x\} -$$

$$- \iota_{\gamma^{-1}(\alpha_x)}\{\alpha_x\wedge\beta_x\} = -\{\iota_{\gamma^{-1}(\alpha_x)}\alpha_x\}\beta_x = -\|\alpha_x\|^2\beta_x\,,$$

where we used the fact that the interior product is a skew-derivation of degree -1. Thus $\sigma_2(\Delta)(\alpha_x)$ is multiplication with $-\|\alpha_x\|^2$, in particular, the Laplacian Δ is elliptic.

An r-form α on M is said to be *harmonic* if $\Delta\alpha=0$. The vector space of all harmonic r-forms is denoted by $H^r(M)$ and we set $H(M) = \bigoplus_{r=0}^{m} H^r(M)$. By (5.1) and (5.6), the vector space

$$P^r(M) = \{\alpha\in D^r(M)\,|\,\nabla\alpha=0\}$$

of *parallel r-forms* on M is a linear subspace of $H^r(M)$, $r=0,\dots,m$. We put $P(M) = \bigoplus_{r=0}^{m} P^r(M)\subset H(M)$. Finally, we set $P_1(M) = \{X\in V(M)\,|\,\nabla X=0\}$; elements of $P_1(M)$ are then said to be *parallel vector fields*. By duality, $P_1(M)\cong P^1(M)$.

A powerful tool in studying the geometric properties of the Laplacian is the following Bochner-Weitzenböck formula.

Theorem 6.2. *For any* $\alpha \in \mathcal{D}^1(M)$,

$$\Delta\alpha = -\text{trace } \nabla^2\alpha + \text{Ric}(\alpha) , \tag{6.3}$$

where trace $\nabla^2\alpha$ *is the (pointwise) trace of the* $C^\infty(M)$*-bilinear map*

$$\nabla^2\alpha : \mathcal{V}(M)\times\mathcal{V}(M) \to \mathcal{D}^1(M)$$

defined by

$$(\nabla^2\alpha)(X,Y) = \nabla_X\nabla_Y\alpha - \nabla_{\nabla_X Y}\alpha , \quad X,Y\in\mathcal{V}(M)$$

and the Ricci endomorphism Ric *is considered to act on* $T^*(M)$ *by duality, i.e., for* $X\in\mathcal{V}(M)$,

$$(\text{Ric}(\alpha))(X) = \alpha(\text{Ric}(X)) .$$

The proof of 6.2 will be postponed to Section 1 of Chapter II, where we verify the Bochner-Weitzenböck formula in a more general setting. (See also Lichnerowicz (1958) pages 1-2.)

Theorem 6.2 yields an explicit expression for the Laplacian of the scalar $\|\alpha\|^2$, $\alpha\in\mathcal{D}^1(M)$, as follows.

Corollary 6.4. *For any* 1-*form* α *on* M ,

$$(\Delta\alpha,\alpha) = \frac{1}{2}\Delta(\|\alpha\|^2) + \|\nabla\alpha\|^2 + (\text{Ric}(\alpha),\alpha) , \tag{6.5}$$

where, with respect to an orthonormal base $\{e^i\}_{i=1}^m \subset T_x(M)$ *at* $x\in M$,

$$\|\nabla\alpha\|^2(x) = \sum_{i,j=1}^m \|(\nabla_{e^i}\alpha)(e^j)\|^2$$

and

$$(\text{Ric}(\alpha),\alpha)_x = \sum_{j=1}^m (\alpha(\text{Ric}(e^j)),\alpha(e^j)) . \tag{6.6}$$

Proof. The Bochner-Weitzenböck formula (6.3) gives

$$(\Delta\alpha,\alpha) = -(\text{trace } \nabla^2\alpha,\alpha) + (\text{Ric}(\alpha),\alpha) .$$

Now, (6.5) follows from the classical formula

$$-(\text{trace } \nabla^2\alpha,\alpha) = \|\nabla\alpha\|^2 + \frac{1}{2}\Delta(\|\alpha\|^2)$$

which can be verified by a straightforward local computation (cf. Lichnerowicz (1958) page 3). ✓

We illustrate the use of the Bochner-Weitzenböck formula in deriving the following classical result of Bochner.

Theorem 6.7. *Let M be a compact oriented Riemannian manifold. If the Ricci tensor field* Ric *as a symmetric linear endomorphism on the fibres of $T(M)$, is positive semidefinite everywhere on M then $H^1(M) = P^1(M)$, or equivalently, for every harmonic* 1-*form ω on M we have $\nabla\omega=0$. If, moreover,* Ric *is positive definite at a point of M then there is no nonzero harmonic* 1-*form on M.*

The proof is preceded by a lemma due to E. Hopf.

Lemma 6.8. *Let M be a compact Riemannian manifold. If $\mu \in C^\infty(M)$ such that $\Delta\mu \geq 0$ everywhere on M then μ is constant.*

Proof. Without loss of generality we may suppose that M is oriented by taking its oriented 2-sheeted covering if necessary. (Here we use the fact that the Laplacian acting on scalars commutes with local isometries. In fact, if $\{E^i\}_{i=1}^m$ is a local orthonormal moving frame (adapted to an orthonormal base $\{e^i\}_{i=1}^m \subset T_x(M)$ at $x \in M$) over a normal coordinate neighbourhood centered at x then, for $\mu \in C^\infty(M)$, we have

$$(\Delta\mu)(x) = (\partial(d\mu))(x) = -\sum_{i=1}^m (\nabla_{e^i}(d\mu))(e^i) =$$
$$= -\sum_{i=1}^m e^i((d\mu)E^i) = -\sum_{i=1}^m e^i(E^i(\mu)) , \tag{6.9}$$

and the properties of the moving frame $\{E^i\}_{i=1}^m$ being preserved

under local isometries, the condition $\Delta\mu \geq 0$ is also preserved when going up to the oriented covering.)
Now, for any scalar $\mu \in C^{\infty}(M)$, by 5.5, we have

$$\int_M \Delta\mu \operatorname{vol}(M,g) = ((\Delta\mu, 1)) = ((\partial(d\mu), 1)) = 0 . \tag{6.10}$$

Assuming $\Delta\mu \geq 0$, by (6.10), it follows that $\Delta\mu$ vanishes identically on M. Applying again (6.10) to the scalar $\frac{1}{2}\mu^2$, the elementary identity

$$\frac{1}{2}\Delta(\mu^2) = \mu \cdot \Delta\mu - \| d\mu \|^2$$

yields

$$0 = \frac{1}{2}\int_M \Delta(\mu^2) \operatorname{vol}(M,g) =$$

$$= \int_M \mu \cdot \Delta\mu \operatorname{vol}(M,g) - \int_M \| d\mu \|^2 \operatorname{vol}(M,g) .$$

As $\Delta\mu = 0$, it follows that $\| d\mu \|^2 = 0$ and hence μ is constant.✓

Proof of Theorem 6.7. By virtue of (6.5), a harmonic 1-form ω on M satisfies the equation

$$\frac{1}{2}\Delta(\|\omega\|^2) + \|\nabla\omega\|^2 + (\operatorname{Ric}(\omega), \omega) = 0 , \tag{6.11}$$

where, with respect to an orthonormal base $\{e^i\}_{i=1}^m \subset T_x(M)$ at $x \in M$,

$$(\operatorname{Ric}(\omega), \omega)_x = \sum_{j=1}^m (\omega(\operatorname{Ric}(e^j)), \omega(e^j)) .$$

Now, the Ricci curvature $\operatorname{Ric}(e^j)$ in the direction e^j decomposes as

$$\operatorname{Ric}(e^j) = \sum_{k=1}^m (\operatorname{Ric})_{jk} e^k ,$$

where, by assumption, the matrix $((\operatorname{Ric})_{jk})_{j,k=1}^m \in M(m,m)$ is positive semidefinite. In particular,

$$(\operatorname{Ric}(\omega), \omega)_x = \sum_{j,k=1}^m (\operatorname{Ric})_{jk} \omega(e^j) \omega(e^k) \geq 0$$

and hence, by (6.11),

$$\Delta(-\|\omega\|^2) = 2\|\nabla\omega\|^2 + 2(\mathrm{Ric}(\omega),\omega) \geq 0 .$$

Then, Hopf's lemma 6.8 applies yielding $\|\omega\|^2$=const. Again by (6.11),

$$\|\nabla\omega\|^2 = 0 \quad \text{and} \quad (\mathrm{Ric}(\omega),\omega) = 0$$

are satisfied everywhere on M. From the first relation we obtain that ω is parallel. Assuming the positive definiteness of the Ricci tensor field Ric at $x \in M$, by the second relation, $\omega_x = 0$. It remains only to show that the parallel 1-form ω vanishes identically on M. For an arbitrary point $y \in M$ choose a curve $c:[0,1] \to M$ with $c(0)=x$ and $c(1)=y$. Since ω is parallel on M, the 1-form $\omega \circ c$ is, by (4.6), a parallel section of $T^*(M)$ along c. As $\omega_{c(0)} = \omega_x = 0$, it follows that $\omega_y = \omega_{c(1)} = 0$. ✓

Now, let M be an m-dimensional *oriented* Riemannian manifold. By virtue of 5.5, for $\alpha,\beta \in \mathcal{D}^r(M)$, $r=0,\ldots,m$, with supp $\alpha \cap$ supp $\beta \subset M$ compact, we have

$$((\Delta\alpha,\beta)) = ((d(\partial\alpha),\beta)) + ((\partial(d\alpha),\beta)) =$$

$$= ((\partial\alpha,\partial\beta)) + ((d\alpha,d\beta)) = ((\alpha,\Delta\beta)) ,$$

i.e. the Laplacian Δ is self-adjoint and nonnegative ($((\Delta\alpha,\alpha)) \geq 0$, $\alpha \in \mathcal{D}^r(M)$). Also, an r-form α on M, with supp $\alpha \subset M$ compact, is harmonic if and only if it is closed and *coclosed*, i.e. $d\alpha=0$ and $\partial\alpha=0$ hold. Thus, for compact M, 1.18 can be specified yielding the classical finiteness and Hodge-de Rham decomposition.

Theorem 6.12. *Let M be an m-dimensional compact oriented Riemannian manifold. Then,* $\mathrm{Spec}(\Delta,M) \subset \mathbf{R}$ *is discrete, the vector space $H^r(M)$ of harmonic r-forms on M is finite dimensional for all $r=0,\ldots,m$. Further, we have the orthogonal decomposition*

$$\mathcal{D}^r(M) = d(\mathcal{D}^{r-1}(M)) \oplus \partial(\mathcal{D}^{r+1}(M)) \oplus H^r(M), \tag{6.13}$$

$$r=0,\ldots,m\ ,$$

or equivalently, for each $\alpha\in\mathcal{D}^r(M)$ *, there exists a unique harmonic* r*-form* ω *such that*

$$\alpha = du + \partial\lambda + \omega\ , \qquad (6.14)$$

for some $u\in\mathcal{D}^{r-1}(M)$ *and* $\lambda\in\mathcal{D}^{r+1}(M)$.

Proof. By definition, $\ker(\Delta|\mathcal{D}^r(M)) = H^r(M)$ and 1.18 applies yielding discreteness of $\mathrm{Spec}(\Delta,M)$ and finite dimensionality of $H^r(M)$. Clearly, the summands on the right hand side of (6.13) are mutually orthogonal, i.e. we need only to show that their direct sum actually spans $\mathcal{D}^r(M)$. But, using (1.19), we have

$$\mathcal{D}^r(M) = \Delta(\mathcal{D}^r(M)) \oplus H^r(M) \subset d(\partial\mathcal{D}^r(M)) \oplus$$

$$\oplus\ \partial(d\mathcal{D}^r(M)) \oplus H^r(M) \subset d(\mathcal{D}^{r-1}(M)) \oplus$$

$$\oplus\ \partial(\mathcal{D}^{r+1}(M)) \oplus H^r(M)\ .\ \checkmark$$

Remark. For M compact, $\mathcal{D}(M)$ is a prehilbert space with respect to the global scalar product (1.13). The space $H(M) = \bigoplus_{r=0}^{m} H^r(M)$ of all harmonic forms on M is, by 6.12, a finite dimensional linear subspace of $\mathcal{D}(M)$ and hence the orthogonal projection

$$H:\mathcal{D}(M) \to H(M) \qquad (6.15)$$

is a continuous operator. For $\alpha\in\mathcal{D}^r(M)$ the harmonic form $H\alpha\in H^r(M)$ (occurring as ω in (6.14)) is said to be the *harmonic part* of α .

The Hodge-de Rham decomposition plays a crucial role in computing the de Rham cohomology groups of an m-dimensional compact oriented Riemannian manifold M since, as the following result asserts, the determination of $H^r_{de\ R}(M)$, $r=0,\ldots,m$, can be reduced to solving the Laplace equation $\Delta\alpha=0$, for $\alpha\in\mathcal{D}^r(M)$.

Theorem 6.16. *Let M be an m-dimensional compact oriented Riemannian manifold. Then, every cohomology class in $H^r_{de\ R}(M;\mathbf{R})$, $r=0,\dots,m$, contains a unique harmonic r-form, in particular,*

$$H^r_{de\ R}(M;\mathbf{R}) \cong H^r(M) ,$$

as vector spaces over $\mathbf{R}$. In particular, all Betti numbers $b_r(M)$, $r=0,\dots,m$, are finite.

Proof. If $\alpha\in\mathcal{D}^r(M)$, $r=0,\dots,m$, is closed and has the Hodge-de Rham decomposition

$$\alpha = du + \partial\lambda + H\alpha , \tag{6.17}$$

then, applying d to both sides, we get $d\partial\lambda=0$. Hence $0 = ((d\partial\lambda,\lambda)) = ((\partial\lambda,\partial\lambda))$, i.e. we have $\partial\lambda=0$ and (6.17) reduces to

$$\alpha = du + H\alpha ,$$

in particular, the harmonic part $H\alpha$ is in the same cohomology class as α. Thus, every cohomology class in $H^r_{de\ R}(M;\mathbf{R})$ contains a harmonic representative. Finally, if $\omega,\omega'\in H^r(M)$ represent the same cohomology class in $H^r_{de\ R}(M;\mathbf{R})$ then

$$\omega - \omega' = du ,$$

for some $u\in\mathcal{D}^{r-1}(M)$. Applying the exterior codifferentiation ∂ to both sides, we obtain $\partial du=0$ and hence $0 = ((\partial du,u)) = ((du,du))$. It follows that $du=0$ and thus $\omega=\omega'$. ✓

7. THE SPECTRUM OF A RIEMANNIAN MANIFOLD

The Laplacian Δ acting on scalars of an m-dimensional compact oriented Riemannian manifold M is of particular interest and deserves special attention here because of later applications. In this case, the corresponding spectrum

$$\mathrm{Spec}(M) = \mathrm{Spec}(\Delta\,|\,C^\infty(M),M)\subset\mathbf{R}$$

is said to be the *spectrum of the Riemannian manifold* M . By 1.18 and the previous section, $\mathrm{Spec}(M)$ is a discrete series of nonnegative numbers and, for each eigenvalue $\lambda \in \mathrm{Spec}(M)$, the corresponding eigenspace $V_\lambda \subset C^\infty(M)$ is finite dimensional. Note also that, by 6.8, the eigenspace V_0 corresponding to the zero eigenvalue is 1-dimensional and consists of the constant functions on M .

For later reference, it will be convenient the express the Laplacian of a scalar $\mu \in C^\infty(M)$ in terms of geodesics of M . Namely, given an orthonormal base $\{e^i\}_{i=1}^m \subset T_x(M)$ at a point $x \in M$, for sufficiently small $\varepsilon > 0$, the geodesics $c^i : (-\varepsilon, \varepsilon) \to M$, $c^i(t) = \exp(te^i)$, $t \in (-\varepsilon, \varepsilon)$, are defined for all $i = 1, \ldots, m$. We claim that

$$(\Delta\mu)(x) = -\sum_{i=1}^m \frac{\partial^2(\mu \circ c^i)}{\partial t^2}(0) . \tag{7.1}$$

Indeed, let $\{E^i\}_{i=1}^m$ be the local orthonormal moving frame adapted to the base $\{e^i\}_{i=1}^m$ (over a normal coordinate neighbourhood of M centered at x). Then, by (6.9), we have

$$(\Delta\mu)(x) = -\sum_{i=1}^m e^i(E^i(\mu)) = -\sum_{i=1}^m \dot{c}^i(0)(E^i(\mu)) =$$

$$= -\sum_{i=1}^m \frac{\partial}{\partial t}\{(E^i(\mu))(c^i(t))\}_{t=0} =$$

$$= -\sum_{i=1}^m \frac{\partial}{\partial t}\{E^i_{c^i(t)}(\mu)\}_{t=0} = -\sum_{i=1}^m \frac{\partial^2(\mu \circ c^i)}{\partial t^2}(0)$$

and (7.1) follows.

Our subsequent applications are mainly concerned with the spectrum of the *Euclidean m-sphere* S^m (i.e. the Riemannian structure on S^m is induced from the Euclidean metric on $\mathbf{R}^{m+1}$ (cf. (4.24)) via the inclusion $S^m \subset \mathbf{R}^{m+1}$), so the rest of this section is devoted to determination of $\mathrm{Spec}(S^m)$ and, for each $\lambda \in \mathrm{Spec}(S^m)$, the eigenspace $V_\lambda \subset C^\infty(S^m)$. To simplify the notations, the restriction of a scalar $\mu \in C^\infty(\mathbf{R}^{m+1})$ to S^m will be denoted by $\tilde{\mu} \in C^\infty(S^m)$. In a similar vein, for any linear subspace

$V \subset C^\infty(\mathbf{R}^{m+1})$, we set

$$\tilde{V} = \{\tilde{\mu} \in C^\infty(S^m) \,|\, \mu \in V\} \ .$$

Our starting point is the following elementary lemma relating the Laplacians $\Delta^{\mathbf{R}^{m+1}}$ and Δ^{S^m}, where, as usual, the superscripts stand for the respective base manifolds.

Lemma 7.2. *For any scalar* $\mu \in C^\infty(\mathbf{R}^{m+1})$,

$$(\Delta^{\mathbf{R}^{m+1}} \mu)^\sim = \Delta^{S^m} \tilde{\mu} - (\frac{\partial^2 \mu}{\partial r^2})^\sim - m(\frac{\partial \mu}{\partial r})^\sim \ , \tag{7.3}$$

where $\frac{\partial}{\partial r}$ *and* $\frac{\partial^2}{\partial r^2}$ *denote the radial differentiations.*

Proof. By the canonical identification $\check{}: T(\mathbf{R}^{m+1}) \to \mathbf{R}^{m+1}$, an orthonormal base $\{e^i\}_{i=1}^m \subset T_x(S^m)$ at a point $x \in S^m$ gives rise to an orthonormal base $\{\check{e}^i\}_{i=0}^m \subset \mathbf{R}^{m+1}$, where $\check{e}^0 = x$. By virtue of (7.1), an elementary calculation yields

$$(\Delta^{\mathbf{R}^{m+1}} \mu)(x) = -\sum_{i=0}^{m} \{\frac{\partial^2 \mu(x+t\check{e}^i)}{\partial t^2}\}_{t=0} =$$

$$= -\frac{\partial^2 \mu}{\partial r^2}(x) - \sum_{i=1}^{m} \{\frac{\partial^2 \mu(x+t\check{e}^i)}{\partial t^2}\}_{t=0} =$$

$$= -\frac{\partial^2 \mu}{\partial r^2}(x) - m\frac{\partial \mu}{\partial r}(x) - \sum_{i=1}^{m} \{\frac{\partial^2 \tilde{\mu}(\cos t \cdot x + \sin t\check{e}^i)}{\partial t^2}\}_{t=0} =$$

$$= -\frac{\partial^2 \mu}{\partial r^2}(x) - m\frac{\partial \mu}{\partial r}(x) + (\Delta^{S^m} \tilde{\mu})(x) \ ,$$

where the last equality holds because, for each $i=1,\ldots,m$, $t \to \cos tx + \sin t\check{e}^i$, $t \in \mathbf{R}$, is the geodesic on S^m with tangent vector e^i at $t=0$. √

Let $P_m^k \subset C^\infty(\mathbf{R}^{m+1})$ denote the vector space of all homogeneous polynomials of degree $k \in \mathbf{Z}_+$ on $\mathbf{R}^{m+1}$. The global scalar product on $\tilde{P}_m^k \subset C^\infty(S^m)$ induces a scalar product $(,)$ on P_m^k (via the isomorphism $P_m^k \cong \tilde{P}_m^k$), i.e. we define

$$(P,Q) = \int_{S^m} \tilde{P}\cdot\tilde{Q}\ \mathrm{vol}(S^m), \quad P,Q\in P^k_m, \tag{7.4}$$

where $\mathrm{vol}(S^m)\in D^m(S^m)$ is the volume form of S^m. Setting

$$H^k_m = \{P\in P^k_m \mid \Delta^{\mathbf{R}^{m+1}} P=0\},$$

by (7.3), for $P\in H^k_m$, we have

$$\Delta^{S^m}\tilde{P} = \left(\frac{\partial^2 P}{\partial r^2}\right)^{\sim} + m\left(\frac{\partial P}{\partial r}\right)^{\sim} = k(k+m-1)\tilde{P},$$

or equivalently, $\tilde{P}\in C^\infty(S^m)$ is an eigenfunction of the Laplacian Δ^{S^m} with eigenvalue $\lambda_k=k(k+m-1)$. In other words, $\tilde{H}^k_m\subset V_{\lambda_k}$, where $\lambda_k=k(k+m-1)\in\mathrm{Spec}(S^m)$. Elements of $\tilde{H}^k_m$ are said to be *spherical harmonics of order* k *on* S^m.

Theorem 7.5. *The spectrum of the Euclidean m-sphere* S^m *is the set of all numbers* $\lambda_k=k(k+m-1)$, $k\in\mathbf{Z}_+$, *and, for each* $k\in\mathbf{Z}_+$, *the eigenspace* V_{λ_k} *equals* $\tilde{H}^k_m$.

The proof of 7.5 is preceded by the following.

Lemma 7.6. *For each* $k\in\mathbf{Z}_+$, *we have the orthogonal decomposition*

$$P^{k+2}_m = H^{k+2}_m \oplus r_o^2\cdot P^k_m, \tag{7.7}$$

where $r_o^2\in P^2_m$ *is given by*

$$r_o^2(x) = \sum_{i=1}^{m+1}(x^i)^2, \quad x=(x^1,\dots,x^{m+1})\in\mathbf{R}^{m+1}.$$

Proof. By elementary calculation, (7.7) is valid for $k=0$. We proceed by induction assuming that (7.7) holds for indices $<k$, in particular,

$$P^k_m = \bigoplus_{\ell=0}^{[\frac{k}{2}]} r_o^{2\ell}\cdot H^{k-2\ell}_m. \tag{7.8}$$

As noted above, $\tilde{H}^{k+2}_m\subset V_{\lambda_{k+2}}$ and, by (7.8), $\tilde{P}^k_m\subset \bigoplus_{\ell=0}^{[\frac{k}{2}]} V_{\lambda_{k-2\ell}}$.

Furthermore, Δ^{S^m} being self-adjoint, $V_{\lambda_{k+2}}$ and $\bigoplus_{\ell=0}^{[\frac{k}{2}]} V_{\lambda_{k-2\ell}}$ are orthogonal with respect to the global scalar product which implies that H_m^{k+2} and $r_0^2 \cdot P_m^k$ are orthogonal with respect to the scalar product (7.4). Thus, it remains to show that if $P \in P_m^{k+2}$ such that $(P, r_0^2 \cdot P_m^k) = 0$ then $\Delta^{\mathbf{R}^{m+1}} P = 0$. Evidently, $\Delta^{\mathbf{R}^{m+1}} P \in P_m^k$ and, by (7.8), to accomplish the proof, we have to verify that

$$\int_{S^m} (\Delta^{\mathbf{R}^{m+1}} P)^{\sim} \cdot \tilde{Q} \operatorname{vol}(S^m) = 0$$

holds for each $Q \in H_m^{k-2\ell}$, $\ell = 0, \ldots, [\frac{k}{2}]$.
Now, by making use of (7.3) and the orthogonality relation $(P, r_0^2 \cdot P_m^k) = 0$, we have

$$\int_{S^m} (\Delta^{\mathbf{R}^{m+1}} P)^{\sim} \cdot \tilde{Q} \operatorname{vol}(S^m) =$$

$$= \int_{S^m} \{\Delta^{S^m} \tilde{P} - (\frac{\partial^2 P}{\partial r^2})^{\sim} - m(\frac{\partial P}{\partial r})^{\sim}\} \tilde{Q} \operatorname{vol}(S^m) =$$

$$= \int_{S^m} \{\Delta^{S^m} \tilde{P} - (k+2)(m+k+1)\tilde{P}\} \tilde{Q} \operatorname{vol}(S^m) =$$

$$= \int_{S^m} (\Delta^{S^m} \tilde{P}) \tilde{Q} \operatorname{vol}(S^m) = \int_{S^m} \tilde{P} (\Delta^{S^m} \tilde{Q}) \operatorname{vol}(S^m) =$$

$$= (k-2\ell)(k-2\ell+m-1) \int_{S^m} \tilde{P}\tilde{Q} \operatorname{vol}(S^m) = 0 \,. \checkmark$$

Proof of Theorem 7.5. By the Stone-Weierstrass theorem (cf. Dieudonné (1974)), $\bigoplus_{k \in \mathbf{Z}_+} \tilde{P}_m^k$ is dense in $C^\infty(S^m)$ (with respect to, say, the global scalar product). By virtue of 7.6 (or rather (7.8)) $\bigoplus_{k \in \mathbf{Z}_+} \tilde{H}_m^k \subset C^\infty(S^m)$ is a dense linear subspace. The rest is clear. ✓

Corollary 7.9. *For each* $k \in \mathbf{Z}_+$,

$$\dim H_m^k = \dim \tilde{H}_m^k = (m+2k-1)\frac{(m+k-2)!}{k!(m-1)!} \quad . \tag{7.10}$$

Proof. As $\dim P_m^k = \binom{m+k}{k}$, by (7.7), we have

$$\dim H_m^k = \dim P_m^k - \dim P_m^{k-2} =$$

$$= \binom{m+k}{k} - \binom{m+k-2}{k-2} = (m+2k-1)\frac{(m+k-2)!}{k!(m-1)!} \quad . \ \checkmark$$

8. ISOMETRIES

Let M be a Riemannian manifold with metric tensor $g=(,)\in$ $\in C^\infty(S^2(T^*(M)))$. Then, $X\in V(M)$ is said to be a *Killing vector field* on M if the local 1-parameter group of local transformations $(\phi_t)_{t\in\mathbf{R}}$ induced by X in a neighbourhood of each point of M consist of local isometries. Using the differential calculus developed in Sections 2-6, in this section we give various characterizations of Killing vector fields.

Theorem 8.1. *Let X be a vector field on a Riemannian manifold M with metric tensor $g=(,)$. Then, the following conditions are mutually equivalent:*

(i) *X is a Killing vector field,*

(ii) $L_X g = 0$,

(iii) *The derivation $A_X = L_X - \nabla_X$ is skew-symmetric with respect to g ,*

(iv) $b(\gamma(X)) = 0$, *where, for $\alpha \in D^1(M)$, the symmetric 2-tensor $b(\alpha)$ on M is defined by*

$$b(\alpha)(Y,Z) = (\nabla_Y \alpha)(Z) + (\nabla_Z \alpha)(Y) \ , \qquad Y,Z\in V(M) \ ,$$

(v) *$\nabla\gamma(X)$ is skew-symmetric, i.e. $\nabla\gamma(X)\in D^2(M)$.*

Proof. First we show the equivalence of (i) and (ii). Denote by $(\phi_t)_{t\in\mathbf{R}}$ the local 1-parameter group of local transformations induced by the vector field X . Let $U\subset M$ be an open set such that ϕ_t is defined on U for $t\in(-\varepsilon,\varepsilon)$, where $\varepsilon>0$.

Introduce the scalar

$$\mu:(-\varepsilon,\varepsilon)\times U\times V(U)\times V(U)\to \mathbf{R}$$

by the formula

$$\mu(t,x,Y,Z) = ((\phi_t)_*Y_{\phi_{-t}(x)},(\phi_t)_*Z_{\phi_{-t}(x)}) -$$

$$-(Y_{\phi_{-t}(x)},Z_{\phi_{-t}(x)}) , \quad |t|<\varepsilon , \quad x\in U , \quad Y,Z\in V(U) .$$

Clearly, ϕ_t is an isometry on U for all $t\in(-\varepsilon,\varepsilon)$ if and only if μ is constant in $t\in(-\varepsilon,\varepsilon)$ for all $(x,Y,Z)\in U\times V(U)\times \times V(U)$. Further, by definition

$$\mu(t+s,x,Y,Z) = \mu(t,\phi_{-s}(x),(\phi_s)_*Y,(\phi_s)_*Z)$$

for all possible arguments on which μ is defined and hence μ is constant in $t\in(-\varepsilon,\varepsilon)$ if and only if

$$\left.\frac{\partial\mu(t,x,Y,Z)}{\partial t}\right|_{t=0} = 0$$

is valid for all $(x,Y,Z)\in U\times V(U)\times V(U)$. Finally, by (2.2), we have

$$\left.\frac{\partial\mu(t,x,Y,Z)}{\partial t}\right|_{t=0} = \lim_{t\to 0}\frac{1}{t}\{(Y,Z)_x-(Y,Z)_{\phi_{-t}(x)}\} -$$

$$-\lim_{t\to 0}\frac{1}{t}\{(Y_x,Z_x)-((\phi_t)_*Y_{\phi_{-t}(x)},(\phi_t)_*Z_{\phi_{-t}(x)})\} =$$

$$= X(Y,Z) - (L_XY,Z) - (Y,L_XZ) = (L_Xg)(Y,Z)$$

and the equivalence of (i) and (ii) follows.

Condition (v) being just a reformulation of (iv), we prove the equivalence of (ii), (iii) and (iv) by showing that, for any vector fields $Y,Z\in V(M)$,

$$(L_Xg)(Y,Z) = -(A_XY,Z) - (Y,A_XZ) = b(\gamma(X))(Y,Z) .$$

Indeed, we have

$$(L_X g)(Y,Z) = X(Y,Z) - (L_X Y,Z) - (Y,L_X Z) =$$

$$= (\nabla_X Y - L_X Y,Z) + (Y,\nabla_X Z - L_X Z) =$$

$$= -(A_X Y,Z) - (Y,A_X Z) = (\nabla_Y X,Z) + (Y,\nabla_Z X) =$$

$$= Y(X,Z) - (X,\nabla_Y Z) + Z(Y,X) - (\nabla_Z Y,X) =$$

$$= Y(\gamma(X)Z) - \gamma(X)(\nabla_Y Z) + Z(\gamma(X)Y) - \gamma(X)(\nabla_Z Y) =$$

$$= (\nabla_Y \gamma(X))(Z) + (\nabla_Z \gamma(X))(Y) = b(\gamma(X))(Y,Z) \ . \ \checkmark$$

Corollary 8.2. *For any Killing vector field* X *on a Riemannian manifold* M *the associated* 1-*form* $\gamma(X)$ *is coclosed, i.e.* $\partial\gamma(X)=0$.

Proof. For $x \in M$, choose a local orthonormal moving frame $\{E^i\}_{i=1}^m$ (adapted to an orthonormal base $\{e^i\}_{i=1}^m \subset T_x(M)$) over a normal coordinate neighbourhood centered at x . By 8.1(v), $\nabla\gamma(X)$ is skew-symmetric and hence

$$(\partial\gamma(X))_x = -\sum_{i=1}^m (\nabla_{e^i}\gamma(X))(e^i) =$$

$$= -\sum_{i=1}^m (\nabla\gamma(X))(e^i,e^i) = 0 \ . \ \checkmark$$

Denote by $I(M)$ the set of all Killing vector fields on the Riemannian manifold M . By virtue of 8.1(ii), $I(M)$ is a vector space and, moreover, 2.3 implies that, for $X,Y \in I(M)$,

$$L_{[X,Y]}g = [L_X,L_Y]g = 0 \ ,$$

i.e. $I(M)$ is a Lie algebra with respect to the usual Lie bracket operation.

As an application of the Bochner-Weitzenböck formula, the next result of Lichnerowicz gives a characterization of the elements of $I(M)$ in terms of a second order differential equation.

Theorem 8.3. *Let M be an m-dimensional Riemannian manifold. If $X \in V(M)$ is a Killing vector field on M then the associated 1-form $\alpha = \gamma(X)$ satisfies the equation*

$$\Delta\alpha - 2\,\mathrm{Ric}(\alpha) + d(\partial\alpha) = 0\ . \tag{8.4}$$

Conversely, if M is compact and oriented, $X \in V(M)$ is a Killing vector field on M if and only if

$$((\Delta\alpha - 2\,\mathrm{Ric}(\alpha) + d(\partial\alpha), \alpha)) = 0\ . \tag{8.5}$$

Proof. We first show that for any 1-form α on M,

$$(\Delta\alpha - 2\,\mathrm{Ric}(\alpha) + d(\partial\alpha))(e^i) = \tag{8.6}$$

$$= -\sum_{j=1}^{m} (\nabla_{e^j} b(\alpha))(e^i, e^j)\ ,$$

where $\{e^i\}_{i=1}^m$ is an orthonormal base at $x \in M$. By 8.1(iv), this will clearly imply the first statement.

Let $\{E^i\}_{i=1}^m$ be the local orthonormal moving frame (adapted to the orthonormal base $\{e^i\}_{i=1}^m \subset T_x(M)$) over a normal coordinate neighbourhood on M. Then, for $i=1,\ldots,m$, we have

$$\sum_{j=1}^{m} (\nabla_{e^j} b(\alpha))(e^i, e^j) = \sum_{j=1}^{m} e^j(b(\alpha)(E^i, E^j)) =$$

$$= \sum_{j=1}^{m} e^j\{(\nabla_{E^j}\alpha)(E^i)\} + \sum_{j=1}^{m} e^j\{(\nabla_{E^i}\alpha)E^j\} =$$

$$= \sum_{j=1}^{m} (\nabla_{e^j}(\nabla_{E^j}\alpha))(e^i) + \sum_{j=1}^{m} e^j E^i(\alpha(E^j)) -$$

$$- \sum_{j=1}^{m} e^j(\alpha(\nabla_{E^i}E^j)) = (\mathrm{trace}\ \nabla^2\alpha)(e^i) +$$

$$+ \sum_{j=1}^{m} e^i E^j(\alpha(E^j)) - \sum_{j=1}^{m} \alpha(\nabla_{e^j}\nabla_{E^i}E^j) =$$

$$= (\text{trace } \nabla^2\alpha)(e^i) + \sum_{j=1}^{m} e^i E^j(\alpha(E^j)) -$$

$$- \sum_{j=1}^{m} \alpha(R(e^j,e^i)e^j) - \sum_{j=1}^{m} \alpha(\nabla_{e^i}\nabla_{E^j}E^j) =$$

$$= (\text{trace } \nabla^2\alpha)(e^i) + \text{Ric}(\alpha)(e^i) +$$

$$+ \sum_{j=1}^{m} \{e^i E^j(\alpha(E^j)) - \alpha(\nabla_{e^i}\nabla_{E^j}E^j)\} =$$

$$= (\text{trace } \nabla^2\alpha)(e^i) + \text{Ric}(\alpha)(e^i) +$$

$$+ \sum_{j=1}^{m} e^i\{E^j(\alpha(E^j)) - \alpha(\nabla_{E^j}E^j)\} =$$

$$= (\text{trace } \nabla^2\alpha)(e^i) + \text{Ric}(\alpha)(e^i) +$$

$$+ \sum_{j=1}^{m} e^i\{(\nabla_{E^j}\alpha)E^j\} = (\text{trace } \nabla^2\alpha)(e^i) +$$

$$+ \text{Ric}(\alpha)(e^i) - (d(\partial\alpha))(e^i) \ .$$

Combining this with (6.3), we obtain (8.6).

Assume now that M is compact and oriented and introduce the scalar $\| b(\alpha) \|^2 \in C^\infty(M)$ by setting

$$\| b(\alpha) \|^2 (x) = \frac{1}{2} \sum_{i,j=1}^{m} \| b(\alpha)(e^i,e^j) \|^2 \ .$$

To prove the second statement, it is enough to show that

$$((\Delta\alpha - 2\,\text{Ric}(\alpha) + d(\partial\alpha), \alpha)) = \int_M \| b(\alpha) \|^2 \,\text{vol}(M,g) \qquad (8.7)$$

holds for $\alpha\in\mathcal{D}^1(M)$. Let $Z\in\mathcal{V}(M)$ be defined by

$$(Y,Z) = (\alpha,b(\alpha)(.,Y)) , \quad Y\in\mathcal{V}(M) .$$

We compute the divergence of Z as

$$(\mathrm{div}\, Z)(x) = \sum_{j=1}^{m} (\nabla_{e^j} Z,e^j) = \sum_{j=1}^{m} e^j(Z,E^j) =$$

$$= \sum_{j=1}^{m} e^j(\alpha,b(\alpha)(.,E^j)) = \sum_{i,j=1}^{m} e^j(\alpha(E^i)\cdot b(\alpha)(E^i,E^j)) =$$

$$= \sum_{i,j=1}^{m} (\nabla_{e^j}\alpha)(e^i)\cdot b(\alpha)(e^i,e^j) +$$

$$+ \sum_{i=1}^{m} \alpha(e^i) \sum_{j=1}^{m} (\nabla_{e^j} b(\alpha))(e^i,e^j) =$$

$$= \| b(\alpha)\|^2(x) + \sum_{i=1}^{m} \alpha(e^i) \sum_{j=1}^{m} (\nabla_{e^j} b(\alpha))(e^i,e^j) .$$

Multiplying (8.6) with $\alpha(e^i)$ and taking a summation over $i=1,\ldots,m$, the divergence of Z above reduces (8.6) to the form

$$(\Delta\alpha-2\,\mathrm{Ric}(\alpha)+d(\partial\alpha),\alpha) = -\mathrm{div}\, Z + \| b(\alpha)\|^2 .$$

Integrating and using Green's theorem (8.7) follows. √

Since a 1-form α associated to a Killing vector field X on M is, by Corollary 8.2, coclosed, equation (8.4) implies

$$\Delta\alpha - 2\,\mathrm{Ric}(\alpha) = 0 ,$$

i.e. α belongs to the kernel of the elliptic operator $\Delta-2\,\mathrm{Ric}\in\mathrm{Diff}_2(T^*(M),T^*(M))$.

Now, take M to be an m-dimensional compact Riemannian manifold. Then, by the above, 1.18 implies that $I(M)$ is a finite dimensional Lie algebra. (Note that orientability of M can be dispensed with by taking, if necessary, its oriented 2-

sheeted covering $\bar{M}$ with locally isometric projection $\pi:\bar{M}\to M$ and lifting the elements of $I(M)$ to that of $I(\bar{M})$.) The group of all isometries $i(M)$ of M with respect to the compact-open topology being a compact (possibly disconnected) Lie group acting on M (cf. Myers-Steenrod (1939)), it follows easily that the Lie algebra of the identity component $i^{\circ}(M)$ of $i(M)$ can be identified with $I(M)$. (In fact, if X is in the Lie algebra of $i^{\circ}(M)$ then the 1-parameter subgroup $t\to\phi_t$, $t\in\mathbf{R}$, determined by X is nothing but a 1-parameter group of isometries of M which induces the Killing vector field on M corresponding to X. See also Kobayashi-Nomizu (1963) page 239.)

To illustrate the use of (8.4) we prove the following.

Theorem 8.8. *Let M be a compact oriented Riemannian manifold with metric tensor $g=(,)$. If the Ricci tensor field* Ric *is negative semidefinite everywhere on M then $I(M)=P_1(M)$, in particular, $I(M)$ is commutative. If, moverover,* Ric *is negative definite at a point of M then the group of isometries $i(M)$ of M is finite.*

Proof. By virtue of 8.2 and 8.3, for $X\in I(M)$ we have

$$((\Delta\gamma(X),\gamma(X))) = 2\int_M \mathrm{Ric}(X,X)\,\mathrm{vol}(M,g) \tag{8.9}$$

since

$$(\mathrm{Ric}(\gamma(X)),\gamma(X)) = (\gamma(\mathrm{Ric}(X)),\gamma(X)) =$$

$$= (\mathrm{Ric}(X),X) = \mathrm{Ric}(X,X)\ .$$

As the Laplacian is nonnegative, the curvature assumption yields

$$((\Delta\gamma(X),\gamma(X))) = 0\ ,$$

in particular, $\gamma(X)\in\mathcal{D}^1(M)$ is closed. Now, by (3.2), for $Y,Z\in$ $\in V(M)$, we have

$$0 = (d\gamma(X))(Y,Z) = Y(\gamma(X)Z) - Z(\gamma(X)Y) - \gamma(X)[Y,Z]) =$$

$$= Y(X,Z) - Z(X,Y) - (X,\nabla_Y Z) + (X,\nabla_Z Y) =$$

$$= (\nabla_Y X,Z) - (\nabla_Z X,Y) = 2(\nabla_Y X,Z) ,$$

where, in the last equality, we used 8.1(iii). Thus, $\nabla X=0$ for $X\in I(M)$ and so

$$[X,X'] = \nabla_X X' - \nabla_{X'} X = 0 , \quad X,X'\in I(M) .$$

Assuming the negative definiteness of the Ricci tensor field at $x\in M$, by (8.9), we obtain that $X\in I(M)$ vanishes on an open neighbourhood of x. Now, by the same reasoning as in the proof of 6.7, the Killing vector field X vanishes everywhere on M .√

PROBLEMS FOR CHAPTER I

1. Given vector bundles ξ and η over a manifold M, prove that a linear map $D:C^\infty(\xi) \to C^\infty(\eta)$ is a first order differential operator (i.e. $D\in\mathrm{Diff}_1(\xi,\eta)$) if and only if, for $\mu\in C^\infty(M)$ and $v\in C^\infty(\xi)$, we have $(D(\mu\cdot v))(x)=0$ at a point $x\in M$ whenever $\mu(x)=0$ and $v_x=0$.
2. Using the definition of Lie differentiation, deduce Corollary 3.7 from Proposition 3.8.
3. Let ξ be a vector bundle over a manifold M with covariant differentiation ∇. Given a curve $c:I\to M$ and a closed interval $[t',t'']\subset I$, $t',t''\in I$, prove that each vector $w_{t'}\in \xi_{c(t')}$ extends uniquely to a parallel section w of ξ along $c|[t',t'']$. (Hint: In local coordinates, the condition $\nabla_{\frac{\partial}{\partial t}} w=0$ yields a system of ordinary differential equations of first order; cf. also Boothby (1975) page 319, or Cheeger-Ebin (1975) pages 2-3.)
4. Let S^n denote the Euclidean n-sphere whose Riemannian metric (,) is induced from that of $\mathbf{R}^{n+1}$ (cf. Example (4.24)) via the inclusion $S^n\subset\mathbf{R}^{n+1}$. Show that, for the Riemannian curvature tensor R of $T(S^n)$, we have

$$R(X,Y)Z = (Y,Z)X - (X,Z)Y , \quad X,Y,Z\in V(S^n) .$$

(Hint: The orthogonal group $O(n+1)$ acts on S^n by isometries carrying each tangent 2-plane to any tangent 2-plane in S^n. Thus, all the sectional curvatures of S^n are equal (to 1), i.e.

$$(R(X,Y)Y,X) = \|X\|^2 \cdot \|Y\|^2 - (X,Y)^2 \,, \quad X,Y \in V(S^n) \,.$$

The required formula follows by multilinear algebra; cf. also Kobayashi and Nomizu (1963) pages 198-203.)

5. Let M be an m-dimensional oriented Riemannian manifold with metric tensor $g=(\,,\,)$. The *Hodge star operator* is a vector bundle isomorphism

$$*: \Lambda^r(T^*(M)) \to \Lambda^{m-r}(T^*(M)) \,, \quad r=0,\ldots,m \,,$$

defined as follows.
Given an oriented orthonormal base $\{e^i\}_{i=1}^m \subset T_x(M)$ at $x \in M$, set

$$*(\gamma(e^{i_1}) \wedge \ldots \wedge \gamma(e^{i_r})) = \varepsilon(i_1,\ldots,i_r)\gamma(e^{j_1}) \wedge \ldots \wedge \gamma(e^{j_{m-r}}) \,,$$

where $1 \le i_1 < \ldots < i_r \le m$ and $1 \le j_1 < \ldots < j_{m-r} \le m$ with $\{i_1,\ldots,i_r, j_1,\ldots,j_{m-r}\} = \{1,\ldots,m\}$ and $\varepsilon(i_1,\ldots,i_r)$ stands for the sign of the permutation $(i_1,\ldots,i_r,j_1,\ldots,j_{m-r})$. Prove that

(i) $**\alpha = (-1)^{r(m-r)}\alpha$, for each $\alpha \in D^r(M)$,

(ii) $\alpha \wedge (*\beta) = (\alpha,\beta)\mathrm{vol}(M,g)$, for each $\alpha,\beta \in D^r(M)$, in particular, $\alpha \wedge (*\beta) = \beta \wedge (*\alpha)$ and $(*\alpha, *\beta) = (\alpha,\beta)$, i.e. $*$ is orthogonal.

(iii) On $D^r(M)$, $r=0,\ldots,m$,

$$d^* = (-1)^r *^{-1} \circ d \circ * \,.$$

(iv) The Laplacian Δ commutes with $*$.

(v) For M compact,

$$b_r(M) = b_{m-r}(M) \,, \quad r=0,\ldots,m \,.$$

6. Using (iii) of Problem 5, show, by a direct computation, that, for a Killing vector field $X \in \mathcal{V}(M)$ on an oriented Riemannian manifold M, the associated from $\gamma(X) \in \mathcal{D}^1(M)$ is coclosed, i.e. $\partial\gamma(X)=0$.

7. Prove that the Lie algebra $I(S^n)$ of the Killing vector fields on the Euclidean n-sphere S^n is isomorphic with the Lie algebra $so(n+1)$ $(\subset M(n+1,n+1))$ of skew-symmetric matrices (with the commutator as the Lie bracket). (Hint: The isomorphism in question associates to $X \in so(n+1)$ the vector field $\check{X} \in \mathcal{V}(S^n)$ whose value at $x \in S^n$ is determined by

$$\check{X}_x = X \cdot x ,$$

where, on the right hand side, the matrix X acts on $x \in \mathbf{R}^{n+1}$ by the usual multiplication.)

8. Show that a vector field X on a Riemannian manifold M is Killing if and only if, for each $Y \in \mathcal{V}(M)$,

$$L_X(\gamma(Y)) = \gamma([X,Y]) .$$

CHAPTER II

Harmonic Maps into Flat Codomains and the Structure of Compact Nonnegatively Ricci Curved Riemannian Manifolds

This chapter begins with developing a differential calculus on the *twisted bundle* $\Lambda(T^*(M))\otimes\xi$, where ξ is a Riemannian-connected vector bundle over a Riemannian manifold M . The classical differential operators introduced in Chapter I generalize here to yield the (twisted) exterior differentiation, exterior codifferentiation and Laplacian (acting on $\Lambda(T^*(M))\otimes\xi$). A general Bochner-Weitzenböck formula is derived yielding, in particular, a proof of Theorem 6.2, Chapter I. In Section 2, a further application of this formula is given to the case $\xi=$ $=f^*(T(N))$, where $f:M\to N$ is a harmonic map between Riemannian manifolds M and N . Since one of the main objects of this book is the study of harmonic maps, in Section 3 the generalized Hodge-de Rham decomposition is exploited yielding Eells-Sampson's existence and Hartman's uniqueness of harmonic maps in a given homotopy class of maps into *flat* codomains. Specializing to maps into flat tori, in Section 4, the pure existence problem is replaced by full classification by constructing the Albanese map with a universal factorizing property. Following the ideas of Lichnerowicz, in Section 5 we refine the analysis of the Albanese map to give a detailed account on the fibre structure of nonnegatively Ricci curved manifolds over the Albanese torus.

1. CALCULUS ON TWISTED TENSOR BUNDLES

Let ξ be a Riemannian-connected vector bundle over an m-dimensional manifold M with fibre metric $(\,,)\in C^\infty(S^2(\xi^*))$ and covariant differentiation $\nabla\in\mathrm{Diff}_1(\xi,\mathrm{Hom}(T(M),\xi))$. The tensor product

$$\Lambda^r(T^*(M)) \otimes \xi\ , \qquad r=0,\dots,m\ ,$$

is said to be the *(twisted) vector bundle of* r*-covectors on* M *with values in* ξ. Let $\mathcal{D}^r(\xi)$ denote the $C^\infty(M)$-module of sections of $\Lambda^r(T^*(M)) \otimes \xi$, $r=0,\dots,m$, and set $\mathcal{D}(\xi) = \bigoplus_{r=0}^{m} \mathcal{D}^r(\xi)$. The exterior multiplication Λ gives $\mathcal{D}(M)$ a (left) $\mathcal{D}(\xi)$-module structure. The elements of $\mathcal{D}^r(\xi)$ are said to be r-*forms on* M *with values in* ξ. ($\sigma \in \mathcal{D}^r(\xi)$ can then be considered as a skew-symmetric multilinear bundle map

$$\sigma : T(M) \oplus \dots \oplus T(M) \to \xi\ ,$$

i.e. for $X_x^1,\dots,X_x^r \in T_x(M)$, $x \in M$, the value $\sigma(X_x^1,\dots,X_x^r)$ is nothing but a vector in the fibre ξ_x.) For $\sigma \in \mathcal{D}^r(\xi)$, $r=0,\dots,m$, we define the *exterior differential* $d\sigma \in \mathcal{D}^{r+1}(\xi)$ of σ by

$$\begin{aligned}(d\sigma)(X^1,\dots,X^{r+1}) &= \\ &= \sum_{i=1}^{r+1} (-1)^{i+1} \nabla_{X^i}(\sigma(X^1,\dots,\hat{X}^i,\dots,X^{r+1})) + \\ &+ \sum_{i<j} (-1)^{i+j} \sigma([X^i,X^j],X^1,\dots,\hat{X}^i,\dots,\hat{X}^j,\dots,X^{r+1})\ ,\end{aligned} \tag{1.1}$$

$$X^1,\dots,X^{r+1} \in \mathcal{V}(M)\ .$$

In particular, for $v \in \mathcal{D}^0(\xi)$ $(=C^\infty(\xi))$, the exterior differential dv reduces to the covariant differential $\nabla v \in \mathcal{D}^1(\xi)$. The linear map $d : \mathcal{D}(\xi) \to \mathcal{D}(\xi)$ sending an r-form σ with values in ξ to its exterior differential $d\sigma$ is said to be the *exterior differentiation* of the twisted bundle $\Lambda(T^*(M)) \otimes \xi$. The next proposition gives an alternative description of the exterior differentiation.

Proposition 1.2. *For any* r*-form* α *on* M *and* $v \in C^\infty(\xi)$,

$$d(\alpha\otimes v) = (d\alpha)\otimes v + (-1)^r \alpha\wedge\nabla v \ . \tag{1.3}$$

Proof. Using the definition I.(3.2) of the exterior differentiation d acting on $\mathcal{D}(M)$, for $X^1,\ldots,X^{r+1}\in V(M)$, we have

$$(d(\alpha\otimes v))(X^1,\ldots,X^{r+1}) =$$

$$= \sum_{i=1}^{r+1} (-1)^{i+1} X^i(\alpha(X^1,\ldots,\hat{X}^i,\ldots,X^{r+1}))\cdot v +$$

$$+ \sum_{i=1}^{r+1} (-1)^{i+1} \alpha(X^1,\ldots,\hat{X}^i,\ldots,X^{r+1})\cdot\nabla_{X^i} v +$$

$$+ \sum_{i<j} (-1)^{i+j} \alpha([X^i,X^j],X^1,\ldots,\hat{X}^i,\ldots,\hat{X}^j,\ldots,X^{r+1})v =$$

$$= ((d\alpha)\otimes v)(X^1,\ldots,X^{r+1}) +$$

$$+ \sum_{i=1}^{r+1} (-1)^{i+1} \alpha(X^1,\ldots,\hat{X}^i,\ldots,X^{r+1})\nabla_{X^i} v =$$

$$= ((d\alpha)\otimes v)(X^1,\ldots,X^{r+1}) + (-1)^r(\alpha\wedge\nabla v)(X^1,\ldots,X^{r+1}) . \ \checkmark$$

Any section of the tensor product $\Lambda^r(T^*(M))\otimes\xi$, $r=0,\ldots,m$, can locally be expressed as a sum of sections of type $\alpha\otimes v$, $\alpha\in\mathcal{D}^r(M)$ and $v\in C^\infty(\xi)$, and hence, d being a local operator, (1.3) and I.1.7 imply that $d\in\mathrm{Diff}_1(\Lambda^r(T^*(M))\otimes\xi\ ,\ \Lambda^{r+1}(T^*(M))\otimes\xi)$, $r=0,\ldots,m$ (cf. also Problem 1 for Chapter I). Furthermore, the sybol

$$\sigma_1(d)(\alpha_x):\Lambda^r(T^*_x(M))\otimes\xi_x \to \Lambda^{r+1}(T^*_x(M))\otimes\xi_x ,$$

$$\alpha_x\in T'_x(M) \ , \quad x\in M \ ,$$

is seen to be exterior multiplication with the covector α_x.

A new phenomenon occurring in the twisted case is that d^2 is, in general, nonzero (and hence a "twisted cohomology theory" cannot be built up in the way as it was sketched in the end of Section 3, Chapter I) but d^2 involves the curvature of the Riemannian-connected vector bundle ξ .

Corollary 1.4. *For any r-form* α *on* M *and* $v \in C^\infty(\xi)$,

$$d^2(\alpha \otimes v) = \alpha \wedge R \cdot v \ , \qquad (1.5)$$

where $R \in \mathcal{D}^2(\mathrm{Hom}(\xi,\xi))$ *is the curvature of* ξ .

Proof. By virtue of (1.3),

$$d^2(\alpha \otimes v) = d\{(d\alpha)\otimes v\} + (-1)^r d\{\alpha \wedge \nabla v\} =$$

$$= (-1)^{r+1}(d\alpha)\wedge \nabla v + (-1)^r (d\alpha)\wedge \nabla v + \alpha \wedge d(\nabla v) = \alpha \wedge d(\nabla v) \ .$$

On the other hand, by making use of (1.1), for $X,Y \in V(M)$, we get

$$(d(\nabla v))(X,Y) = \nabla_X((\nabla v)Y) - \nabla_Y((\nabla v)X) - (\nabla v)([X,Y]) =$$

$$= \nabla_X \nabla_Y v - \nabla_Y \nabla_X v - \nabla_{[X,Y]} v = R(X,Y)\cdot v \ . \ \checkmark$$

Now, let M be an m-dimensional Riemannian manifold with metric tensor $g=(,) \in C^\infty(S^2(T^*(M)))$ and ξ a Riemannian-connected vector bundle over M with fibre metric $(,) \in C^\infty(S^2(\xi^*))$ and orthogonal covariant differentiation ∇ . Then, as it was explained in I.4, the bundle of covectors $\Lambda(T^*(M))$ on M becomes a Riemannian-connected vector bundle over M with respect to the canonical fibre metric $(,)$ and the Levi-Civita covariant differentiation ∇ . As ξ is Riemannian-connected by assumption, the tensor product $\Lambda(T^*(M))\otimes\xi$ inherits a Riemannian-connected vector bundle structure as well. (In fact, the fibre metric $(,)$ and the orthogonal covariant differentiation ∇ on the twisted bundle $\Lambda(T^*(M))\otimes\xi$ are characterized by

$$(\alpha\otimes v \ , \ \alpha'\otimes v') = (\alpha,\alpha')(v,v')$$

and

$$\nabla_X(\alpha\otimes v) = (\nabla_X \alpha)\otimes v + \alpha\otimes(\nabla_X v) \ ,$$

$\alpha,\alpha' \in \mathcal{D}(M)$, $v,v' \in C^\infty(\xi)$, $X \in V(M)$, respectively.)

We introduce the *exterior codifferentiation*

$$\partial \in \mathrm{Diff}_1(\Lambda^{r+1}(T^*(M))\otimes\xi\ ,\ \Lambda^r(T^*(M))\otimes\xi)\ ,\quad r=0,\ldots,m-1\ ,$$

as

$$(\partial\sigma)_x = -\sum_{i=1}^m (\iota_{e^i}\circ\nabla_{e^i})\sigma\ ,\quad \sigma\in\mathcal{D}^{r+1}(\xi)\ ,\quad x\in M\ , \tag{1.6}$$

where $\{e^i\}_{i=1}^m \subset T_x(M)$ is an orthonormal base and ι is the canonical extension of the interior product to $\mathcal{D}(\xi)$. (On $\mathcal{D}^0(\xi) = C^\infty(\xi)$, we set $\partial=0$.) The computation of the symbol $\sigma_1(\partial)$ given in I.5 applies verbatim to the twisted case yielding that

$$\sigma_1(\partial)(\alpha_x):\Lambda^{r+1}(T_x^*(M))\otimes\xi_x \to \Lambda^r(T_x^*(M))\otimes\xi_x\ ,$$

$$\alpha_x\in T_x'(M)\ ,\quad x\in M\ ,$$

is the interior product $-\iota_{\gamma^{-1}(\alpha_x)}$. Again, by the same proof as that of I.5.5. it follows that ∂ is nothing but the adjoint of d, provided that M is oriented.

We define the *Laplacian* $\Delta\in\mathrm{Diff}_2(\Lambda^r(T^*(M))\otimes\xi\ ,\ \Lambda^r(T^*(M))\otimes\xi)$, $r=0,\ldots,m$, by

$$\Delta = d\circ\partial + \partial\circ d\ . \tag{1.7}$$

The computation of $\sigma_2(\Delta)$ for the ordinary Laplacian applies verbatim to the twisted case and we obtain that

$$\sigma_2(\Delta)(\alpha_x):\Lambda^r(T_x^*(M))\otimes\xi_x \to \Lambda^r(T_x^*(M))\otimes\xi_x\ ,$$

$$\alpha_x\in T_x'(M)\ ,\quad x\in M\ ,$$

is multiplication with $-\|\alpha_x\|^2$; in particular, Δ is elliptic. An r-form σ on M with values in ξ is said to be *harmonic* if $\Delta\sigma=0$. The vector space of all harmonic r-forms on M with values in ξ is denoted by $H^r(\xi)$ and we set $H(\xi) = \bigoplus_{r=0}^m H^r(\xi)$.

By (1.1) and (1.6) the vector space

$$P^r(\xi) = \{\sigma \in \mathcal{D}^r(\xi) \mid \nabla\sigma = 0\}$$

of parallel r-forms on M with values in ξ is a linear subspace of $H^r(\xi)$, $r=0,\ldots,m$. We set $P(\xi) = \bigoplus_{r=0}^{m} P^r(\xi)$.

In what follows, to study the geometric properties of the Laplacian Δ acting on $\Lambda(T^*(M))\otimes\xi$, we present a general Bochner-Weitzenböck formula.

Let ξ be a Riemannian-connected vector bundle over a Riemannian manifold M with orthogonal covariant differentiation ∇. Notice first that, for fixed $v \in C^\infty(\xi)$, the bilinear map

$$\nabla^2 v : V(M)\times V(M) \to C^\infty(\xi)$$

defined by

$$(\nabla^2 v)(X,Y) = \nabla_X\nabla_Y v - \nabla_{\nabla_X Y} v\,, \quad X,Y \in V(M)\,, \tag{1.8}$$

is, in fact, $C^\infty(M)$-linear in each argument. Hence, $\nabla^2 v$ gives rise to a bilinear form on each of the tangent spaces $T_x(M)$, $x \in M$, with values in ξ_x. We then introduce the differential operator

$$\operatorname{trace} \nabla^2 \in \operatorname{Diff}_2(\xi,\xi)$$

as

$$(\operatorname{trace} \nabla^2 v)(x) = \{\text{trace of the bilinear form } \nabla^2 v$$

$$\text{on } T_x(M)\}\,,$$

where $x \in M$ and $v \in C^\infty(\xi)$. With respect to a local orthonormal moving frame $\{E^i\}_{i=1}^m$ (adapted to an orthonormal base $\{e^i\}_{i=1}^m \subset T_x(M)$) over a normal coordinate neighbourhood centered at x,

$$(\operatorname{trace} \nabla^2 v)(x) = \sum_{i=1}^{m} \nabla_{e^i}(\nabla_{E^i} v)\,, \quad v \in C^\infty(\xi)\,. \tag{1.9}$$

Our first result interprets the Laplacian Δ, restricted to $\mathcal{D}^0(\xi)=C^\infty(\xi)$, in terms of the covariant differentiation ∇ on ξ.

Proposition 1.10. *For any section v of ξ (considered also as a 0-form with values in ξ)*

$$\Delta v = -\text{trace } \nabla^2 v \; ; \tag{1.11}$$

in particular, the differential operator trace ∇^2 *is elliptic.*

Proof. Using the local orthonormal moving frame $\{E^i\}_{i=1}^m$ around $x \in M$ occurring in (1.9), by (1.6), we have

$$(\Delta v)_x = (\partial(dv))_x = -\sum_{i=1}^m (\iota_{e^i}(\nabla_{e^i}(dv)) =$$

$$= -\sum_{i=1}^m (\nabla_{e^i}(dv))(e^i) = -\sum_{i=1}^m \nabla_{e^i}((dv)(E^i)) =$$

$$= -\sum_{i=1}^m \nabla_{e^i}(\nabla_{E^i} v) = -(\text{trace } \nabla^2 v)_x \; . \; \checkmark$$

The Bochner-Weitzenböck formula which follows generalizes I.(6.3) to 1-forms on M with values in the vector bundle ξ.

Theorem 1.12. *For any* $\sigma \in \mathcal{D}^1(\xi)$

$$\Delta\sigma = -\text{trace } \nabla^2\sigma + R(\sigma) \; , \tag{1.13}$$

where ∇ *stands for the orthogonal covariant differentiation on* $\Lambda(T^*(M)) \otimes \xi$ *and* $R(\sigma) \in \mathcal{D}^1(\xi)$ *, with respect to an orthonormal base* $\{e^i\}_{i=1}^m \subset T_x(M)$ *at* $x \in M$ *, is given by*

$$R(\sigma)(X_x) = \sigma(\text{Ric}^M(X_x)) +$$

$$+ \sum_{i=1}^m R^\xi(e^i, X_x)(\sigma(e^i)) \; , \quad X_x \in T_x(M) \; .$$

Proof. For $x \in M$, let $\{E^i\}_{i=1}^m$ be a local orthonormal moving frame (adapted to an orthonormal base $\{e^i\}_{i=1}^m \subset T_x(M)$) over a normal coordinate neighbourhood centered at $x \in M$. By (1.6) and

$\nabla_{e^i}E^j=0$, $i,j=1,\ldots,m$, for fixed $k=1,\ldots,m$, we first compute

$$((\partial\circ d)\sigma)(e^k) = -\sum_{i=1}^{m} (\nabla_{e^i}(d\sigma))(e^i,e^k) =$$

$$= -\sum_{i=1}^{m} \nabla_{e^i}((d\sigma)(E^i,E^k)) = -\sum_{i=1}^{m} \nabla_{e^i}\{\nabla_{E^i}(\sigma(E^k)) -$$

$$- \nabla_{E^k}(\sigma(E^i)) - \sigma([E^i,E^k])\} = -\sum_{i=1}^{m} \nabla_{e^i}\nabla_{E^i}(\sigma(E^k)) +$$

$$+ \sum_{i=1}^{m} \nabla_{e^i}\nabla_{E^k}(\sigma(E^i)) + \sum_{i=1}^{m} \nabla_{e^i}(\sigma([E^i,E^k])) .$$

As ∇ is torsionfree on $T(M)$, we have $[E^i,E^k]_x = \nabla_{e^i}E^k - \nabla_{e^k}E^i = 0$, $i,k=1,\ldots,m$, and hence

$$\nabla_{e^i}(\sigma([E^i,E^k])) = \sigma(\nabla_{e^i}[E^i,E^k]) =$$

$$= \sigma(\nabla_{e^i}\nabla_{E^i}E^k) - \sigma(\nabla_{e^i}\nabla_{E^k}E^i) .$$

Secondly, by (1.6),

$$((d\circ\partial)\sigma)(e^k) = -\sum_{i=1}^{m} d\{(\nabla_{E^i}\sigma)(E^i)\}(e^k) =$$

$$= -\sum_{i=1}^{m} \nabla_{e^k}\{(\nabla_{E^i}\sigma)(E^i)\} = -\sum_{i=1}^{m} \nabla_{e^k}\nabla_{E^i}(\sigma(E^i)) +$$

$$+ \sum_{i=1}^{m} \nabla_{e^k}\{\sigma(\nabla_{E^i}E^i)\} = -\sum_{i=1}^{m} \nabla_{e^k}\nabla_{E^i}(\sigma(E^i)) +$$

$$+ \sum_{i=1}^{m} \sigma(\nabla_{e^k}\nabla_{E^i}E^i) .$$

Summarizing, we obtain

$$(\Delta\sigma)(e^k) = -\sum_{i=1}^{m} \{\nabla_{e^i}\nabla_{E^i}(\sigma(E^k)) - \sigma(\nabla_{e^i}\nabla_{E^i}E^k)\} +$$

$$+ \sum_{i=1}^{m} \{\nabla_{e^i}\nabla_{E^k} - \nabla_{e^k}\nabla_{E^i}\}(\sigma(E^i)) -$$

$$- \sum_{i=1}^{m} \sigma((\nabla_{e^i}\nabla_{E^k} - \nabla_{e^k}\nabla_{E^i})(E^i)) =$$

$$= -(\text{trace } \nabla^2\sigma)(e^k) + \sigma(\text{Ric}^M(e^k)) +$$

$$+ \sum_{i=1}^{m} R^{\xi}(e^i, e^k)(\sigma(e^i)) ,$$

since

$$(\text{trace } \nabla^2\sigma)(e^k) = \sum_{i=1}^{m} (\nabla_{e^i}\nabla_{E^i}\sigma)(e^k) =$$

$$= \sum_{i=1}^{m} \nabla_{e^i}\{(\nabla_{E^i}\sigma)(E^k)\} = \sum_{i=1}^{m} \{\nabla_{e^i}\nabla_{E^i}(\sigma(E^k)) -$$

$$- \nabla_{e^i}(\sigma(\nabla_{E^i}E^k))\} = \sum_{i=1}^{m} \{\nabla_{e^i}\nabla_{E^i}(\sigma(E^k)) - \sigma(\nabla_{e^i}\nabla_{E^i}E^k)\}. \checkmark$$

Remark. For an r-form σ with values in ξ , $r \geq 1$, we have the same formula (1.13) with $R(\sigma) \in D^r(\xi)$ defined by

$$R(\sigma)(X_x^1, \ldots, X_x^r) =$$

$$= - \sum_{i=1}^{m} \sum_{j=1}^{r} (-1)^j (R(e^i, X_x^j)\sigma)(e^i, X_x^1, \ldots, \hat{X}_x^j, \ldots, X_x^r) ,$$

where $X_x^1, \ldots, X_x^r \in T_x(M)$, $x \in M$; $\{e^i\}_{i=1}^m \subset T_x(M)$ is an orthonormal base and, for $X, Y, Y^1, \ldots, Y^r \in V(M)$,

$$(R(X,Y)\sigma)(Y^1, \ldots, Y^r) = R^{\xi}(X,Y)(\sigma(Y^1, \ldots, Y^r)) -$$

$$- \sum_{k=1}^{m} \sigma(Y^1,\ldots,Y^{k-1},R^M(X,Y)Y^k,Y^{k+1},\ldots,Y^r) \; .$$

(For a proof, as well as a general reference of this section, see Eells and Lemaire (1983).)

In the same way as we derived I.6.4 from I.6.2, we obtain an expression for the Laplacian of the scalar $\|\sigma\|^2$, $\sigma\in\mathcal{D}^1(\xi)$, as follows.

Corollary 1.14. *For any* 1-*form* σ *on* M *with values in* ξ ,

$$(\Delta\sigma,\sigma) = \frac{1}{2}\Delta(\|\sigma\|^2) + \|\nabla\sigma\|^2 + (R(\sigma),\sigma) \tag{1.15}$$

where, with respect to an orthonormal base $\{e^i\}_{i=1}^m \subset T_x(M)$ *at* $x\in M$,

$$\|\nabla\sigma\|^2 = \sum_{i,j=1}^{m} \|(\nabla_{e^i}\sigma)(e^j)\|^2$$

and

$$(R(\sigma),\sigma)_x = \sum_{j=1}^{m} (\sigma(\mathrm{Ric}^M(e^j)),\sigma(e^j)) + \sum_{i,j=1}^{m} (R^\xi(e^i,e^j)(\sigma(e^i)),\sigma(e^j)) \; . \; \surd \tag{1.16}$$

Let M be an m-dimensional *oriented* Riemannian manifold. Then, ∂ being the adjoint of the exterior differentiation, it follows, in exactly the same way as in I.6, that the Laplacian acting on $\Lambda(T^*(M))\otimes\xi$ is self-adjoint and nonnegative. Further, for $\sigma\in\mathcal{D}^r(\xi)$, $r=0,\ldots,m$, with supp $\sigma\subset M$ compact, σ is harmonic if and only if $d\sigma=0$ and $\partial\sigma=0$.

Assume now that M is compact and oriented. Then, Δ being an elliptic and self-adjoint differential operator on $\Lambda(T^*(M))\otimes$ $\otimes\xi$, the finiteness theorem and the Hodge-de Rham decomposition I.1.18 imply that the vector space $H^r(\xi)$, $r=0,\ldots,m$, of harmonic r-forms on M with values in ξ , is finite dimensional and

$$\mathcal{D}^r(\xi) = \Delta(\mathcal{D}^r(\xi))\oplus H^r(\xi) \; , \quad r=0,\ldots,m \; , \tag{1.17}$$

is an orthogonal direct sum. (Note, however, that the vector spaces $d(\mathcal{D}^{r-1}(\xi))$ and $\partial(\mathcal{D}^{r+1}(\xi))$ are not orthogonal to each other, unless $R^{\xi}=0$; cf. 1.4.)

2. GENERALITIES ON HARMONIC MAPS

Given a map $f:M\to N$ between manifolds M and N of dimensions m and n, respectively, denote by τ^f the pull-back of the tangent bundle $T(N)$ via f, i.e. set $\tau^f=f^*(T(N))$. A section v of τ^f is said to be a *vector field along* f. Equivalently, v is a map $v:M\to T(N)$ such that the diagram

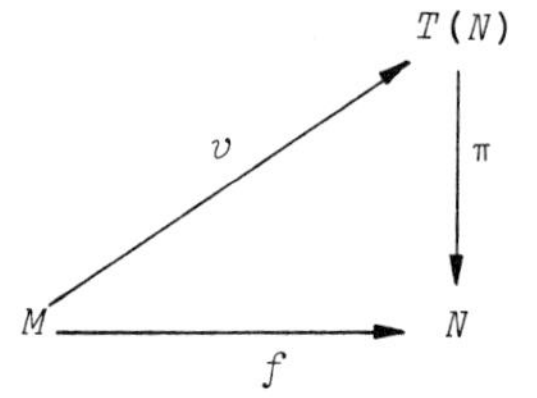

commutes, where $\pi:T(N)\to N$ is the canonical projection. The differential f_* of f sends a vector field X on M to a vector field $f_*(X)$ along f, i.e. it can be considered as a section of the vector bundle $\mathrm{Hom}(T(M),\tau^f) = T^*(M)\otimes\tau^f$, or, in other words, f_* is a 1-form on M with values in τ^f, $f_*\in \mathcal{D}^1(\tau^f)$.

Assume now that the manifold N is endowed with a Riemannian metric $h=(,)\in C^\infty(S^2(T^*(N)))$. Then, by I.4.15, τ^f becomes a Riemannian-connected vector bundle with respect to the fibre metric $(,)$ induced from the metric tensor h, and the pull-back ∇ of the Levi-Civita covariant differentiation of the Riemannian manifold N. Then, by (1.1), the exterior differentiation d of the twisted bundle $\Lambda(T^*(M))\otimes\tau^f$ is defined.

Proposition 2.1. $df_*=0$.

Proof. We work out, in local coordinates, the exterior differential $d\sigma$ of a 1-form σ vith values in τ^f. Let (U,ϕ), $\phi=(x^1,\dots,x^m)$, and (V,ψ), $\psi=(y^1,\dots,y^n)$, be coordinate neighbourhoods on M and N, $f(U)\subset V$, with associated base fields (resp., duals) $\{{}^M\partial_i\}_{i=1}^m \subset V(U)$ (resp., $\{dx^i\}_{i=1}^m \subset \mathcal{D}^1(U)$) and $\{{}^N\partial_r\}_{r=1}^n \subset V(V)$ (resp., $\{dy^r\}_{r=1}^n \subset \mathcal{D}^1(V)$) . Then, over $U, \sigma \in \mathcal{D}^1(\tau^f)$ is expressed by

$$\sigma = \sum_{i=1}^{m} \sum_{r=1}^{n} \sigma_i^r \, dx^i \otimes {}^N\partial_r \circ f \, , \quad \sigma_i^r \in C^\infty(U) \, , \qquad i=1,\ldots,m \; ; \quad r=1,\ldots,n \, . \tag{2.2}$$

In particular,

$$f_* = \sum_{i=1}^{m} \sum_{r=1}^{n} {}^M\partial_i(f^r)\, dx^i \otimes {}^N\partial_r \circ f \, , \tag{2.3}$$

where $f^r = y^r \circ (f|U) \in C^\infty(U)$, $r=1,\ldots,n$. As $[{}^M\partial_i, {}^M\partial_j]=0$, for $i,j=1,\ldots,m$, we have

$$(d\sigma)({}^M\partial_i, {}^M\partial_j) = \nabla_{{}^M\partial_i}(\sigma({}^M\partial_j)) - \nabla_{{}^M\partial_j}(\sigma({}^M\partial_i)) =$$

$$= \nabla_{{}^M\partial_i}\{\sum_{r=1}^{n} \sigma_j^r \, {}^N\partial_r \circ f\} - \nabla_{{}^M\partial_j}\{\sum_{r=1}^{n} \sigma_i^r \, {}^N\partial_r \circ f\} =$$

$$= \sum_{r=1}^{n} \{({}^M\partial_i(\sigma_j^r) - {}^M\partial_j(\sigma_i^r))\, {}^N\partial_r \circ f +$$

$$+ \sigma_j^r \, \nabla_{f_*({}^M\partial_i)} {}^N\partial_r - \sigma_i^r \, \nabla_{f_*({}^M\partial_j)} {}^N\partial_r\} \, .$$

Further, letting ${}^N\Gamma_{rs}^t$, $r,s,t=1,\ldots,n$, denote the Christoffel symbols on V , I.(4.22), by (2.3),

$$\nabla_{f_*({}^M\partial_i)} {}^N\partial_r = \sum_{s=1}^{n} {}^M\partial_i(f^s)(\nabla_{{}^N\partial_s} {}^N\partial_r) \circ f =$$

$$= \sum_{s=1}^{n} {}^M\partial_i(f^s)\, {}^N\Gamma_{rs}^t \circ f \cdot {}^N\partial_t \circ f$$

and thus

$$d\sigma = \sum_{i,j=1}^{m} \sum_{t=1}^{n} \{ {}^{M}\partial_i(\sigma_j^t) + \sum_{r,s=1}^{n} {}^{N}\Gamma_{rs}^t \circ f \cdot {}^{M}\partial_i(f^r) \cdot \sigma_j^s \} dx^i \wedge dx^j \otimes {}^{N}\partial_t \circ f \, . \tag{2.4}$$

Setting $\sigma = f_*$, i.e. locally $\sigma_i^r = {}^{M}\partial_i(f^r)$, $i=1,\ldots,m$; $r=1,\ldots$ $\ldots,n$, (2.4) reduces to

$$df_* = \sum_{i,j=1}^{m} \sum_{t=1}^{n} \{ {}^{M}\partial_i {}^{M}\partial_j(f^t) + \sum_{r,s=1}^{n} {}^{N}\Gamma_{rs}^t \circ f \cdot {}^{M}\partial_i(f^r) \cdot {}^{M}\partial_j(f^s) \} dx^i \wedge dx^j \otimes {}^{N}\partial_t \circ f \, .$$

As the Levi-Civita covariant differentiation on N is torsion-free, ${}^{N}\Gamma_{rs}^t = {}^{N}\Gamma_{sr}^t$, $r,s,t=1,\ldots,n$, I.(4.23), the coefficient of $dx^i \wedge dx^j$ in the bracket $\{\ldots\}$ is symmetric in the indices i and j. √

Remark. By similar computations as in the proof above, for $v \in C^\infty(\tau^f)$,

$$dv = \nabla v = \sum_{i=1}^{m} \sum_{t=1}^{n} \{ {}^{M}\partial_i(v^t) + \sum_{r,s=1}^{n} {}^{N}\Gamma_{rs}^t \circ f \cdot {}^{M}\partial_i(f^r) \cdot v^s \} dx^i \otimes {}^{N}\partial_t \circ f, \tag{2.5}$$

where v is locally expressed by

$$v = \sum_{r=1}^{n} v^r \cdot {}^{N}\partial_r \circ f \, , \quad v^r \in C^\infty(U) \, , \; r=1,\ldots,n \, . \tag{2.6}$$

Let $f: M \to N$ be a map between Riemannian manifolds M and N with metric tensors $g=(,) \in C^\infty(S^2(T^*(M)))$ and $h=(,) \in$ $\in C^\infty(S^2(T^*(N)))$, respectively. As it has already been mentioned in Section 1, the twisted vector bundle $\Lambda(T^*(M)) \otimes \tau^f$ is Riemannian-connected with respect to the canonical (tensor product) fibre metric (,) and covariant differentiation ∇ induced from the Levi-Civita covariant differentiations on M and N. Then, for any vector fields X and Y on M, the *second fundamental form* $\beta(f)$ of f associates to the pair (X,Y) the section $\beta(f)(X,Y)$ of τ^f defined by

$$\beta(f)(X,Y) = (\nabla_X f_*)(Y) ,$$

or equivalently, in a more concise form, $\beta(f)$ is a section of the vector bundle $T_2^0(M)\otimes\tau^f$ given by $\beta(f)=\nabla f_*$.

Proposition 2.7. *The second fundamental form* $\beta(f)$ *of a map* $f:M\to N$ *between Riemannian manifolds is symmetric, i.e.* $\beta(f)\in$ $\in C^\infty(S^2(T^*(M))\otimes\tau^f)$.

Proof. By virtue of (1.1) and 2.1, for $X,Y\in V(M)$, we have

$$0 = (df_*)(X,Y) = \nabla_X(f_*(Y)) - \nabla_Y(f_*(X)) - f_*([X,Y]) =$$

$$= \{\nabla_X(f_*(Y))-f_*(\nabla_X Y)\} - \{\nabla_Y(f_*(X))-f_*(\nabla_Y X)\} =$$

$$= (\nabla_X f_*)(Y) - (\nabla_Y f_*)(X) = \beta(f)(X,Y) - \beta(f)(Y,X) . \checkmark$$

The second fundamental from $\beta(f)$ can also be considered as a symmetric bilinear form on each of the tangent spaces $T_x(M)$ with values in $\tau_x^f=(T_{f(x)}(N))$, $x\in M$. Taking traces (pointwise) with respect to the metric tensor g on M, trace $\beta(f)$ gives rise to a vector field along f and, by the very definition (1.6) of the exterior codifferentiation ∂ on the twisted bundle $\Lambda(T^*(M))\otimes\tau^f$,

$$\partial f_* = -\text{trace}\ \beta(f) .$$

A map $f:M\to N$ between Riemannian manifolds M and N is said to be *harmonic* if $\partial f_*=0$, or equivalently, if the trace of its second fundamental form $\beta(f)$ vanishes. By 2.1, if $f:M\to N$ is a harmonic map then its differential $f_*\in D^1(\tau^f)$ is a harmonic 1-form on M with values in τ^f and the converse also holds provided that M is compact and oriented. (Indeed, if $\Delta f_*=0$ then, the Laplacian being self-adjoint, we have

$$0 = ((\Delta f_*, f_*)) = ((\ (d\circ\partial+\partial\circ d)f_*, f_*)) =$$

$$= ((\partial f_*, \partial f_*)) + ((df_*, df_*)) ,$$

and hence $\partial f_*=0$.) The map $f:M\to N$ is said to be *totally geodesic* if f_* is a parallel 1-form on M with values in τ^f , i.e. if $f_*\in P^1(\tau f)$, or equivalently, the second fundamental form $\beta(f)$ vanishes identically. A totally geodesic map is obviously harmonic.

Remark. The equation $\partial f_*=0$ of harmonicity is, locally, a second order semilinear elliptic differential system and hence, applying classical regularity, it follows that any map $f:M\to N$ of class C^2 satisfying $\partial f_*=0$ is automatically C^∞ (cf. E. Hopf (1931) page 210) and, moreover, if M and N are analytic Riemannian manifolds then f is analytic as well (Petrovsky (1939) page 4).

Example 2.8. Setting $M=S^1$, the condition of harmonicity of a map $f:S^1\to N$ translates into the equation of geodesics

$$\nabla_\partial(f_*(\partial)) = 0 ,$$

where $\partial\in V(S^1)$ is a unit base field. Thus, $f:S^1\to N$ is harmonic (and totally geodesic) if and only if f maps S^1 onto a closed geodesic of N , linearly.

As in I.6.7, the Bochner-Weitzenböck formula (1.15), for the twisted bundle $\Lambda(T^*(M))\otimes\tau^f$, implies the next result of Eells-Sampson (1964).

Theorem 2.9. *Let $f:M\to N$ be a harmonic map of a compact oriented Riemannian manifold M into a Riemannian manifold N . If the Ricci tensor field Ric^M is positive semidefinite everywhere on M and all the sectional curvatures of N are non-positive then $f:M\to N$ is a totally geodesic map. If, moreover, Ric^M is positive definite at a point of M then f is a constant map.*

Proof. By virtue of the Bochner-Weitzenböck formula (1.15), the harmonic 1-form f_* on M with values in τ^f satisfies the equation

$$\frac{1}{2}\Delta(\| f_*\|^2) + \| \nabla f_*\|^2 + (R(f_*),f_*) = 0, \qquad (2.10)$$

where, with respect to an orthonormal base $\{e^i\}_{i=1}^m \subset T_x(M)$ at $x \in M$, by I.(4.10),

$$(R(f_*),f_*)_x = \sum_{j=1}^m (f_*(\mathrm{Ric}^M(e^j)),f_*(e^j)) +$$

$$+ \sum_{i,j=1}^m (R^{\tau^f}(e^i,e^j)(f_*(e^i)),f_*(e^j)) =$$

$$= \sum_{i,j=1}^m (\mathrm{Ric}^M)_{ij}(f_*(e^i),f_*(e^j)) -$$

$$- \sum_{i=1}^m (R^N(f_*(e^i),f_*(e^j))(f_*(e^j)),f_*(e^i)) .$$

The first sum on the right hand side is, by hypothesis, non-negative. As for the second, note that

$$-(R^N(f_*(e^i),f_*(e^j))f_*(e^j),f_*(e^i)) , \quad i,j=1,\ldots,m ,$$

is either zero or, in case $f_*(e^i)$ and $f_*(e^j)$ are linearly independent in $T_{f(x)}(N)$, equals

$$-\kappa^N(f_*(e^i),f_*(e^j))\{\| f_*(e^i)\|^2 \cdot$$

$$\cdot \| f_*(e^j)\|^2 -(f_*(e^i),f_*(e^j))^2\} \geq 0 ,$$

where κ^N stands for the sectional curvature on N. Thus, $(R(f_*),f_*) \geq 0$ and, by (2.10),

$$\Delta(-\| f_*\|^2) = 2\| \nabla f_*\|^2 + 2(R(f_*),f_*) \geq 0 .$$

Applying Hopf's lemma I.6.8, we obtain that $\|f_*\|^2$ is a constant. Again by (2.10),

$$\|\nabla f_*\|^2 = 0 \quad \text{and} \quad (R(f_*),f_*) = 0 ,$$

are valid everywhere on M. From the first relation we obtain that $f_* \in P^1(\tau^f)$, i.e. the map f is totally geodesic. Assuming the positive definiteness of the Ricci tensor field on M at $x \in M$, by the second relation, $f_*=0$ at $x \in M$. As f_* is parallel, we obtain $f_*=0$ everywhere on M, i.e. the map f is constant (cf. proof of I.6.7.). √

The totally geodesic maps can geometrically be characterized by the following.

Theorem 2.11. *If $f:M \to N$ is a totally geodesic map then its differential f_* maps each parallel vector field along each curve $c:[0,1] \to M$ into a parallel vector field along the curve $f \circ c$ of N. In particular, f maps geodesics of M onto geodesics of N, linearly, and rank of f is constant.*

Proof. Let $t \to X_t$, $t \in [0,1]$, be a parallel vector field along c. Then, for $t \in [0,1]$ fixed,

$$0 = \beta(f)(\dot{c}(t),X_t) = (\nabla f_*)(\dot{c}(t),X_t) =$$

$$= (\nabla_{\dot{c}(t)} f_*)(X_t) = \nabla_{\frac{\partial}{\partial t}} (f_*(X_t)) - f_*(\nabla_{\frac{\partial}{\partial t}} X_t) =$$

$$= \nabla_{\frac{\partial}{\partial t}} (f_*(X_t))$$

and the first statement follows. In particular, if c is a geodesic segment of M then the vector field $t \to f_*(\dot{c}(t)) = (f \circ c)^{\bullet}(t)$, $t \in [0,1]$, is also parallel along $f \circ c$, i.e. the curve $f \circ c:[0,1] \to N$ is also a geodesic segment of N. As for the rank of f, fix $x,y \in M$, and choose a curve $c:[0,1] \to M$ with $c(0)=x$ and $c(1)=y$. For $X_0 \in \ker f_* \cap T_x(M)$, extend X_0 to a parallel vector field $t \to X_t$, $t \in [0,1]$, along c. Then, by the above, $t \to f_*(X_t)$, $t \in [0,1]$, is a parallel vector field along $f \circ c$ and hence $f_*(X_0)=0$ implies $f_*(X_1)=0$, i.e. $X_1 \in \ker f_* \cap T_y(M)$. We obtain that parallel translation along c carries $\ker f_* \cap T_x(M)$ isomorphically onto $\ker f_* \cap T_y(M)$, in particular, $(\text{rank } f)(x) = (\text{rank } f)(y)$. √

Remark. The totally geodesic maps can, in fact, be characterized by the property of carrying geodesics onto geodesics, linearly (cf. Vilms (1970)). Harmonic maps, however, have a much richer variety of possible behaviour, as the examples below show.

Example 2.12. Take $N=\mathbf{R}^n$ with the Euclidean metric (,) and let $\check{}:T(\mathbf{R}^n) \to \mathbf{R}^n$ denote the canonical identification (cf. I.4.24). Given a map $\bar{f}:M \to \mathbf{R}^n$, a vector field $v:M \to T(\mathbf{R}^n)$ along $\bar{f}$ gives rise to a vector function $\check{v}:M \to \mathbf{R}^n$. By I.(4.6) and I.(4.25), the pull-back of the Levi-Civita covariant differentiation on $T(\mathbf{R}^n)$ via $\bar{f}$ is characterized by

$$(\nabla_X v)\check{} = X(\check{v}) \ , \quad X \in V(M) \ , \quad v \in C^\infty(\tau^{\bar{f}}) \ , \tag{2.13}$$

where X , on the right hand side, acts on the vector function $\check{v}$ componentwise. Fix $x \in M$ and choose a local orthonormal moving frame $\{E^i\}_{i=1}^m$ (adapted to an orthonormal base $\{e^i\}_{i=1}^m \subset T_x(M)$) over a normal coordinate neighbourhood centered at x. Then, by I.(6.9) and (2.13),

$$(\partial \bar{f}_*)\check{}_x = - \sum_{i=1}^m (\nabla_{e^i}(\bar{f}_*(E^i)))\check{} = - \sum_{i=1}^m e^i((\bar{f}_*(E^i))\check{}) =$$

$$= - \sum_{i=1}^m e^i(E^i(\bar{f})) = (\Delta \bar{f})(x) \ , \tag{2.14}$$

since, for $X \in V(M)$, we have $(\bar{f}_*(X))\check{} = X(\bar{f})$. If follows that $\bar{f}:M \to \mathbf{R}^n$ is a harmonic map if and only if its components are harmonic functions on M . In particular, if M is compact then Hopf's lemma I.6.8 implies that any harmonic map $\bar{f}:M \to \mathbf{R}^n$ is constant.

These considerations extend to the case of harmonic maps into complete flat Riemannian manifolds. Indeed, first take M to be a manifold with base point $o \in M$. Then the *universal covering*

$\tilde{M}$ of M, represented by the homotopy classes of curves starting from the point $o \in M$, carries a manifold structure such that the end-point map $\pi_M : \tilde{M} \to M$ is a covering projection, in particular, a local diffeomorphism. (For details on covering spaces, see Steenrod (1951).) If M is Riemannian with metric tensor $g \in C^\infty(S^2(T^*(M)))$ then we endow the universal covering manifold $\tilde{M}$ with the pull-back metric $\tilde{g} = \pi_M^*(g) \in C^\infty(S^2(T^*(\tilde{M})))$ and hence $\pi_M : \tilde{M} \to M$ becomes a local isometry.
Secondly, given a map $f : M \to N$ of a Riemannian manifold M with base point $o \in M$ into a Riemannian manifold N with base point $f(o) \in N$, we have the commutative diagram

$$\begin{array}{ccc} \tilde{M} & \xrightarrow{\ \tilde{f}\ } & \tilde{N} \\ \downarrow{\scriptstyle \pi_M} & & \downarrow{\scriptstyle \pi_N} \\ M & \longrightarrow & N \end{array} \qquad (2.15)$$

where the *universal lift* $\tilde{f} : \tilde{M} \to \tilde{N}$ is the map induced by f on homotopy. Now, if $f : M \to N$ is harmonic then, π_M and π_N being local isometries, $\tilde{f}$ is also a harmonic map. Finally, if N is complete and *flat* $(R^N = 0)$ Riemannian manifold then the universal covering $\tilde{N}$ is known to be isometric to $\mathbf{R}^n$, $n = \dim N$ endowed with the Euclidean metric (cf. Kobayashi and Nomizu (1963) page 212). Summarizing, we obtain that a map $f : M \to N$ into a complete flat Riemannian manifold N is harmonic if and only if the components of the universal lift $\tilde{f} : \tilde{M} \to \tilde{N}$ are harmonic functions on $\tilde{M}$.

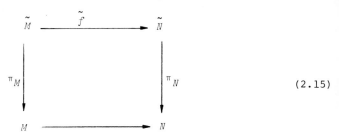

Example 2.16. Let $f : M \to N$ be an *immersion* between manifolds M and N (i.e. $f_* : T_x(M) \to T_{f(x)}(N)$ is injective for all $x \in M$). If N is endowed with a Riemannian metric $h \in C^\infty(S^2(T^*(N)))$ then the domain M inherits, as a Riemannian structure, the pull-back metric tensor $g = f^*(h) \in C^\infty(S^2(T^*(M)))$ and $f : M \to N$ becomes an isometric immersion. $T(M)$ is an (isometric) direct summand of the Riemannian-connected vector bundle τ^f which,

then, splits into the orthogonal direct sum of Euclidean vector bundles

$$\tau^f \cong T(M) \oplus \nu^f ,$$

where ν^f stands for the *normal bundle of* f . Denoting by

$$^{\top}:\tau^f \to T(M)$$

and

$$^{\perp}:\tau^f \to \nu^f$$

the tangential and orthogonal projections, respectively, for $X \in V(M)$, we set

$$\nabla^{\top}_X u = (\nabla_X u)^{\top} , \quad u \in V(M) , \tag{2.17}$$

and

$$\nabla^{\perp}_X v = (\nabla_X v)^{\perp} , \quad v \in C^{\infty}(\nu^f) . \tag{2.18}$$

Clearly, $\nabla^{\top} \in \mathrm{Diff}_1(T(M) , \mathrm{Hom}(T(M),T(M))$ and $\nabla^{\perp} \in \mathrm{Diff}_1(\nu^f, \mathrm{Hom}(T(M),\nu^f))$ are covariant differentiations on $T(M)$ and ν^f, respectively. Moreover, orthogonality of ∇ on τ^f translates into orthogonality of $\nabla^{\top}$ and $\nabla^{\perp}$ and hence $T(M)$ and ν^f are Riemannian-connected vector bundles. Finally, I.(4.21) implies that $\nabla^{\top}$ is torsionfree and, by uniqueness in I.4.20, this means that $\nabla^{\top}$ is nothing but the Levi-Civita covariant differentiation on $T(M)$ with respect to the pull-back metric g on M . (In fact, the question being local, M can be considered, via f , as a submanifold of N and thus, for $X,Y \in V(M)$, the Lie bracket $[X,Y]$ being a vector field on M , I.(4.21) entails

$$\nabla^{\top}_X Y - \nabla^{\top}_Y X = [X,Y] , \quad X,Y \in V(M) .)$$

In particular, for $X,Y \in (M)$, the second fundamental form satisfies

$$\beta(f)(X_x,Y_x) = (\nabla_{X_x} f_*)(Y_x) = \nabla_{X_x}(f_*(Y)) - \\ - f_*(\nabla_{X_x} Y) = (\nabla_{X_x}(f_*(Y)))^{\perp} \in \nu^f_x , \quad x \in M . \tag{2.19}$$

Combining this with 2.7, it follows that $\beta(f)$ is a symmetric bilinear form on each of the tangent spaces $T_x(M)$ with values in ν_x^f, $x\in M$, or equivalently,

$$\beta(f)\in C^\infty(S^2(T^*(M))\otimes\nu^f) .$$

Taking traces, the section $\mu(f) = \frac{1}{m}$ trace $\beta(f)$, $m=\dim M$, of ν^f is said to be the *mean curvature* of f and if $\mu(f)=0$ we say that f is a *minimal (isometric) immersion*. Hence, we have the following.

Proposition 2.20. *An isometric immersion $f:M\to N$ is minimal if and only if f is harmonic.*

In particular, any *Riemannian* (i.e. isometric) *covering* is harmonic. If $f:M\to N$ is an isometric minimal embedding then M is said to be a *minimal submanifold* of N. Finally, if $f:M\to N$ is an isometric totally geodesic embedding then M is called a *totally geodesic submanifold* of N.

Example 2.21. Take $N=S^n$ to be the *Euclidean n-sphere*. A map $f:M\to S^n$ of a Riemannian manifold of dimension m into S^n induces, via $S^n\subset\mathbf{R}^{n+1}$, a vector function $\bar{f}:M\to\mathbf{R}^{n+1}$ with $(\bar{f},\bar{f})=1$, where $(,)$ stands for the Euclidean scalar product on $\mathbf{R}^{n+1}$. The (restriction of the) canonical identification $\check{}:T(S^n)\to\mathbf{R}^{n+1}$, I.4.24, establishes a 1-1 correspondence between the vector fields v along f and the vector functions $\check{v}:M\to\mathbf{R}^{n+1}$ satisfying the linear constraint $(\bar{f},\check{v})=0$. The covariant differentiation ∇_X in the direction of a vector field X on M can then be characterized by

$$(\nabla_X v)\check{} = X(\check{v}) - (X(\check{v}),\bar{f})\bar{f} , \quad v\in C^\infty(\tau^f) . \tag{2.22}$$

Indeed, again by the inclusion $S^n\subset\mathbf{R}^{n+1}$, a vector field $v:M\to T(S^n)$ along f defines a vector field $\bar{v}:M\to T(\mathbf{R}^{n+1})$ along $\bar{f}:M\to\mathbf{R}^{n+1}$. Hence, by (2.13) and 2.16 (applied to the inclusion $S^n\subset\mathbf{R}^{n+1}$),

$$\begin{aligned}(\nabla_X v)\check{} &= ((\nabla_X\bar{v})^{\top})\check{} = (\nabla_X\bar{v})\check{} - ((\nabla_X\bar{v})\check{},\bar{f})\bar{f} = \\ &= X(\check{v}) - (X(\check{v}),\bar{f})\bar{f}\end{aligned}$$

and (2.22) follows.

Now, we reformulate the condition of harmonicity of f in terms of $\bar{f}$. For fixed $x \in M$, choosing a local orthonormal moving frame $\{E^i\}_{i=1}^m$ (adapted to an orthonormal base $\{e^i\}_{i=1}^m \subset T_x(M)$) over a normal coordinate neighbourhood, by I.(6.9) and (2.22),

$$-(\partial f_*)_x^{\vee} = \sum_{i=1}^m (\nabla_{e^i}(f_*(E^i)))^{\vee} =$$

$$= \sum_{i=1}^m e^i((f_*(E^i))^{\vee}) - \sum_{i=1}^m (e^i((f_*(E^i))^{\vee}), \bar{f}(x))\bar{f}(x) =$$

$$= \sum_{i=1}^m e^i(E^i(\bar{f})) - \sum_{i=1}^m (e^i(E^i(\bar{f})), \bar{f}(x))\bar{f}(x) =$$

$$= -(\Delta\bar{f})(x) + ((\Delta\bar{f})(x), \bar{f}(x))\bar{f}(x) \quad .$$

It follows that $f: M \to S^n$ is a harmonic map if and only if the vector function $\Delta\bar{f}$ is proportional to $\bar{f}$. (For the converse, if $\Delta\bar{f} = \mu\cdot\bar{f}$ holds for some scalar μ on M then $\mu = (\Delta\bar{f}, \bar{f})$ and harmonicity of f follows.)

Example 2.23. Let $\bar{f}: M \to \mathbf{R}^{n+1}$ be an isometric immersion of a Riemannian manifold M of dimension m into $\mathbf{R}^{n+1}$ endowed with the Euclidean metric. By (2.14),

$$\Delta\bar{f} = -m\cdot\mu(\bar{f})^{\vee} , \tag{2.24}$$

where $\mu(\bar{f}) \in C^\infty(\nu^{\bar{f}})$ stands for the mean curvature of $\bar{f}$. The following result of Takahashi (1966) indicates how to build up minimal isometric immersions into spheres.

Theorem 2.25. *Let $\bar{f}: M \to \mathbf{R}^{n+1}$ be an isometric immersion of a Riemannian manifold M into $\mathbf{R}^{n+1}$. Assume that*

$$\Delta\bar{f} = \lambda\cdot\bar{f} \tag{2.26}$$

holds for $(0\neq)\lambda \in \mathbf{R}$. *Then* $\lambda > 0$ *and* $\mathrm{im}(\bar{f}) \subset r\cdot S^n$, *with* $r = (\frac{m}{\lambda})^{\frac{1}{2}}$, *such that the induced map* $f: M \to r\cdot S^n$ *is minimal.*

Proof. The scalar $(\bar{f},\bar{f})$ is constant on M since, by making use of (2.24) and (2.26), for $X_x \in T_x(M)$, $x \in M$, we have

$$X_x(\bar{f},\bar{f}) = 2(X_x(\bar{f}),\bar{f}(x)) = 2((\bar{f}_*(X_x))^\vee,\bar{f}(x)) =$$

$$= \frac{2}{\lambda}((\bar{f}_*(X_x))^\vee,(\Delta\bar{f})(x)) = -\frac{2m}{\lambda}((f_*(X_x))^\vee,(\mu(\bar{f})_x)^\vee) =$$

$$= -\frac{2m}{\lambda}(f_*(X_x),\mu(\bar{f})_x) = 0 \ .$$

Hence, $(\bar{f},\bar{f})=r^2$, for some constant $r>0$, yielding $\mathrm{im}(\bar{f})\subset r\cdot S^n$. Denote by $f: M\to r\cdot S^n$ the isometric immersion induced by $\bar{f}$. To compute the mean curvature $\mu(f)$ in terms of $\mu(\bar{f})$, let $\top$ and $\perp$ be the tangential and orthogonal projections, respectively, corresponding to the isometric immersion $r\cdot S^n \subset \mathbf{R}^{n+1}$ (cf. 2.16), while, for a vector field $v: M\to T(r\cdot S^n)$ along $f: M\to r\cdot S^n$, let $v^\perp_f \in C^\infty(\nu^f)$ denote the component of v *in the normal bundle of* f . For fixed $x\in M$, choosing a local orthonormal moving frame $\{E^i\}_{i=1}^m$ (adapted to an orthonormal base $\{e^i\}_{i=1}^m \subset T_x(M)$) over a normal coordinate neighbourhood of M centered at $x\in M$, we have

$$\mu(\bar{f})^\vee_x = \frac{1}{m}\{\sum_{i=1}^m (\nabla_{e^i}(\bar{f}_*(E^i)))^\perp\}^\vee +$$

$$+ \frac{1}{m}\{\sum_{i=1}^m ((\nabla_{e^i}(\bar{f}_*(E^i)))^\top)^\perp_f\}^\vee =$$

$$= \frac{1}{r^2}(\mu(\bar{f})^\vee_x,\bar{f}(x))\bar{f}(x) + \frac{1}{m}\{\sum_{i=1}^m (\nabla_{e^i}(f_*(E^i)))^\perp_f\}^\vee =$$

$$= \frac{1}{r^2}(\mu(\bar{f})^\vee_x,\bar{f}(x))\bar{f}(x) + \mu(f)^\vee_x \ .$$

As $\mu(\bar{f})^\vee$ is proportional to $\bar{f}$,

$$\mu(f)^\vee = \mu(\bar{f})^\vee - \frac{1}{r^2}(\mu(\bar{f})^\vee,\bar{f})\bar{f} = 0 \ ,$$

i.e. f is minimal. Finally,

$$\lambda\bar{f}(x) = (\Delta\bar{f})(x) = -m\mu(\bar{f})^{\vee}_{x} =$$

$$= -\sum_{i=1}^{m}((\nabla_{e^i}(\bar{f}_*(E^i)))^{\perp})^{\vee} = -\sum_{i=1}^{m}((\nabla_{e^i}(\bar{f}_*(E^i)))^{\vee} ,$$

$$\frac{\bar{f}(x)}{r})\frac{\bar{f}(x)}{r} = -\frac{1}{r^2}\sum_{i=1}^{m}\{e^i((\bar{f}_*(E^i))^{\vee},\bar{f})\}\bar{f}(x) +$$

$$+\frac{1}{r^2}\sum_{i=1}^{m}(\bar{f}_*(e^i),\bar{f}_*(e^i))\bar{f}(x) = \frac{m}{r^2}\bar{f}(x) ,$$

since $((\bar{f}_*(E^i))^{\vee},\bar{f}) = (E^i(\bar{f}),\bar{f}) = \frac{1}{2}E^i(\bar{f},\bar{f}) = 0$. Thus, $\lambda=\frac{m}{r^2}>0$.✓

Example 2.27. The Hopf-Whitehead construction which follows produces a large variety of harmonic maps between spheres. Recall that a bilinear map $\pi:\mathbf{R}^{m+1}\times\mathbf{R}^{m+1}\to\mathbf{R}^n$ satisfying

$$\|\pi(x,y)\| = \|x\|\cdot\|y\| , \quad x,y\in\mathbf{R}^{m+1} ,$$

is said to be an *orthogonal multiplication*. (For details and examples, see Husemoller (1966).) Given an orthogonal multiplication $\pi:\mathbf{R}^{m+1}\times\mathbf{R}^{m+1}\to\mathbf{R}^n$, define a map

$$h:\mathbf{R}^{2m+2}\to\mathbf{R}^{n+1}$$

by setting

$$h(x,y) = (\|x\|^2-\|y\|^2, \ 2\pi(x,y)) , \quad x,y\in\mathbf{R}^{m+1} .$$

Then, for $x,y\in\mathbf{R}^{m+1}$ with $\|x\|^2+\|y\|^2=1$, we have

$$\|h(x,y)\|^2 = (\|x\|^2-\|y\|^2)^2+4\|\pi(x,y)\|^2 =$$

$$= (\|x\|^2-\|y\|^2)^2 + 4\|x\|^2\|y\|^2 =$$

$$= (\|x\|^2+\|y\|^2)^2 = 1 ,$$

or equivalently, h restricts to a map

$$f:S^{2m+1}\to S^n .$$

Furthermore, $\Delta^{\mathbf{R}^{2m+2}}h=0$ since

$$\Delta^{\mathbf{R}^{2m+2}}(\|x\|^2-\|y\|^2)=$$

$$=\Delta^{\mathbf{R}^{2m+2}}((x^1)^2+\ldots+(x^{m+1})^2-(y^1)^2-\ldots-(y^{m+1})^2)=0 ,$$

$x=(x^1,\ldots,x^{m+1})$, $y=(y^1,\ldots,y^{m+1})\in\mathbf{R}^{m+1}$, and, π being linear in both arguments, $\Delta^{\mathbf{R}^{2m+2}}\pi=0$. By making use of I.(7.3), it follows that

$$\Delta\bar{f}=2(m+1)\bar{f} \tag{2.28}$$

and 2.21 implies that the map $f:S^{2m+1}\to S^n$ is harmonic. In particular, if we take π to be the complex, quaternionic and Cayley multiplications the corresponding maps $f:S^3\to S^2$, $f:S^7\to S^4$ and $f:S^{15}\to S^8$ are just the *Hopf-bundles*. By the above, these are all harmonic and, by making use of 2.11, easily seen to be non-totally geodesic maps.

Example 2.29. Assume that $f:M'\times M''\to N$ is a map from the product $M'\times M''$ of Riemannian manifolds M' and M'' into a Riemannian manifold N such that $f(.,x''):M'\to N$ and $f(x',.):M''\to N$ are harmonic maps for all $x'\in M'$ and $x''\in M''$. Then, with obvious notations,

$$\partial f_\star=\partial' f_\star+\partial'' f_\star$$

and hence f is harmonic. In particular, if G is a compact Lie group, endowed with a biinvariant Riemannian metric, then the multiplication map $\pi:G\times G\to G$, being an isometry in each variable separately, is harmonic.

We close this section by discussing some composition properties of harmonic maps. Recall first that a *surjective* map $f:M\to N$ between manifolds M and N of dimensions m and n , respectively, is said to be a *submersion* if its rank equals n everywhere on M, i.e. if the differential f_* maps each of the tangent spaces $T_x(M)$ onto $T_{f(x)}(N)$, $x \in M$. As an application of the inverse function theorem, it follows that, for $y \in N$, the *fibre* $f^{-1}(y) \subset M$ is a (possibly disconnected) closed (regular) submanifold of M. (Moreover, by a result of Ehresmann (1947) if all the fibres $f^{-1}(y) \subset M$, $y \in N$, are compact then $f: M \to N$ is necessarily the projection of a locally trivial fibre bundle.) Taking M and N to be Riemannian manifolds with metric tensors $g = (\,,\,) \in C^\infty(S^2(T^*(M)))$ and $h = (\,,\,) \in C^\infty(S^2(T^*(N)))$, respectively, to a submersion $f: M \to N$ there is associated a distribution H on M (called the *horizontal distribution of* f) by setting $H(x)$, $x \in M$, to be the orthogonal complement of $T_x(f^{-1}(f(x)))$ in $T_x(M)$ with respect to g. The submersion $f: M \to N$ is said to be *Riemannian* if $f_*|H(x): H(x) \to T_{f(x)}(N)$ is a linear isometry for all $x \in M$.

The next result of Eells-Sampson (1964) gives a geometric description of harmonic Riemannian submersions.

Theorem 2.30. *A Riemannian submersion $f: M \to N$ is harmonic if and only if all the fibres are minimal submanifolds of M.*

Proof. Fix $x \in M$ with $y = f(x) \in N$ and choose a local orthonormal moving frame $\{F^i\}_{i=1}^n$ (adapted to an orthonormal base at y) over a normal coordinate neighbourhood of N centered at y. By the local product structure of M, there exists an orthonormal moving frame $\{E^i\}_{i=1}^m$ on a neighbourhood of x in M such that $f_*(E^i) = F^i \circ f$, $i = 1, \ldots, n$, and $f_*(E^i) = 0$, $i = n+1, \ldots$ $\ldots, m$. (Note that $(\nabla_{F^i} F^j)(y) = 0$, $i, j = 1, \ldots, n$, but $\nabla_{E^i} E^j$, $i, j = 1, \ldots, m$, do not vanish, in general, at x.) Then, setting $e^i = E^i_x$, $i = 1, \ldots, m$, we have

$$-(\partial f_*)_x = \sum_{i=1}^{m} (\nabla_{e^i} f_*)(e^i) =$$

$$= \sum_{i=1}^{n} (\nabla_{e^i} f_*)(e^i) + \sum_{i=n+1}^{m} (\nabla_{e^i} f_*)(e^i) .$$

We claim that the first sum on the right hand side is always zero. Indeed,

$$\sum_{i=1}^{n} (\nabla_{e^i} f_*)(e^i) = \sum_{i=1}^{n} \nabla_{e^i}(f_*(E^i)) - \sum_{i=1}^{n} f_*(\nabla_{e^i} E^i)$$

and, fixing $i=1,\ldots,n$,

$$\nabla_{e^i}(f_*(E^i)) = \nabla_{e^i}(F^i \circ f) = \nabla_{f_*(e^i)} F^i = (\nabla_{F^i} F^i)(y) = 0 .$$

As the vector fields E^i and F^i are f-related, for $i,j=1,\ldots,n$, $f_*([E^i,E^j]_x) = [F^i,F^j]_y = (\nabla_{F^i} F^j)(y) - (\nabla_{F^j} F^i)(y) = 0$, (cf. Kobayashi and Nomizu (1963) page 10), and we have

$$(\nabla_{e^i} E^i, e^j) = -(e^i, \nabla_{e^i} E^j) = -(e^i, [E^i,E^j]_x) -$$

$$- (e^i, \nabla_{e^j} E^i) = -(e^i, \nabla_{e^j} E^i) = -\frac{1}{2} e^j (\| E^i \|^2) = 0 .$$

Equivalently,

$$f_*(\nabla_{e^i} E^i) = 0 , \quad i=1,\ldots,n ,$$

and the claim follows.

Now

$$(\partial f_*)_x = - \sum_{i=n+1}^{m} (\nabla_{e^i} f_*)(e^i) = - \sum_{i=n+1}^{m} \nabla_{e^i}(f_*(E^i)) +$$

$$+ \sum_{i=n+1}^{m} f_*(\nabla_{e^i} E^i) = f_*(\sum_{i=n+1}^{m} \nabla_{e^i} E^i)$$

since $f_*(E^i)=0$, for $i=n+1,\ldots,m$. It follows that $(\partial f_*)_x=0$ if and only if $\sum_{i=n+1}^{m} \nabla_{e^i} E^i$ is tangent to the fibre $f^{-1}(y)$. On the other hand, the horizontal component of the vector $\sum_{i=n+1}^{m} \nabla_{e^i} E^i$ is nothing but m times the mean curvature of the submanifold $f^{-1}(y) \subset M$ at x . √

Remark 1. According to a result of Hermann (1969), if $f:M \to N$ is a Riemannian submersion then the unique horizontal lift of a geodesic segment $c:[0,1] \to N$ starting from $c(0)=f(x)$, $x \in M$, is a geodesic segment starting from $x \in M$. The original proof of Eells and Sampson exploited this fact to derive 2.30.

Remark 2. A Riemannian submersion $f:M \to N$ is a totally geodesic map if and only if the fibres of f are totally geodesic submanifolds and the horizontal distribution H is integrable (cf. Vilms (1970)). For example, the Hopf-map $f:S^3 \to S^2$ is well-known to be a Riemannian submersion with totally geodesic fibres. (In fact, the fibres are great circles of S^3.) According to 2.27 (or 2.30) f is harmonic but, as noted in 2.27, non totally geodesic and hence the horizontal distribution H of dimension 2 on S^3 is nonintegrable.

Remark 3. There is a large supply of harmonic Riemannian submersions $f:M \to S^n$ provided by a method due to Baird and Fuglede (cf. Baird (1983)).

To start with the composition properties of harmonic maps observe first that, as the Levi-Civita covariant differentiation commutes with the isometries, for a harmonic map $f:M \to N$ between Riemannian manifolds M and N, the composition $b \circ f \circ a : M \to N$, with $a \in i(M)$ and $b \in i(N)$, is also harmonic. The following result of R.T. Smith generalizes this observation to harmonic Riemannian submersions.

Theorem 2.31. *Let $f:M \to N$ be a harmonic Riemannian submersion. Then, for any map $f':N \to N'$ between Riemannian manifolds,*

$$\partial(f' \circ f)_* = (\partial f'_*) \circ f \, , \tag{2.32}$$

in particular, f' is harmonic if and only if $f' \circ f$ is harmonic.

Proof. Choose local orthonormal moving frames $\{E^i\}_{i=1}^m$ and $\{F^i\}_{i=1}^n$ around $x \in M$ and $y=f(x) \in N$, respectively, as in the proof of 2.30. Then,

$$-(\partial(f'\circ f)_*)_x = \sum_{i=1}^{m} \nabla_{e^i}((f'\circ f)_* E^i) - \sum_{i=1}^{m} (f'\circ f)_*(\nabla_{e^i} E^i).$$

The second sum on the right hand side, by the proof of 2.30, vanishes. As for the first, we have

$$\sum_{i=1}^{m} \nabla_{e^i}((f'\circ f)_* E^i) = \sum_{i=1}^{m} \nabla_{e^i}\{f'_*(f_*(E^i))\} =$$

$$= \sum_{i=1}^{n} \nabla_{e^i}\{f'_*(F^i\circ f)\} = \sum_{i=1}^{n} \nabla_{f_*(e^i)} \{f'_*(F^i)\} =$$

$$= \sum_{i=1}^{n} \nabla_{F^i_y}(f'_*(F^i)) = \sum_{i=1}^{n} (\nabla_{F^i_y} f'_*)(F^i_y) = -(\partial f'_*)_y \;. \checkmark$$

Remark 2.31. can also be verified by making use of the general composition formula

$$\partial(f'\circ f)_* = f'_*(\partial f_*) + \text{trace } \beta(f')(f_*, f_*) \;, \qquad (2.33)$$

valid for any maps $f:M\to N$ and $f':N\to N'$ between Riemannian manifolds (cf.Eells and Lemaire (1983)).In particular, from (2.33) it follows that a harmonic map followed by a totally geodesic map is always harmonic. Note, however, that, as for compositions of harmonic maps, we shall only use Smith's theorem.

By (2.14), for any scalar $\mu\in C^\infty(N)$, the vector function $(\partial\mu_*)\check{}$ reduces to $\Delta\mu$ and from 2.31 we deduce:

Corollary 2.34. *Let $f:M\to N$ be a harmonic Riemannian submersion. Then, for $\mu\in C^\infty(N)$,*

$$\Delta(\mu\circ f) = (\Delta\mu)\circ f \;. \checkmark \qquad (2.35)$$

Remark 1. Based on classical local analysis, a direct proof of 2.34 is given in Wallach (1972) (see also O'Neil (1966)).

Remark 2. Harmonic maps satisfying the more general property

$$\Delta(\mu \circ f) = \lambda^2 \cdot (\Delta\mu) \circ f \,, \tag{2.36}$$

(with a fixed scalar $\lambda \in C^\infty(M)$) for all functions μ defined on open sets of N, were introduced in Fuglede (1978) and (1979). In particular, a nonconstant map f satisfying (2.36) is open, harmonic and horizontally conformal. Note that 2.31 remains valid if the harmonic Riemannian submersion $f:M \to N$ is replaced by such maps (cf. Eells and Lemaire (1983) page 43).

We close this section by showing that the composition of harmonic maps does not, in general, result a new harmonic map.

Example 2.37. The quotient of the 2-plane $\mathbf{R}^2$ modulo the lattice $(2\pi\mathbf{Z})\times(2\pi\mathbf{Z}) \subset \mathbf{R}^2$ is the 2-torus $\mathbf{T}^2$. Scalars on $\mathbf{T}^2$ are then identified with doubly periodic functions on $\mathbf{R}^2$ with period $(2\pi, 2\pi)$. Hence, the map

$$f:\mathbf{T}^2 \to S^3$$

given by

$$\bar{f}(\phi,\psi) = \frac{1}{\sqrt{2}}(\cos\phi, \sin\phi, \cos\psi, \sin\psi), \quad \phi,\psi \in \mathbf{R} \,,$$

is a well-defined embedding of $\mathbf{T}^2$ into the Euclidean 3-sphere. Moreover, the base fields associated to the (global) coordinate neighbourhood $(\mathbf{R}^2, id_{\mathbf{R}^2})$ project down to $\mathbf{T}^2$ via the universal covering $\pi_{\mathbf{T}^2}:\mathbf{R}^2 \to \mathbf{T}^2$ yielding globally defined vector fields $\frac{\partial}{\partial\phi}, \frac{\partial}{\partial\psi} \in V(\mathbf{T}^2)$. Taking on $\mathbf{T}^2$ the flat metric (inherited from that of $\mathbf{R}^2$ via $\pi_{\mathbf{T}^2}$) the Laplacian Δ on $C^\infty(\mathbf{T}^2)$ is expressed as

$$\Delta = -\left(\frac{\partial}{\partial\phi}\right)^2 - \left(\frac{\partial}{\partial\psi}\right)^2 \tag{2.38}$$

(cf. I.4.24). Hence,

$$\Delta\bar{f} = \bar{f}$$

and, by 2.21, $f:\mathbf{T}^2 \to S^3$ is a harmonic map. On the other hand, the closed geodesic $c:S^1 \to T^2$, $c(\phi)=(\phi,0)$, $\phi \in R$, is by 2.8, harmonic while the composition $f \circ c:S^1 \to S^3$ is not. In particular, according to 2.11, $f:T^2 \to S^3$ is non-totally geodesic.

3. EXISTENCE AND UNIQUENESS OF HARMONIC MAPS INTO FLAT CODOMAINS

We begin here the systematic study of the structure of harmonic maps into flat codomains which is the principal object of the remaining sections of this chapter. Due to the curvature assumption on the codomain, this is essentially the only case when the classical Hodge-de Rham theory applies. The existence theorem which follows provides us, in the present situation, abundant harmonic maps.

Theorem 3.1. *Any map $f:M \to N$ of a compact oriented Riemannian manifold M into a flat complete Riemannian manifold N is homotopic to a harmonic map.*

This result, first published by Fuller (1954)(see also Eells and Sampson (1964)), is a prototype of theorems known to compose the so called *existence theory* for harmonic maps. Based on variational calculus (treated here in Chapter III), Eells and Sampson posed the following existence question:
Given Riemannian manifolds M and N, does there exist a harmonic map in each homotopy class of maps $M \to N$?

Though the study of the possible solutions of this problem far exceeds the scope of this book, we make here a few historical comments. As we know from 2.8, for $M=S^1$, harmonic maps $f:S^1 \to N$ are nothing but the closed geodesics of N, using Hilbert's method, the answer to the existence question is affirmative in this case, provided that N is compact. Restricting ourselves to maps between compact Riemannian manifolds, the first general existence theorem, based on heat equation technique was achieved by Eells and Sampson (1964) giving an affirmative answer in case when the sectional curvatures of N are nonpositive. (For a different proof, see Uhlenbeck (1970).) By direct method of calculus of variations, Lemaire solved the existence problem affirmatively for maps from surfaces into Riemannian manifolds N with trivial second homotopy group (cf. Lemaire (1975) and (1978)); the same result was also obtained by Sacks and Uhlenbeck (1977) using different methods. As for counterexamples, by a fine complex analysis, Eells and Wood (1976) have shown that there is no harmonic map of Brouwer degree 1 from the flat 2-torus into the Euclidean 2-sphere. Finally, for maps between spheres, we shall give a proof of R.T. Smith's solution of the existence problem in Chapter III. (Note also that, more recently, Schoen and Uhlenbeck (1982) and Giaquinta and Giusti (1982) have developed a powerful regularity theory for harmonic maps yielding joint treatment of some existence theorems which seemed to belong to entirely different domains.)

Turning to the proof of 3.1, for later purposes, we start with a more general setting than necessary. Let $f:M\to N$ be a map between Riemannian manifolds M and N. By a *variation* of f we mean a (smooth) 1-parameter family of maps

$$f_t:M\to N\ ,\quad t\in\mathbf{R}\ ,$$

satisfying $f_0=f$. Equivalently, a variation $t\to f_t$, $t\in\mathbf{R}$, of f can be thought of as a map

$$\phi:M\times\mathbf{R}\to N$$

with $\phi(x,t)=f_t(x)$, for $x\in M$ and $t\in\mathbf{R}$.
Given a variation $t\to f_t$, $t\in\mathbf{R}$, for fixed interval $[t',t'']\subset\mathbf{R}$, the parallel translations

$$\tau^{t'}_{t''}:T_{f_{t'}(x)}(N)\to T_{f_{t''}(x)}(N)\ ,\qquad x\in M\ ,$$

along the curves $t\to f_t(x)$, $t\in[t',t'']$, $x\in M$, define a fibre metric preserving isomorphism

$$\tau^{t'}_{t''}:\tau^{f_{t'}}\to\tau^{f_{t''}} \tag{3.2}$$

between the corresponding Riemannian-connected pull-back bundles $\tau^{f_{t'}}=(f_{t'})^*(T(N))$ and $\tau^{f_{t''}}=(f_{t''})^*(T(N))$. (Here and in what follows, we use the notations and results of I.4.) It extends to an isomorphism

$$id_{\Lambda(T^*(M))} \otimes \tau^{t'}_{t''} : \Lambda(T^*(M)) \otimes \tau^{f_{t'}} \to \Lambda(T^*(M)) \otimes \tau^{f_{t''}} \tag{3.3}$$

which is also denoted by $\tau^{t'}_{t''}$ and we omit 0's in f_0, $\tau^{t'}_0$, etc. Moreover, we set $(\tau^{t'}_{t''})^{-1} = \tau^{t''}_{t'}$, $t', t'' \in \mathbf{R}$. The differentials $(f_t)_* \in \mathcal{D}^1(\tau^{f_t})$, $t \in \mathbf{R}$, can be conveniently described by considering the 1-parameter family

$$t \to P(t) = \tau^t (f_t)_* \in \mathcal{D}^1(\tau^f), \quad t \in \mathbf{R}, \tag{3.4}$$

of 1-forms on M with values in τ^f.
(The main advantage of using (3.4) instead of the 1-parameter family of differentials $t \to (f_t)_* \in \mathcal{D}^1(\tau^{f_t})$, $t \in \mathbf{R}$, lies in the fact that we have to work in the fixed vector bundle $\Lambda(T^*(M)) \otimes \otimes \tau^f$ rather than in the 1-parameter family of bundles $t \to \to \Lambda(T^*(M)) \otimes \tau^{f_t}$, $t \in \mathbf{R}$.)

Proposition 3.5. $\left.\dfrac{\partial P(t)}{\partial t}\right|_{t=0} = dv$, *where* $v = \left.\dfrac{\partial f_t}{\partial t}\right|_{t=0} \in C^\infty(\tau^f)$.

Proof. Consider the map $\phi : M \times \mathbf{R} \to N$ defined by $\phi(x,t) = f_t(x)$, $x \in M$, $t \in \mathbf{R}$. Any vector field X on M extends canonically to $M \times \mathbf{R}$ and is denoted by the same symbol. Letting $\frac{\partial}{\partial t} \in V(M \times \mathbf{R})$ denote the unit vector field tangent to $\{x\} \times \mathbf{R}$, $x \in M$, we have $[X, \frac{\partial}{\partial t}] = \nabla_X \frac{\partial}{\partial t} - \nabla_{\frac{\partial}{\partial t}} X = 0$ and hence, by the symmetry of the second fundamental form $\beta(\phi)$, 2.7,

$$\nabla_{\frac{\partial}{\partial t}} (\phi_*(X)) = \nabla_X (\phi_*(\tfrac{\partial}{\partial t})) .$$

At $(x,0) \in M \times \mathbf{R}$, the left hand side restricts to

$$\left.\nabla_{\frac{\partial}{\partial t}}\right|_{t=0} \phi_*(X_{(x,t)}) = \lim_{t \to 0} \frac{1}{t}\{\tau^t(\phi_*(X_{(x,t)})) - \phi_*(X_x)\} =$$

$$= \lim_{t \to 0} \frac{1}{t}\{\tau^t (f_t)_*(X_x) - f_*(X_x)\} = \lim_{t \to 0} \frac{1}{t}\{P(t)X_x - P(0)X_x\} =$$

$$= \left. \frac{\partial P(t) X_x}{\partial t} \right|_{t=0}$$

while, for the right hand side, we obtain

$$\nabla_{X_x} (\phi_* (\left. \frac{\partial}{\partial t} \right|_{t=0})) = \nabla_{X_x} v = (dv)(X_x) \ . \ \checkmark$$

Remark. The proof above was indicated by Sealey. A different argument based on the Nash embedding applied to N is given in Tóth (1981) page 392.

Now we specialize our investigations in two ways. First, we assume that N is complete and *locally symmetric*. (Recall that a Riemannian manifold N is said to be locally symmetric if its Riemannian curvature tensor R commutes with parallel translations, i.e. given a curve $c:I \to N$, for $[t',t''] \subset I$, we have

$$\tau_{t''}^{t'} (R(X_{t'}, Y_{t'}) Z_{t'}) = R(\tau_{t''}^{t'} (X_{t'}) ,$$

$$\tau_{t''}^{t'} (Y_{t'})) \tau_{t''}^{t'} (Z_{t'}) ,$$

for all $X_{t'}, Y_{t'}, Z_{t'} \in T_{c(t')}(N)$.) Secondly, we shall be concerned with *geodesic variations* $t \to f_t$, $t \in \mathbf{R}$, of the given map $f:M \to N$, i.e. we assume that $t \to f_t(x)$, $t \in \mathbf{R}$, is a (linearly parametrized) geodesic for all $x \in M$. A variation of this type is completely determined by the vector field $v = \left. \frac{\partial f_t}{\partial t} \right|_{t=0} \in C^\infty(\tau^f)$ along f, namely, $f_t = \exp \circ (tv)$, $t \in \mathbf{R}$, where $\exp: T(N) \to N$ is the exponential map (cf. I.4 and also note that, by completeness of N, exp is defined on the whole of $T(N)$.) In this case, to stress the dependence of (3.3) and (3.4) on v, whenever convenient, we apply the notations $\tau_{t''}^{t'}[v] = \tau_{t''}^{t'}$ and $P_v(t) = P(t)$, respectively. In what follows, we derive a Jacobi equation for the 1-parameter family

$$t \to P_v(t) \in \mathcal{D}^1(\tau^f) , \quad t \in \mathbf{R} ,$$

defined by a variation $t \to f_t = \exp \circ (tv)$, $t \in \mathbf{R}$, $v \in C^\infty(\tau^f)$.

Lemma 3.6. *Given a geodesic variation* $t \to f_t = \exp\circ(tv)$, $t \in \mathbf{R}$, *of a map* $f: M \to N$ *into a complete locally symmetric Riemannian manifold* N , *for any vector field* w *along* f , *we have*

$$(\tau^{t_o}[v] \circ \nabla_X \circ \tau_{t_o}[v])(w) - \nabla_X w =$$

$$= -R\left(\int_0^{t_o} P_v(s) X ds, v\right) w \ , \quad t_o \in \mathbf{R} \ , \quad X \in V(M) \ , \tag{3.7}$$

where R *is the Riemannian curvature tensor of* N .

Proof. At $x \in M$ the left and right hand sides of (3.7) define curves $c_1: \mathbf{R} \to T_{f(x)}(N)$ and $c_2: \mathbf{R} \to T_{f(x)}(N)$, respectively. Obviously, $c_1(0) = c_2(0)$ and so we need only to show that $\dot{c}_1(t_o) = \dot{c}_2(t_o)$ holds for all $t_o \in \mathbf{R}$. By the notations introduced above,

$$\dot{c}_1(t_o) = \lim_{t \to 0} \frac{1}{t}\{(\tau^{t_o+t}[v] \circ \nabla_{X_x} \circ \tau_{t_o+t}[v](w) -$$

$$-(\tau^{t_o}[v] \circ \nabla_{X_x} \circ \tau_{t_o}[v])(w)\} =$$

$$= \tau^{t_o}[v] \lim_{t \to 0} \frac{1}{t}\{(\tau^{t_o+t}_{to}[v] \circ \nabla_{X_x} \circ \tau^{t_o}_{t_o+t}[v])(\tau_{t_o}[v](w)) -$$

$$-\nabla_{X_x}(\tau_{t_o}[v](w))\}$$

and, making use of local symmetry of N ,

$$\dot{c}_2(t_o) = -R(P_v(t_o)X_x, v_x)w_x =$$

$$= -\tau^{t_o}[v] R((f_{t_o})_* X_x, \tau_{t_o}[v](v_x))(\tau_{t_o}[v](w_x)) \ .$$

Thus, it remains to show that

$$\lim_{t \to 0} \frac{1}{t}\{(\tau^{t_o+t}_{t_o}[v] \circ \nabla_{X_x} \circ \tau^{t_o}_{t_o+t}[v])(\tau_t[v](w)) -$$

$$-\nabla_{X_x}(\tau_{t_o}[v](w))\} = -R((f_{t_o})_* X_x ,$$

$$(3.8)$$

$$\tau_{t_o}[v](v_x))(\tau_{t_o}[v](w_x)) .$$

Now, we are in the position to apply I.4.11 with $C:(-\varepsilon,\varepsilon)\times\times(-\varepsilon,\varepsilon)\to N$, $C(s,t) = f_{t_o+t}(c(s))$, $s,t\in(-\varepsilon,\varepsilon)$, $\varepsilon>0$, where $c:(-\varepsilon,\varepsilon)\to M$ is a curve with $\dot{c}(0)=X_x$ and $w_s = (\tau_{t_o}[v](w))_{c(s)}\in T_{C(s,0)}(N)$, $s\in(-\varepsilon,\varepsilon)$. Then, as v is a geodesic variation, we obtain $C_*(\frac{\partial}{\partial s}\Big|_{(0,0)}) = (f_{t_o})_* X_x$ and $C_*(\frac{\partial}{\partial t}\Big|_{(0,0)}) = \tau_{t_o}[v]v_x$, $C(0,0) = f_{t_o}(x)$, where $\frac{\partial}{\partial s},\frac{\partial}{\partial t}\in$ $\in V((-\varepsilon,\varepsilon)\times(-\varepsilon,\varepsilon))$ denote the canonical coordinate fields, respectively. Hence, the right hand side of I. (4.12) takes the form

$$-R((f_{t_o})_* X_x , \tau_{t_o}[v]v_x)(\tau_{t_o}[v](w_x))$$

and the left hand side reduces to that of (3.8), since, for fixed $s\in(-\varepsilon,\varepsilon)$,

$$\tau^{t_o+t'}_{t_o+t''}[v] = \tau^{t'}_{t''}(s) , \quad t',t''\in\mathbf{R} .\surd$$

Remark. When the codomain N is flat $(R=0)$ and we do not require the variation $t\to f_t$, $t\in\mathbf{R}$, to be geodesic, the same proof applies yielding that the parallel translations $\tau^{t'}_{t''}$, t' , $t''\in\mathbf{R}$, commute with the covariant differentiations of the respective bundles $\tau^{f_{t'}}$ and $\tau^{f_{t''}}$.

For fixed $t_o\in\mathbf{R}$, applying 3.5 to $f_{t_o}:M\to N$ and $\frac{\partial f_{t_o+t}}{\partial t}\Big|_{t=0} = \tau_{t_o}[v](v)\in C^\infty(\tau^{f_{t_o}})$, for $X\in V(M)$, we have

$$\nabla_X(\tau_{t_o}[v](v)) = \frac{\partial}{\partial t}\{\tau^t[\tau_{t_o}[v](v)](f_{t_o+t})_* X\}_{t=0} =$$

$$= \frac{\partial}{\partial t}\{\tau^{t_o+t}_{t_o}[v](f_{t_o+t})_* X\}_{t=0} =$$

$$= \tau_{t_o}[v]\frac{\partial}{\partial t}\{\tau^{t_o+t}[v](f_{t_o+t})_*X\}_{t=o} =$$

$$= \tau_{t_o}[v]\left(\frac{\partial P_v(t)X}{\partial t}\right)_{t=t_o} .$$

Hence, setting $w=v$ in (3.7),

$$\left.\frac{\partial P_v(t)X}{\partial t}\right|_{t=t_o} - \nabla_X v =$$

$$= -R\left(\int_o^{t_o} P_v(s)Xds, v\right)v \ , \ t_o\in\mathbf{R} . \tag{3.9}$$

Differentiating at $t_o\in\mathbf{R}$ and omitting $X\in V(M)$ and the index o , we obtain the *Jacobi equation*

$$\frac{\partial^2 P_v(t)}{\partial t^2} + R(P_v(t),v)v = 0 \tag{3.10}$$

of the geodesic variation $t \to f_t = \exp\circ(tv)$, $t\in\mathbf{R}$. By (3.9), the two initial data for (3.10) are

$$P_v(0) = f_* \quad \text{and} \quad \left.\frac{\partial P_v(t)}{\partial t}\right|_{t=o} = dv . \tag{3.11}$$

The Jacobi equation (3.10) with the initial data (3.11) can be thought of as an initial value problem for the variable $P_v(t)$, $t\in\mathbf{R}$, which has a unique solution given by a convergent Taylor series in $t\in\mathbf{R}$.

Our next purpose is to express harmonicity of the deformed maps $f_t:M\to N$, $t\in\mathbf{R}$, in terms of the 1-parameter famliy $t \to$ $\to P_v(t)\in\mathcal{D}^1(\tau^f)$, $t\in\mathbf{R}$. In fact, by (3.7), we have

$$\tau^t[v](\partial(f_t)_*) = -\tau^t[v](\text{trace}\{\nabla(f_t)_*\}) =$$

$$= -\text{trace}\{(\tau^t[v]\circ\nabla\circ\tau_t[v])P_v(t)\} =$$

$$= -\text{trace}\{\nabla P_v(t)\} + \text{trace}\{R(\int_o^t P_v(s)ds, v)P_v(t)\} ,$$

i.e. the map $f_t : M \to N$ is harmonic if and only if

$$\psi(v,t) = \partial P_v(t) + \mathrm{trace}\{R(\int_0^t P_v(s)ds, v)P_v(t)\} = 0 \,. \qquad (3.12)$$

Proof of Theorem 3.1. Let $f:M \to N$ be a map from a compact oriented Riemannian manifold M into a complete flat Riemannian manifold N. As the Riemannian curvature R of N vanishes, by (1.5), we have

$$d^2 = 0 \qquad (3.13)$$

on the twisted bundle $\Lambda(T^*(M)) \otimes \tau^f$. The proof of I.6.12 applies verbatim yielding the refined Hodge-de Rham decomposition

$$\mathcal{D}(\tau^f) = d(\mathcal{D}(\tau^f)) \oplus \partial(\mathcal{D}(\tau^f)) \oplus \mathcal{H}(\tau^f) \,. \qquad (3.14)$$

In particular, there exists a harmonic 1-form ω on M with values in τ^f such that

$$f_* = dv + \partial\lambda + \omega$$

holds for some $v \in \mathcal{D}^0(\tau^f)(=C^\infty(\tau^f))$ and $\lambda \in \mathcal{D}^2(\tau^f)$. By virtue of 2.1, $0 = df_* = d\partial\lambda$ and hence $0 = ((d\partial\lambda, \lambda)) = ((\partial\lambda, \partial\lambda))$, i.e. the decomposition of f_* above reduces to

$$f_* = dv + \omega \,. \qquad (3.15)$$

Consider now the geodesic variation $t \to f_t = \exp \circ (tv)$, $t \in \mathbf{R}$, of f given by the vector field v along f. Then, as $R=0$, the unique solution of the initial value problem (3.10) with (3.11) takes the form

$$P_v(t) = f_* + t\,dv$$

and substituting this into (3.12) and using (3.15), we have

$$\psi(v,t) = \partial f_* + t\partial dv = \partial(dv+\omega) + t\partial dv = (1+t)\partial dv \,.$$

It follows that $f_{-1} = \exp\circ(-v):M\to N$ is a harmonic map homotopic to f . √

Remark. For a different proof of 3.1, see Nagano and Smyth (1975)

By 3.1, each homotopy class of maps from a compact oriented Riemannian manifold M into a complete flat Riemannian manifold N contains a harmonic representative. Next, we investigate to what extent this representative is unique, i.e. we deal with the uniqueness problem of harmonic maps.

Theorem 3.16. *Let $f,f':M\to N$ be harmonic maps of a compact oriented Riemannian manifold M into a complete flat Riemannian manifold N. If f and f' are homotopic then they are also homotopic through harmonic maps, i.e. there exists a (smooth) 1-parameter family of harmonic maps*

$$h_t:M\to N\ ,\quad t\in[0,1]\ ,$$

such that $h_0=f$ and $h_1=f'$.

Proof. Choose a (smooth) 1-parameter family of maps

$$f_t:M\to N\ ,\quad t\in[0,1]\ ,$$

with $f_0=f$, $f_1=f'$. Given $t\in[0,1]$, by the proof of 3.1, the Hodge-de Rham decomposition of the differential $(f_t)_*\in\mathcal{D}^1(\tau^{f_t})$ reduces to

$$(f_t)_* = dv^t + \omega^t\ , \tag{3.17}$$

where $\omega^t\in H^1(\tau^{f_t})$ is the (uniquely defined) harmonic part of $(f_t)_*$ and $v^t\in C^\infty(\tau^{f_t})$. In general, $v^t\in C^\infty(\tau^{f_t})$, $t\in[0,1]$, is unique only up to an additive parallel section of τ^{f_t} . But by fixing a base point $o\in M$ and requiring

$$v_o^t = 0\ , \tag{3.18}$$

the vector field v^t along f_t occurring in (3.17) is unique; in particular, $v^0=v^1=0$. (Here we use the fact that a parallel section of τ^{f_t} vanishing at $o\in M$ is identically zero on M; cf. the proof of I.6.7.) Due to the curvature constraint $R=0$, from Remark after 3.6, it follows that the parallel translations $\tau^{t'}_{t''}:\Lambda(T^*(M))\otimes\tau^{f_{t'}} \to \Lambda(T^*(M))\otimes\tau^{f_{t''}}$, $t',t''\in[0,1]$, commute with the covariant differentiations ∇ of the respective bundles $\Lambda(T^*(M))\otimes\tau^{f_{t'}}$ and $\Lambda(T^*(M))\otimes\tau^{f_{t''}}$. As the exterior differentiation d and codifferentiation ∂ (and hence the Laplacian Δ) can be expressed via ∇ (cf. (1.3) and (1.6)), they also commute with $\tau^{t'}_{t''}$, $t',t''\in[0,1]$. In particular,

$$P(t) = d\tau^t(v^t) + \tau^t(\omega^t)$$

is nothing but the Hodge-de Rham decomposition of the 1-form $P(t)=\tau^t(f_t)_*$ with values in $\tau^f(=\tau^{f_0})$. We claim that the 1-parameter families

$$t \to \tau^t(\omega^t) \quad\text{and}\quad t\to\tau^t(v^t)\ ,\ t\in[0,1]\ ,$$

are smooth in t. More generally, given a smooth 1-parameter family $t\to\sigma(t)\in\mathcal{D}^1(\tau^f)$, with $d\sigma(t)=0$, $t\in[0,1]$, we shall prove that $t\to v(t)\in C^\infty(\tau^f)$, with $v(t)_o=0$, $t\in[0,1]$, and $t\to\omega(t)\in H^1(\tau^f)$, $t\in[0,1]$, occurring in the Hodge-de Rham decomposition

$$\sigma(t) = dv(t) + \omega(t)\ ,\quad t\in[0,1]\ , \tag{3.19}$$

are also smooth in t.
As $H^1(\tau^f)\subset\mathcal{D}^1(\tau^f)$ is a finite dimensional linear subspace, the orthogonal projection $H:\mathcal{D}^1(\tau^f)\to H^1(\tau^f)$ (with respect to the prehilbert structure on $\mathcal{D}^1(\tau^f)$ given by the global scalar product) is a continuous linear operator so that

$$t\to H(\sigma(t)) = \omega(t) \quad\text{and}\quad t\to dv(t)\ ,\ t\in[0,1]\ ,$$

are continuous in t. Setting

$$\Omega = \{x\in M \mid t\to v(t)_x\ ,\ t\in[0,1]\ ,\ \text{is continuous in } t\}$$

we show that $\Omega=M$ holds. Indeed, if $x\in\Omega$ then, taking a coordinate neighbourhood (U,ϕ) around x with vanishing Christoffel symbols (cf.I.4.24), by (2.5) and (2.6), we have

$$dv(t) = \sum_{i=1}^{m}\sum_{r=1}^{n} {}^{M}\partial_i v^r(t)\cdot dx^i\otimes{}^{N}\partial_r\circ f ,$$

where

$$v = \sum_{r=1}^{n} v^r(t)\cdot{}^{N}\partial_r\circ f , \quad v^r(t)\in C^\infty(U) , \quad t\in[0,1] , \quad r=1,\ldots,n,$$

with $m=\dim M$ and $n=\dim N$. By hypothesis, $t\to v^r(t)_x$, $t\in[0,1]$, is continuous in t at x and the partial derivative $t\to {}^{M}\partial_i v^r(t)$, $t\in[0,1]$, is continuous in t *on the whole of* U, for all $i=1,\ldots,m$ and $r=1,\ldots,n$. It follows that all the scalars $t\to v^r(t)$, $t\in[0,1]$, $r=1,\ldots,n$, are continuous in t at each point of U. (In fact, given $y\in U$, choose a curve $c:[0,1]\to U$ with $c(0)=x$ and $c(1)=y$. Then, using the decomposition

$$\dot{c} = \sum_{i=1}^{m} a^i\cdot{}^{M}\partial_i\circ c , \quad a^i\in C^\infty([0,1]) , \quad i=1,\ldots,m ,$$

we have

$$v^r(t)_y = v^r(t)_x + \int_0^1 \frac{\partial v^r(t)_{c(s)}}{\partial s} ds =$$

$$= v^r(t)_x + \int_0^1 \dot{c}(s)\{v^r(t)\} ds =$$

$$= v^r(t)_x + \sum_{i=1}^{m}\int_0^1 a^i(s)\{{}^{M}\partial_i v^r(t)\}(c(s))\, ds ,$$

which is continuous in t.) Hence, $\Omega\subset M$ is open (and nonvoid since $o\in\Omega$) and a standard continuation argument shows that actually $\Omega=M$, or equivalently, $t\to v(t)$, $t\in[0,1]$, is continuous in t *throughout* M.

Now, apply the argument above to the 1-parameter family

$$t \to \frac{\partial\sigma(t)}{\partial t}\in D^1(\tau^f) , \quad t\in[0,1] ,$$

with Hodge-de Rham decomposition

$$\frac{\partial\sigma(t)}{\partial t} = dv'(t) + \omega'(t) , \tag{3.20}$$

where $v'(t)\in C^\infty(\tau^f)$, $v'(t)_o=0$, and $\omega'(t)\in H^1(\tau^f)$, $t\in[0,1]$. (Note that $d(\frac{\partial\sigma(t)}{\partial t}) = \frac{\partial}{\partial t}\{d\sigma(t)\} = 0$, $t\in[0,1]$, and hence $\frac{\partial\sigma(t)}{\partial t}$ has the decomposition (3.20) above.) It follows that

$$t \to \omega'(t) \quad \text{and} \quad t \to v'(t) , \ t\in[0,1],$$

are continuous in t . Integrating (3.20), we have

$$\sigma(t) - \sigma(0) = \int_0^t \frac{\partial\sigma(s)}{\partial s}ds = \int_0^t \{dv'(s)\}ds +$$

$$+ \int_0^t \omega'(s)ds = d\{\int_0^t v'(s)ds\} + \int_0^t \omega'(s)ds$$

which is the Hodge-de Rham decomposition of $\sigma(t)-\sigma(0)$ since $\int_0^t \omega'(s)ds\in H^1(\tau^f)$. (In fact, the integration takes place in the (finite dimensional) vector space $H^1(\tau^f)$.) Thus, by uniqueness,

$$\int_0^t \omega'(s)ds = \omega(t) - \omega(0) \quad \text{and}$$

$$\int_0^t v'(s)ds = v(t) - v(0) , \quad t\in[0,1] ,$$

and we obtain that

$$t \to \omega(t) \quad \text{and} \quad t \to v(t), \ t\in[0,1] ,$$

are of class C^1 in t . Now, by an easy induction, the claim follows. In particular, the 1-parameter family

$$t \to v^t\in C^\infty(\tau^{f_t}) , \quad t\in[0,1] ,$$

is smooth in t . Setting

$$h_t = \exp\circ(-v^t) \ , \quad t\in[0,1] \ ,$$

the proof of 3.1 applies yielding that $h_t: M \to N$ is a harmonic map for each $t\in[0,1]$. As $v_0=0$ and $v_1=0$ we obtain $h_0=f_0$ and $h_1=f_1$. √

Remark. 3.16, in a more general form, is due to Hartman (1967). Namely, applying the heat equation technique, he showed that any two homotopic harmonic maps $f,f': M \to N$ of a compact Riemannian manifold into a complete Riemannian manifold N with nonpositive sectional curvatures are homotopic through harmonic maps. (Obviously, the curvature assumption cannot be dropped, e.g. take $f: S^1 \to S^2$ to be the canonical inclusion and $f': S^1 \to S^2$ a constant map.) Moreover, if N is locally symmetric then, according to a result of Sunada (1979), the universal lifts

$$\tilde{f}, \tilde{f}': \tilde{M} \to \tilde{N}$$

differ by performing an isometry of $\tilde{N}$, i.e. there exists $a\in I(\tilde{N})$ such that $\tilde{f}'=a\circ\tilde{f}$. Finally, the full description of the space of harmonic maps $M \to N$ in a given homotopy class was achieved by Schoen and Yau (1979), namely, under the hypothesis that M is compact and N is complete real analytic with nonpositive sectional curvatures, they proved that this space can be identified with a compact immersed totally geodesic submanifold of N

4. THE ALBANESE MAP

Theorem 3.1 of the previous section establishes the existence of a harmonic map in a given homotopy class of maps from a compact oriented Riemannian manifold into a complete flat Riemannian manifold but the proof does not involve a constructive method to build up such harmonic maps. Following the ideas of Lichnerowicz in this section we give an effective way of describing all harmonic maps into flat tori.

Let M denote an m-dimensional compact oriented Riemannian manifold with metric tensor $g=(,)\in C^\infty(S^2(T^*(M)))$. Fixing a base point $o\in M$, we consider the universal covering manifold $\tilde{M}$ represented by the homotopy classes of curves in M starting at o . Then, $\tilde{o}\in\tilde{M}$ will denote the class of 0-homotopic loops and we endow $\tilde{M}$ with the pull-back metric $\tilde{g}=(,)\in C^\infty(S^2(T^*(\tilde{M})))$ induced by the universal covering projection $\pi_M:\tilde{M}\to M$ (cf. 2.12). Let $H^1(M)^*$ denote the dual of the (finite dimensional) vector space of harmonic 1-forms $H^1(M)$ on M (endowed with the global scalar product as a Euclidean structure) and transport the scalar product of $H^1(M)$ to $H^1(M)^*$. With respect to the induced metric tensor, $H^1(M)^*$ becomes a flat Riemannian manifold (cf. I.4.24).

For $\tilde{x}\in\tilde{M}$, we define a linear functional $\tilde{J}(\tilde{x})\in H^1(M)^*$ on $H^1(M)$ as follows.

The pull-back $\pi_M^*(\omega)$ of a harmonic 1-form ω on M via $\pi_M:$ $:\tilde{M}\to M$ is, by I.3.8, closed and hence, applying I.3.14, it follows that the equation

$$du = \pi_M^*(\omega) \tag{4.1}$$

can be solved yielding a scalar $u\in C^\infty(\tilde{M})$ determined uniquely up to an additive constant. Setting

$$\tilde{J}(\tilde{x})[\omega] = (\tilde{J}(\tilde{x}),\omega) = u(\tilde{x}) - u(\tilde{o}) , \tag{4.2}$$

we obtain a well-defined map

$$\tilde{J}:\tilde{M} \to H^1(M)^* .$$

Furthermore, a scalar u , satisfying (4.1), is automatically harmonic since

$$\Delta u = \partial(du) = \partial(\pi_M^*(\omega)) = \pi_M^*(\partial\omega) = 0$$

and thus, the components of $\tilde{J}$ (with respect to a base in $H^1(M)$) being harmonic functions on $\tilde{M}$, by 2.12, we obtain that $\tilde{J}:\tilde{M}\to H^1(M)^*$ is a harmonic map.

Proposition 4.3. *The map* $\tilde{J}:\tilde{M}\to H^1(M)^*$ *is full, i.e. the image* $im(\tilde{J})\subset H^1(M)^*$ *is not contained in any (affine) hyperplane of* $H^1(M)^*$.

Proof. Assume the existence of a harmonic 1-form ω on M such that $\tilde{J}(\tilde{x})[\omega]=0$ holds for all $\tilde{x}\in\tilde{M}$. Then, setting $du==\pi_M^*(\omega)$, $u\in C^\infty(\tilde{M})$, by (4.2), the scalar u is constant on $\tilde{M}$. Equivalently, $du=0$ and hence $\omega=0$.

Now, we digress to explain some concepts in the de Rham isomorphism theorem to be used subsequently. (For details, see Warner (1970).) For each $r\in\mathbf{Z}_+$ a *smooth singular r-simplex in* M is a (smooth) map from the standard r-simplex $\Delta^r\subset\mathbf{R}^{r+1}$ into M . Let $S_r(M;\mathbf{R})$ denote the (real) vector space generated by the smooth singular r-simplices in M . Then, an element of $S_r(M;\mathbf{R})$ (called a *smooth singular r-chain in* M) is nothing but a finite linear combination of smooth singular r-simplices in M with real coefficients. For $r\in\mathbf{Z}$ with $r<0$, we set $S_r(M;\mathbf{R})=\{0\}$. Just as in continuous singular homology theory, we can define the *boundary operator*

$$\partial:S_{r+1}(M;\mathbf{R})\to S_r(M;\mathbf{R})\ ,\quad r\in\mathbf{Z}\ ,$$

having the property $\partial^2=0$. A smooth singular r-chain z is said to be an *r-cycle* if $\partial z=0$. It is called an *(r+1)-boundary* if there exists a smooth singular $(r+1)$-chain z' such that $z=\partial z'$. Since $\partial^2=0$, every $(r+1)$-boundary is an r-cycle. The quotient of the vector space of r-cycles modulo the linear subspace of $(r+1)$-boundaries is said to be the *r-th real smooth singular homology* $H_r(M;\mathbf{R})$ of M . Using integer coefficients, the $\mathbf{Z}$-modules $S_r(M;\mathbf{Z})$, $r\in\mathbf{Z}$, of *smooth singular r-chains with coefficients in* $\mathbf{Z}$ (and hence the *integral smooth singular homology modules* $H_r(M;\mathbf{Z})$, $r\in\mathbf{Z}$) are defined analogously and the inclusions $S_r(M;\mathbf{Z})\to S_r(M;\mathbf{R})$, $r\in\mathbf{Z}$, give rise to homomorphisms

$$\rho_r:H_r(M;\mathbf{Z})\to H_r(M;\mathbf{R})\ ,\quad r\in\mathbf{Z}\ ,$$

of $\mathbf{Z}$-modules, where the $\mathbf{Z}$-module structure on $H_r(M;\mathbf{R})$ is given by restriction from $\mathbf{R}$ to $\mathbf{Z}$. As M is compact, $H_r(M;\mathbf{Z})$

is known to be finitely generated (Spanier (1966) page 298) and hence $im\,\rho_r \subset H_r(M;\mathbf{R})$, $r \in \mathbf{Z}$, is a discrete subgroup. Moreover, by the *universal coefficient theorem* (Spanier (1966) page 222)

$$H_r(M;\mathbf{Z}) \otimes_{\mathbf{Z}} \mathbf{R} = H_r(M;\mathbf{R}) \ , \qquad r \in \mathbf{Z} \ ,$$

or equivalently, $im\,\rho_r \subset H_r(M;\mathbf{R})$ spans $H_r(M;\mathbf{R})$ over $\mathbf{R}$. Thus, $im\,\rho_r \subset H_r(M;\mathbf{R})$ is a lattice.

If z is an r-cycle representing a real smooth singular homology class $\{z\} \in H_r(M;\mathbf{R})$ and α a closed r-form on M representing a de Rham cohomology class $\{\alpha\} \in H^r_{de\ R}(M;\mathbf{R})$ then the integral

$$\int_z \alpha$$

(defined by linearity and pulling back α to the standard simplex Δ^r via the smooth singular r-simplices in z) only depends on the equivalence classes $\{z\}$ and $\{\alpha\}$ and hence it defines a linear map

$$\int : H_r(M;\mathbf{R}) \to (H^r_{de\ R}(M;\mathbf{R}))^* = H^r(M)^* \ , \quad r \in \mathbf{Z} \ , \tag{4.4}$$

where, by I.6.16, we identify $H^r_{de\ R}(M;\mathbf{R})$ with $H^r(M)$. The de Rham isomorphism theorem asserts that this is an isomorphism. (For a proof based on sheaf theory, see Warner (1970).) In particular, the subgroup $D = im(\int \circ \rho_1) \subset H^1(M)^*$ is a lattice and, dividing out, we obtain the *Albanese torus*

$$A(M) = H^1(M)^*/D \tag{4.5}$$

of M (with base point $\{D\}$). Its dimension is, by I.6.16, nothing but the first Betti number $b_1(M)$ of M . We take on $A(M)$ the flat metric inherited from that of $H^1(M)^*$ via the canonical projection

$$p_M : H^1(M)^* \to A(M) \ .$$

The fundamental group $\pi_1(M,o)$ acts (on the right) on the universal covering manifold $\tilde{M}$ by means of which $\pi_M : \tilde{M} \to M$

becomes the projection of a principal bundle with structure group $\pi_1(M,o)$ (cf. Steenrod (1951)).

Proposition 4.6. *For* $\sigma \in \pi_1(M,o)$,

$$\tilde{J}(\tilde{x}\sigma) - \tilde{J}(\tilde{x}) \in D , \tag{4.7}$$

for all $\tilde{x} \in \tilde{M}$.

Proof. Represent the homotopy class $\tilde{x} \in \tilde{M}$ by a curve $c:[0,1] \to M$ with $c(0)=o$, i.e. set $\tilde{x}=\{c\}$. Then, $\tilde{x}\sigma=\{c\cdot s\}$, where σ is represented by a loop s in M at o and $\cdot$ stands for the product of curves. Define

$$\tilde{z}:[0,1] \to \tilde{M}$$

by

$$\tilde{z}(t) = \begin{cases} \{c_{1-2t}\} , & 0 \le t \le \frac{1}{2} \\ \{(c\cdot s)_{2t-1}\} , & \frac{1}{2} < t \le 1 , \end{cases}$$

where, for any curve $c':[0,1] \to M$ and $t \in [0,1]$, we set $c'_t : [0,1] \to M$, $c'_t(t')=c'(tt')$, $t' \in [0,1]$. Then $\tilde{z}(0)=\tilde{x}$ and $\tilde{z}(1)=\tilde{x}\sigma$ and hence, for each harmonic 1-form ω on M with $du=\pi_M^*(\omega)$, $u \in C^\infty(\tilde{M})$, we have

$$(\tilde{J}(\tilde{x}\sigma)-\tilde{J}(\tilde{x}))[\omega] = u(\tilde{x}\sigma) - u(\tilde{o}) - u(\tilde{x}) + u(\tilde{o}) =$$

$$= u(\tilde{x}\sigma) - u(\tilde{x}) = \int_{\tilde{z}} du = \int_{\tilde{z}} \pi_M^*(\omega) = \int_{\pi_M \circ \tilde{z}} \omega ,$$

where we used the Stokes' theorem. On the other hand, $\pi_M \circ \tilde{z}$ and $c\cdot s\cdot c^{-1}$ represent the same class in $H_1(M;\mathbb{Z})$, i.e.

$$\tilde{J}(\tilde{x}\sigma) - \tilde{J}(\tilde{x}) = \int_{\pi_M \circ \tilde{z}} = \int_{c\cdot s\cdot c^{-1}} \in D . \checkmark$$

By (4.7), there is a unique map $J:M \to A(M)$ making the diagram

$$\begin{array}{ccc} \tilde{M} & \xrightarrow{\tilde{J}} & H^1(M)^* \\ \Big\downarrow{\scriptstyle \pi_M} & & \Big\downarrow{\scriptstyle p_M} \\ M & \xrightarrow{J} & A(M) \end{array} \tag{4.8}$$

commutative. The vertical projections being local isometries, it follows that J is a harmonic map. We call $J: M \to A(M)$ the *Albanese map of* M.

Theorem 4.9. Let *$f: M \to N$ be a base point preserving map between compact oriented Riemannian manifolds M and N with base points $o \in M$ and $f(o) \in N$, respectively. Assume that f pulls back harmonic 1-forms on N to harmonic 1-forms on M. Then there is a unique totally geodesic map $A(f): A(M) \to A(N)$ making the diagram*

$$\begin{array}{ccc} M & \xrightarrow{f} & N \\ \Big\downarrow{\scriptstyle J_M} & & \Big\downarrow{\scriptstyle J_N} \\ A(M) & \xrightarrow{A(f)} & A(N) \end{array} \tag{4.10}$$

commutative, where the vertical arrows denote the respective Albanese maps.

Proof. A base point preserving map $A: A(M) \to A(N)$ between the Albanese tori $A(M)$ and $A(N)$ is totally geodesic if and only if its universal covering $\tilde{A}: H^1(M)^* \to H^1(N)^*$ (cf. (2.15)) is linear. (In fact, using the notations of I.4.24,

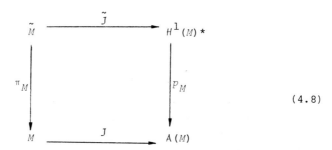

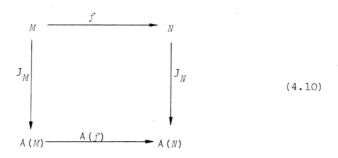

$$((\nabla\tilde{A}_*)(\partial_i,\partial_j))\check{} = \partial_i(A_*(\partial_j))\check{} = \partial_i\partial_j\tilde{A} ,$$

$$i,j=1,\ldots,p = b_1(M) ,$$

where $\{\partial_i\}_{i=1}^{p} \subset V(H^1(M)^*)$ denote the base fields associated to a linear isometry $H^1(M)^* \cong \mathbf{R}^p$.) On the other hand, a linear map $\tilde{A}:H^1(M)^* \to H^1(N)^*$ is completely determined by its action on the full subset $im(\tilde{J}_M) \subset H^1(M)^*$, 4.3, and the uniqueness follows.

As for the existence, the pull-back $f^*:\mathcal{D}^1(N) \to \mathcal{D}^1(M)$, by hypothesis, restricts to a linear map $f^*:H^1(N) \to H^1(M)$ and we first claim that the diagram

$$\begin{array}{ccc} \tilde{M} & \xrightarrow{\ \tilde{f}\ } & \tilde{N} \\ \Big\downarrow{\scriptstyle \tilde{J}_M} & & \Big\downarrow{\scriptstyle \tilde{J}_N} \\ H^1(M)^* & \xrightarrow{\ {}^t(f^*)\ } & H^1(N)^* \end{array}$$

commutes, where $\tilde{f}:\tilde{M} \to \tilde{N}$ is the universal covering map of f sending $\tilde{o}\in\tilde{M}$ to the class $f(o)\tilde{}=\tilde{f}(\tilde{o})\in\tilde{N}$ of 0-homotopic loops at $f(o)$. Indeed, for each harmonic 1-form ω on N with $du=\pi_N^*(\omega)$, $u\in C^\infty(\tilde{N})$, by I.3.8 and (2.15), we have

$$\pi_M^*(f^*(\omega)) = (f\circ\pi_M)^*(\omega) = (\pi_N\circ\tilde{f})^*(\omega) =$$

$$= \tilde{f}^*(\pi_N^*(\omega)) = \tilde{f}^*(du) = d(\tilde{f}^*(u)) = d(u\circ\tilde{f}) ,$$

and hence, for $\tilde{x}\in\tilde{M}$,

$$(({}^t(f^*)\circ\tilde{J}_M)(\tilde{x}))[\omega] = (({}^t(f^*)\circ\tilde{J}_M)(\tilde{x}),\omega) =$$

$$= (\tilde{J}_M(\tilde{x}),f^*(\omega)) = (u\circ\tilde{f})(\tilde{x}) - (u\circ\tilde{f})(\tilde{o}) =$$

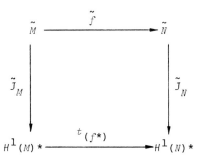

$$= u(\tilde{f}(\tilde{x})) - u(f(o)^{\sim}) = (\tilde{J}_N(\tilde{f}(\tilde{x})),\omega) =$$

$$= ((\tilde{J}_N \circ \tilde{f})(\tilde{x}))[\omega] .$$

Secondly, let $D_M \subset H^1(M)^*$ and $D_N \subset H^1(N)^*$ denote the lattices corresponding to the $\mathbf{Z}$-module homomorphisms

$$H_1(M;\mathbf{Z}) \to H_1(M;\mathbf{R}) \quad \text{and} \quad H_1(N;\mathbf{Z}) \to H_1(N;\mathbf{R}) ,$$

respectively. If z is a 1-cycle representing an integral homology class of M then, for each harmonic 1-form ω on M,

$$({}^t(f^*) \circ (\textstyle\int_z))[\omega] = ({}^t(f^*) \circ (\textstyle\int_z), \omega) =$$

$$= (\textstyle\int_z, f^*(\omega)) = \displaystyle\int_z f^*(\omega) = \int_{f_*(z)} \omega ,$$

where $f_*: H_1(M;\mathbf{Z}) \to H_1(N;\mathbf{Z})$ is the homomorphism induced by f on homology. Equivalently, ${}^t(f^*)(D_M) \subset D_N$, and hence the linear map ${}^t(f^*): H^1(M)^* \to H^1(N)^*$ projects down yielding a totally geodesic map $A(f): A(M) \to A(N)$ such that the diagram

$$\begin{array}{ccc} H^1(M)^* & \xrightarrow{\ {}^t(f^*)\ } & H^1(N)^* \\ \downarrow p_M & & \downarrow p_N \\ A(M) & \xrightarrow{\ A(f)\ } & A(N) \end{array}$$

is commutative, where the vertical arrows are the respective canonical projections. Summarizing, all sides and the upper face of the cube

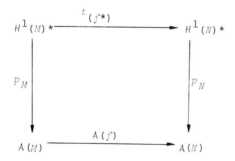

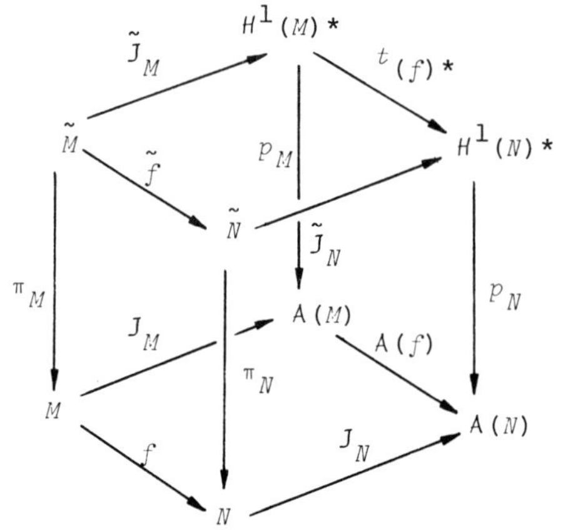

(4.11)

commute and, $\pi_M:\tilde{M}\to M$ being onto, the bottom face also commutes. √

In what follows, we consider specific situations in which the hypothesis of 4.9 is satisfied.

Lemma 4.12. *If* $f:M\to N$ *is a harmonic map between Riemannian manifolds* M *and* N *then* $f^*(P^1(N))\subset H^1(M)$ *, or, in other words,* f *pulls back parallel* 1-*forms on* N *to harmonic* 1-*forms on* M .

Proof. Setting $\omega\in P^1(N)$, by I.(5.6), ω is closed and, applying I.3.8, it follows that $\alpha=f^*(\omega)$ is a closed 1-form on M . Furthermore, choosing a local orthonormal moving frame $\{E^i\}_{i=1}^m$, $m=\dim M$, on M , we have

$$-\partial\alpha = \sum_{i=1}^{m} (\nabla_{E^i}\alpha)(E^i) = \sum_{i=1}^{m} E^i(\alpha(E^i)) - \sum_{i=1}^{m} \alpha(\nabla_{E^i}E^i) =$$

$$= \sum_{i=1}^{m} E^i(\omega(f_*(E^i))) - \sum_{i=1}^{m} \omega(f_*(\nabla_{E^i}E^i)) =$$

$$= \sum_{i=1}^{m} \omega(\nabla_{E^i}(f_*(E^i))) - \sum_{i=1}^{m} \omega(f_*(\nabla_{E^i}E^i)) =$$

$$= \omega\left(\sum_{i=1}^{m} (\nabla_{E^i} f_*)(E^i)\right) = \omega(\mathrm{trace}\{\nabla f_*\}) = 0 \ . \ \checkmark$$

As the next example shows, 4.12 cannot be generalized to r-forms, $r \geq 2$.

Example 4.13. Let $\alpha \in C^\infty(\mathbf{R})$ be the (periodic) solution of the pendulum equation

$$\ddot{x} + \sin x \cdot \cos x = 0 \tag{4.14}$$

with initial data $\alpha(0)=0$ and $\dot{\alpha}(0)=a$, where $(0<)|a|<1$. Denote by Θ the quotient of the Euclidean 2-plane $\mathbf{R}^2$ modulo the lattice $(\omega\mathbf{Z})\times(2\pi\mathbf{Z})\subset\mathbf{R}^2$, where $\omega>0$ is the period of α . Then Θ is a flat 2-torus and the *Delaunay map*

$$f:\Theta \to S^2$$

given by

$$\bar{f}(\phi,\psi) = (-\cos\psi\cdot\cos\alpha(\phi), \quad -\sin\psi\cdot\cos\alpha(\phi), \sin\alpha(\phi)) \ , \quad \phi,\psi\in\mathbf{R} \ , \tag{4.15}$$

is well-defined. Its image is seen to cover a band around an equator of S^2 ; in particular, except for two closed curves, f has rank 2 on Θ . As, by 2.37,

$$\Delta = -\left(\frac{\partial}{\partial\phi}\right)^2 - \left(\frac{\partial}{\partial\psi}\right)^2$$

an easy computation shows that

$$\Delta\bar{f} = (\dot{\alpha}^2+\cos^2\alpha)\bar{f} \ . \tag{4.16}$$

According to 2.21, the Delaunay map $f:\Theta\to S^2$ is harmonic. (Note that, following Calabi, the Delaunay map can also be interpreted as the Gauss map of an embedded surface in $\mathbf{R}^3$ with constant mean curvature, see Delaunay (1841 and Eells (1978)). As parallel translation along any curve of a Riemannian manifold M with metric tensor g maps, by I.4.17, every orthonormal frame into an orthonormal frame and preserves the orientation, the volume form $\mathrm{vol}(M,g)$ is parallel on M (cf. I.4.8); in particular,

$$\mathrm{vol}(S^2)\in P^2(S^2) \quad \text{and} \quad \mathrm{vol}(\Theta)\in P^2(\Theta) .$$

Setting

$$f^*(\mathrm{vol}(S^2)) = \mu\cdot\mathrm{vol}(\Theta) , \quad \mu\in C^\infty(\Theta) , \tag{4.17}$$

and *assuming* $f^*(\mathrm{vol}(S^2))\in H^2(\Theta)$, by (4.17), we have

$$0 = \partial(f^*(\mathrm{vol}(S^2))) = -\mathrm{trace}\ \nabla\{f^*(\mathrm{vol}(S^2))\} =$$

$$= -\mathrm{trace}\ \nabla\{\mu\cdot\mathrm{vol}(\Theta)\} = -\mathrm{trace}\{(d\mu)\otimes\mathrm{vol}(\Theta)\} =$$

$$= -\mathrm{vol}(\Theta)(\gamma^{-1}(d\mu),.) .$$

The volume form being nondegenerate, we obtain that $d\mu=0$, i.e. μ is constant. Hence, by (4.17), f has constant rank on Θ which is impossible. Thus, the pull-back of $\mathrm{vol}(S^2)\in$ $\in P^2(S^2)$ via f is not harmonic.

Remark. Maps $f:M\to N$ between Riemannian manifolds M and N satisfying

$$\partial\circ f^* = f^*\circ\partial$$

on 1-forms on N (respectively, on r-forms on N for some $r\geq 2$) have been classified by Watson (1975), namely, he showed that f is then a Riemannian harmonic (respectively, totally geodesic) submersion.

On a compact oriented Riemannian manifold with everywhere positive semidefinite Ricci tensor field each harmonic 1-form is, by I.6.7, parallel and, as a special instance of 4.9, we obtain from 4.12 the following.

Theorem 4.18. *Let $f:M\to N$ be a base point preserving harmonic map between compact oriented Riemannian manifolds M and N*

with base points $o \in M$ *and* $f(o) \in N$ *, respectively, and assume that* Ric^N *is positive semidefinite everywhere on* N *. Then there exists a unique totally geodesic map* $A(f):A(M) \to A(N)$ *such that the diagram*

$$\begin{array}{ccc} M & \xrightarrow{\quad f \quad} & N \\ \Big\downarrow J_M & & \Big\downarrow J_N \\ A(M) & \xrightarrow{\quad A(f) \quad} & A(N) \end{array} \qquad (4.19)$$

commutes. In particular, if $N=\Theta$ *is a flat torus, any harmonic map* $f:M \to \Theta$ *is the composition of the Albanese map* $J_M:M \to A(M)$ *and a (uniquely determined) totally geodesic map* $\upsilon:A(M) \to \Theta$ *.*

Proof. The first assertion being clear, we take $N=\Theta$ to be a flat torus. Then, as $T^*(\Theta)$ is spanned by everywhere linearly independent harmonic (in fact, parallel) 1-forms, a rethinking of the construction of the Albanese map will then yield that $\tilde{J}_\Theta:\tilde{\Theta} \to H^1(\Theta)^*$ is a totally geodesic diffeomorphism between Euclidean vector spaces, i.e. $J_\Theta:\Theta \to A(\Theta)$ is a totally geodesic diffeomorphism between the flat tori Θ and $A(\Theta)$ as well. The rest is clear. √

Remark 1. The second assertion of 4.18, due to Lichnerowicz (1971), will be called the *factorization theorem.*

Remark 2. A map $f:M \to \Theta$ of a compact oriented Riemannian manifold M into a flat torus Θ is harmonic if and only if f pulls back harmonic 1-forms on Θ to harmonic 1-forms on M (cf. Nagano and Smyth (1975)). In fact, by I.6.7 and 4.12, harmonic maps enjoy this property, and conversely, by the computations given in the proof of 4.12, a map $f:M \to \Theta$ with this property satisfies

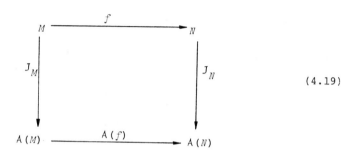

$$\omega(\mathrm{trace}\{\nabla f_*\}) = 0$$

for each harmonic 1-form ω on Θ. As $T^*(\Theta)$ is spanned by pointwise linearly independent harmonic 1-forms, it follows that $\mathrm{trace}\{\nabla f_*\}=0$, i.e. the map $f:M\to\Theta$ is harmonic.

In case of toral codomains a sharpening of the Hartman's uniqueness, 3.16, (and also the Sunada's rigidity (cf. Remark after 3.16)) is given as follows.

Proposition 4.20. *If $f,f':M\to\Theta$ are homotopic harmonic maps of a compact oriented Riemannian manifold M into a flat torus Θ then f and f' differ by a translation of Θ.*

Proof. Fixing a base point $o\in M$, by performing a translation of Θ if necessary, we may put $f(o)=f'(o)$. By I.3.10, $f^*=f'^*$ on cohomology. Hence, if $\omega\in H^1(\Theta)$ is the (unique) representative of a de Rham cohomology class in $H^1_{de\ R}(\Theta;\mathbf{R})$ then $\alpha= =f^*(\omega)-f'^*(\omega)$ is an exact 1-form on M. On the other hand, by Remark 2 above, $\alpha\in H^1(M)$ and hence vanishes, i.e.

$$f^*(\omega) = f'^*(\omega) .$$

Setting

$$\Omega = \{x\in M | f(x) = f'(x)\} ,$$

as the elements of $H^1(\Theta)$ span $T^*(\Theta)$ (pointwise), it follows that $\Omega\subset M$ is open (and nonempty, since $o\in\Omega$) and a standard continuation argument shows that actually $\Omega=M$. √

Remark. The proof above is taken from Nagano and Smyth (1975). They also give a description of compact oriented Riemannian manifolds M admitting harmonic immersions into flat tori Θ. (In this case, the Albanese map $J:M\to A(M)$ has to be an immersion as well.)

The next result is a simple homotopical consequence of the harmonic map theory developed so far and is usually proved by straightforward homotopy argument.

Corollary 4.21. *Let M be a compact oriented manifold. Then*

$b_1(M)=0$ if and only if any map $f:M\to\Theta$ into a torus Θ is homotopic to a constant map.

Proof. Take M to be Riemannian by fixing a metric tensor on M. If $b_1(M)=0$ then the Albanese torus $A(M)$, being of dimension $b_1(M)$, reduces to a point. Given a map $f:M\to\Theta$ into a flat torus Θ, by Hartman 's uniqueness, it is homotopic to a harmonic map $f':M\to\Theta$. By the factorization theorem, f' factors through the Albanese map $J:M\to A(M)$ which is constant. Thus, f' is itself constant. Conversely, the Albanese map $J:M\to A(M)$ is homotopic to a constant map and hence, by 4.20, it is itself constant. As the universal covering map $\tilde{J}:\tilde{M}\to$ $\to H^1(M)^*$ is full, 4.3, the Albanese torus reduces to a point, i.e. $\dim A(M) = b_1(M) = 0$. ✓

Throughout the rest of this section, assuming M to be an m-dimensional compact oriented Riemannian manifold with base point o and first Betti number $p=b_1(M)$, we study the geometric properties of the Albanese map $J:M\to A(M)$. Let $i^0(M)$ denote the identity component of the compact Lie group $i(M)$ of all isometries of M (cf. I.8). Its Lie algebra is then identified with the Lie algebra $I(M)$ of all Killing vector fields on M. Similarly, we consider the compact Lie group $i(A(M))$ of isometries with identity component $i^0(A(M))$ and the corresponding Lie algebra $I(A(M))$ of Killing vector fields on the Albanese torus. By I.8.8, each element $Y\in I(A(M))$ is parallel on $A(M)$. As the canonical projection $p_M:H^1(M)^*\to$ $\to A(M)$ is a local isometry, Y can be lifted up yielding a parallel vector field $\tilde{Y}$ on $H^1(M)^*$, or equivalently, $\tilde{Y}$ is the uniquely determined vector field on $H^1(M)^*$ satisfying

$$(p_M)_*\tilde{Y} = Y\circ p_M \, . \tag{4.22}$$

Now, on the Euclidean vector space $H^1(M)^*$ a parallel vector field $\tilde{Y}$ is necessarily *uniform*, i.e. $\tilde{Y}^{\vee}:H^1(M)^*\to H^1(M)^*$ is a constant map, as can be proved by differentiating the components of $\tilde{Y}$ with respect to the base fields $\{\partial_i\}_{i=1}^p$ associated to a linear isomorphism $H^1(M)^*\cong\mathbf{R}^p$ (cf. I.4.24). Denoting by $(\tilde{\upsilon}_t)_{t\in\mathbf{R}}$ and $(\upsilon_t)_{t\in\mathbf{R}}$ the 1-parameter groups of isometries induced by $\tilde{Y}$ and Y, respectively, (4.22) is easily seen to imply

$$p_M \circ \tilde{\upsilon}_t = \upsilon_t \circ p_M\,, \quad t \in \mathbf{R}\,. \tag{4.23}$$

$\tilde{Y}$ being uniform, each $\tilde{\upsilon}_t$ is a translation of $H^1(M)^*$. It follows that, for $Y \in I(A(M))$, the 1-parameter group $(\upsilon_t)_{t\in\mathbf{R}}$ of isometries induced by Y consists of translations of the Albanese torus $A(M)$. Thus, $i^{\circ}(A(M))$ coincides with the compact Lie group of all translations of $A(M)$, in particular, $i^{\circ}(A(M)) \cong A(M)$, as Lie groups.

If $a \in i(M)$ is an isometry then $J \circ a : M \to A(M)$ is, by 2.31, a harmonic map and hence the factorization theorem 4.18 implies the existence of a (uniquely determined) totally geodesic map $\upsilon : A(M) \to A(M)$ such that the diagram

$$\begin{array}{ccc} M & \xrightarrow{\quad a \quad} & M \\ {\scriptstyle J}\downarrow & & \downarrow{\scriptstyle J} \\ A(M) & \xrightarrow{\quad \upsilon \quad} & A(M) \end{array} \tag{4.24}$$

commutes. Using $a^{-1} \in i(M)$, instead of $a \in i(M)$, uniqueness implies that υ is a totally geodesic diffeomorphism. Now, applying this argument to each element of a 1-parameter subgroup $(\phi_t)_{t\in\mathbf{R}} \subset i^{\circ}(M)$, we obtain the existence of a 1-parameter family $(\upsilon_t)_{t\in\mathbf{R}}$ of totally geodesic diffeomorphisms of the Albanese torus $A(M)$ such that, for each $t \in \mathbf{R}$, the diagram

$$\begin{array}{ccc} M & \xrightarrow{\quad \phi_t \quad} & M \\ {\scriptstyle J}\downarrow & & \downarrow{\scriptstyle J} \\ A(M) & \xrightarrow{\quad \upsilon_t \quad} & A(M) \end{array} \tag{4.25}$$

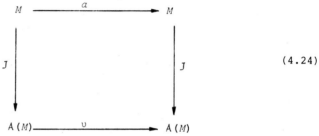

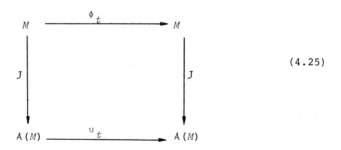

commutes. Again by uniqueness $\upsilon_0=1$ (= identity of $A(M)$) and

$$\upsilon_{t'}\circ\upsilon_t = \upsilon_{t+t'} \quad , \quad t,t'\in\mathbf{R} \ . \tag{4.26}$$

By commutativity of (4.25), $t\to\upsilon_t(y)$, $t\in\mathbf{R}$, is continuous at each point y of $\mathrm{im}(J)\subset A(M)$ and hence, the universal covering map of υ_t being linear (cf. the proof of 4.9), it follows from 4.3 that $t\to\upsilon_t(y)$, $t\in\mathbf{R}$, is continuous for all $y\in A(M)$. Thus, $(\upsilon_t)_{t\in\mathbf{R}}$ is a 1-parameter group of totally geodesic diffeomorphisms of $A(M)$. On the other hand, the identity component of the Lie group of all totally geodesic diffeomorphisms of a compact Riemannian manifold M is well-known to coincide with those of isometries (cf. Kobayashi and Nomizu (1963) page 244) and hence $(\upsilon_t)_{t\in\mathbf{R}}\subset i^o(A(M))$ is a 1-parameter subgroup consisting of isometries of $A(M)$.
To each isometry $a\in i^o(M)$ we now associate a translation $\hat{j}(a)\in i^o(A(M))$ defined as follows. Choose $X\in I(M)$ such that

$$\exp(X) = a \tag{4.27}$$

and consider the 1-parameter group $(\exp(tX)=\phi_t)_{t\in\mathbf{R}}\subset i^o(M)$ of isometries induced by X. (Here, $\exp: I(M)\to i^o(M)$ denotes the exponential map of the Lie group $i^o(M)$ which, due to compactness of $i^o(M)$, is surjective. Hence, $X\in I(M)$ satisfying (4.27) exists; cf. Hochschild (1965).) The construction above applies associating to $(\phi_t)_{t\in\mathbf{R}}$ a 1-parameter group $(\upsilon_t)_{t\in\mathbf{R}}$ of translations of $A(M)$ such that the diagram (4.25) commutes for each $t\in\mathbf{R}$, and υ_t is uniquely determined by ϕ_t. Setting

$$\hat{j}(a) = \upsilon_1\in i^o(A(M))$$

we obtain a well-defined map

$$\hat{j}: i^o(M) \to i^o(A(M))$$

and the commutativity of the diagram (4.25), for $t=1$, translates into

$$J\circ a = \hat{j}(a)\circ J \ , \quad a\in i^o(M) \ . \tag{4.28}$$

Equation (4.28) and the fact that $i^{\circ}(A(M))$ consists of translations imply immediately that $\hat{j}$ is continuous. Furthermore, as $\hat{j}(a)$ is uniquely determined by $a \in i^{\circ}(M)$ via (4.28), $\hat{j}: i^{\circ}(M) \to i^{\circ}(A(M))$ is a (continuous and hence smooth) homomorphism of Lie groups. On the Lie algebra level, $\hat{j}$ induces a homomorphism

$$\hat{J}: I(M) \to I(A(M))$$

of Lie algebras. To compute the effect of $\hat{J}$ on a Killing vector field X of M, let $(\phi_t)_{t\in\mathbf{R}} \subset i^{\circ}(M)$ denote the 1-parameter group of isometries induced by X. Then, by (4.28),

$$J \circ \phi_t = \hat{j}(\phi_t) \circ J \ , \quad t \in \mathbf{R} \ , \tag{4.29}$$

and, differentiating at $t=0$,

$$J_*(X) = \hat{J}(X) \circ J \ . \tag{4.30}$$

Remark. Following Lichnerowicz, one can construct $\hat{j}$ without the use of the surjectivity of the exponential map. Indeed, first define $\hat{j}$ on a sufficiently small open neighbourhood U of $1 \in i^{\circ}(M)$ satisfying (4.28) and then set

$$\hat{j}(a_1 \cdot \ldots \cdot a_k) = \hat{j}(a_1) \cdot \ldots \cdot \hat{j}(a_k) \ ,$$

where $a_i \in U$, $i=1,\ldots,k$. As $i^{\circ}(M) = \bigcup_{k=1}^{\infty} U^k$, this defines $\hat{j}$ everywhere on $i^{\circ}(M)$ and an argument similar to the one above yields that $\hat{j}: i^{\circ}(M) \to i^{\circ}(A(M))$ is a homomorphism of Lie groups.

By virtue of (4.30) the kernel $K(M) \subset I(M)$ of the homomorphism $\hat{J}: I(M) \to I(A(M))$ is an ideal of $I(M)$ consisting of those Killing vector fields X on M for which

$$J_*(X) = 0 \ .$$

Proposition 4.31. *The derived Lie algebra* $I(M)'(=[I(M), I(M)])$ *is contained in* $K(M)$ *; in particular,* $I(M)/K(M)$ *is commutative.*

Proof. By (4.30), for each $X \in I(M)$, the Killing vector fields X and $\hat{J}(X)$ are J-related. Hence, for $X, Y \in I(M)$, the Lie brackets $[X,Y]$ and $[\hat{J}(X), \hat{J}(Y)](=0)$ are J-related, i.e. $J_*([X,Y])=0$. ✓

The kernel $K(M) \subset i^0(M)$ of the homomorphism $\hat{j}: i^0(M) \to i^0(A(M))$ is a (possibly disconnected) closed invariant subgroup of $i^0(M)$ and $K(M)$ is the Lie algebra of the identity component $K^0(M)$ of $K(M)$. (Here and in what follows, we use, without explicit references, standard results of Lie group theory; cf. Hochschild (1965) or Dieudonné (1974).) Let $Z^0 \subset i^0(M)$ denote the identity component of the centre Z of $i^0(M)$. Then Z^0 is a closed invariant subgroup of $i^0(M)$ and its Lie algebra Z is nothing but the centre of $I(M)$. Moreover, Z^0, being compact and commutative, is a toroidal subgroup in $i^0(M)$. The Lie algebra of the (identity component of the) closed invariant subgroup $Z^0 \cap K^0(M)$ is the ideal $Z \cap K(M)$. Choose a base $\{z^i\}_{i=1}^r$ of $Z \cap K(M)$ and let $z^{r+1}, \ldots, z^{r+q} \in Z$ such that $\{z^i\}_{i=1}^{r+q}$ is a base of Z. Fixing a norm $\|\cdot\|$ on the vector space Z, there exists $\varepsilon > 0$ such that if $\{x^i\}_{i=1}^{r+q} \subset Z$ is a system of vectors with $\| z^i - x^i \| < \varepsilon$, $i=1,\ldots,r+q$, then $\{x^i\}_{i=1}^{r+q}$ is also a base of Z. Applying the Kronecker's approximation theorem to the toroidal subgroup $Z^0 \subset i^0(M)$ (cf. Dieudonné (1974) page 188) we obtain the existence of vectors $y^1, \ldots, y^q \in Z$ with $\| z^{r+j} - y^j \| < \varepsilon$, $j=1,\ldots,q$, such that $\exp(\mathbf{R} \cdot y^j) \subset Z^0$ is closed in Z^0. By the above, $\{z^1, \ldots, z^r; y^1, \ldots, y^q\}$ is a base of Z. Letting $P \subset I(M)$ to denote the linear span of $\{y^j\}_{j=1}^q$ in $I(M)$, as $P \subset Z$, it follows that P is an ideal in $I(M)$. Furthermore, as the kernel $K(M)$ of the homomorphism $\hat{J}: I(M) \to I(A(M))$ contains, by 4.31, the derived Lie algebra $I(M)'$, we have $I(M) = Z + K(M)$ and hence, by construction,

$$I(M) = P \oplus K(M) \tag{4.32}$$

is a direct sum decomposition of the Lie algebra $I(M)$ into ideals. Let $P \subset i^0(M)$ denote the connected subgroup with Lie algebra $P \subset I(M)$. As $P \subset Z$, the subgroup P is central, i.e.

$P \subset Z^{\circ}$. Moreover, $P' = \prod_{j=1}^{q} \exp(\mathbf{R}Y^{j}) \subset P$ is a closed subgroup in P and, as $\dim P' = \dim P = q$, P' is relatively open in P , i.e. actually $P' = P$. In particular, $P \subset i^{\circ}(M)$ is a closed (central) subgroup. The intersection

$$H = P \cap K(M)$$

is then a finite central subgroup of $i^{\circ}(M)$ since the corresponding Lie algebra $P \cap K(M)$ is trivial. Equivalently, H is the kernel of the restriction

$$\hat{j}|P : P \to i^{\circ}(\mathrm{A}(M))$$

and setting $\hat{j}(P) = Q$, it follows that $\hat{j}|P$ defines an epimorphism

$$\hat{j}_{\circ} : P \to Q \,, \tag{4.33}$$

with finite kernel H , between the toroidal Lie groups P and Q . On the other hand, P (resp., Q) acts on M (resp., on $\mathrm{A}(M)$) by isometries. For fixed $x \in M$, by (4.28), the restriction of the Albanese map J to the orbit $P(x) = \{a(x) \in M \mid a \in P\}$ of P through x maps onto the orbit $Q(y) = \{\upsilon(y) \in \mathrm{A}(M) \mid \upsilon \in Q\}$ of Q through $y = J(x) \in \mathrm{A}(M)$. We claim that

$$J|P(x) : P(x) \to Q(y) \,, \quad y = J(x) \,, \tag{4.34}$$

is a finite covering. In fact, the subgroup H is discrete in P and hence there exists an open neighbourhood U of 1 in P with

$$(U^{-1} \cdot U) \cap H = \{1\} \,. \tag{4.35}$$

Then, the restriction $\hat{j}_{\circ}|U$ maps U diffeomorphically onto the open neighbourhood $U_{\mathrm{A}} = \hat{j}_{\circ}(U)$ of 1 in Q . The action of the translation group $i^{\circ}(\mathrm{A}(M))$ (and hence that of Q) being free on $\mathrm{A}(M)$, the image $U_{\mathrm{A}}(y) \subset \mathrm{A}(M)$ is an open neighbourhood of y in the orbit $Q(y)$. Then,

$$(J|P(x))^{-1}(U_A(y)) = \cup\{U(z) \,|\, z \in P(x) \text{ with } J(z)=y\}$$

and, by (4.35),

$$U(z) \cap U(z') = \emptyset$$

whenever $z, z' \in P(x)$, $z \neq z'$, with $J(z)=J(z')=y$. Hence $J|P(x)$ is a finite covering and the claim follows. (Clearly, the multiplicity of this covering equals the index of the isotropy subgroup $P_x = \{a \in P \,|\, a(x)=x\} \subset P$ at x in H.) We now come to the principal result of this section, the so-called *fibration theorem*, as follows.

Theorem 4.36. *Let M be a compact oriented Riemannian manifold (with base point $o \in M$) and assume that the rank of the Albanese map $J: M \to A(M)$ does not exceed the codimension of the ideal $K(M)$ in $I(M)$. Then* codim $K(M) = b_1(M)$ *and*

$$J: M \to A(M)$$

is the (harmonic) projection of a (locally trivial) fibre bundle over the Albanese torus $A(M)$ with connected fibres and finite commutative structure group.

Proof. By the arguments above, for $x \in M$ with $y=J(x)$, we have

$$\text{codim } K(M) = \dim P = \dim P = \dim Q = \dim Q(y) \leq$$

$$\leq (\text{rank } J)(x)$$

and hence, by hypothesis,

$$\text{rank } J = \text{codim } K(M) \text{ (=const.)}$$

and J maps the whole of M onto the single orbit $Q(y) \subset A(M)$. Q, being a subgroup of $i^o(A(M))$, consists of translations of $A(M)$ and so the universal covering $\tilde{J}: \tilde{M} \to H^1(M)^*$ maps the whole of $\tilde{M}$ onto an affine subspace of $H^1(M)^*$. By fullness,

4.3, we deduce that im $\tilde{J}=H^1(M)^*$, or equivalently, $J:M \to A(M)$ has everywhere maximal rank$=b_1(M)$ $(=\dim A(M)=\operatorname{codim} K(M))$. In particular, $Q=i^{\circ}(A(M))$ and, for each $y \in A(M)$, the fibre $J^{-1}(y)$ is a (possibly disconnected) submanifold of M. To prove the local triviality, let $U \subset P$ and $U_A \subset Q$ be open neighbourhoods of the respective identity elements such that $\hat{j}_o$ maps U diffeomorphically onto U_A. If $y \in A(M)$ then $U_A(y)$ is an open neighbourhood of y in $A(M)$. Define

$$h:J^{-1}(U_A(y)) \to U_A(y) \times J^{-1}(y)$$

as follows.
For $z \in J^{-1}(U_A(y))$, we have $J(z)=\upsilon(y)$ for some $\upsilon \in U_A$. Set

$$h(z) = (J(z), a^{-1}(z)),$$

where $a \in U$ is the *unique* element with $\hat{j}_o(a)=\upsilon$. Now, $J(z) \in U_A(y)$ and, by (4.28),

$$J(a^{-1}(z)) = \hat{j}_o(a^{-1})(J(z)) = \hat{j}_o(a)^{-1}(\upsilon(y)) =$$

$$= (\hat{j}_o(a)^{-1} \circ \upsilon)(y) = y,$$

or equivalently, $a^{-1}(z) \in J^{-1}(y)$, i.e. the map h is well-defined. A routine computation shows that h is a diffeomorphism preserving the fibre structures, i.e. a bundle map.
Fixing $y, y' \in A(M)$ with $\upsilon(y)=y'$, $\upsilon \in i^{\circ}(A(M))$, there exists $a \in P$ such that $\hat{j}_o(a)=\upsilon$. Then, by (4.28), the isometry a maps the fibre $J^{-1}(y)$ diffeomorphically onto the fibre $J^{-1}(y')$ and so as a structure group of the fibre bundle

$$J:M \to A(M)$$

the (maximal) subgroup of P mapping one (and hence every) fibre onto itself is an appropriate choice. On the other hand, for $a \in P$, we have

$$a(J^{-1}(y)) \subset J^{-1}(y), \quad \text{for some } y \in A(M),$$

if and only if

$$J(a(z)) = \hat{j}_o(a)(y) = y \ , \text{ for all } z \in J^{-1}(y) \ .$$

Since $i^o(A(M))$ acts freely on $A(M)$, we obtain that $\ker \hat{j}_o = H$ is a structure group of the fibre bundle $J:M \to A(M)$. To prove the connectedness of the fibres, first define an equivalence relation $\sim$ on M by setting $x \sim x'$, $x,x' \in M$, if x and x' belong to the same connected component of a fibre of $J:M \to A(M)$. Then, the Albanese map J factorizes through the canonical projection

$$\pi : N \to A(M) \ ,$$

where $N = M/\sim$ stands for the quotient space, i.e. there is a unique (continuous) map $\bar{J}:M \to N$ such that the diagram

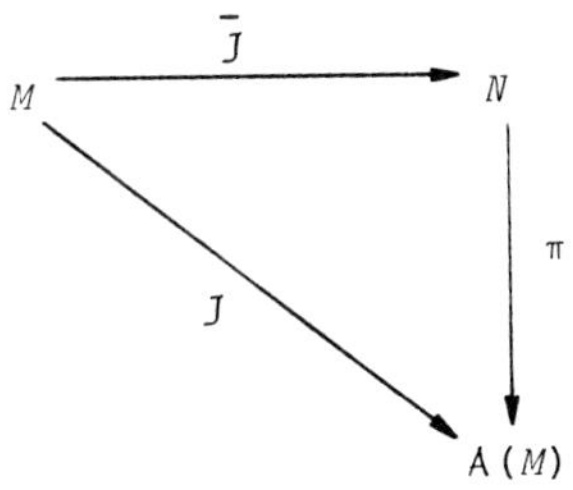

(4.37)

commutes. Clearly, it is enough to show that π is actually a diffeomorphism. For each $y \in A(M)$ as above, the same procedure applied to the restriction $J|J^{-1}(U_A(y)):J^{-1}(U_A(y)) \to U_A(y)$ shows that N can be endowed with a (smooth) manifold structure such that $\bar{J}$ and π are smooth maps with $\pi:N \to A(M)$ a finite covering projection. Taking on N the Riemannian metric induced (by π) from that of $A(M)$, the covering projection $\pi:N \to A(M)$ becomes a local isometry. In particular, $\bar{J}:M \to N$ is also a harmonic map and N, being a finite isometric covering of a Euclidean torus, is itself a Euclidean torus. This being so, we apply the factorization theorem 4.18 yielding the existence of a totally geodesic map $\upsilon : A(M) \to N$ with $\bar{J} = \upsilon \circ J$.

Thus, by (4.37), we have

$$J = \pi \circ \bar{J} = (\pi \circ \upsilon) \circ J \quad .$$

Since $\pi \circ \upsilon : A(M) \to A(M)$ is a totally geodesic map acting as the identity on $\operatorname{im} J$, 4.3 implies

$$\pi \circ \upsilon = id_{A(M)} ,$$

i.e. the map $\pi : N \to A(M)$ is a diffeomorphism. ✓

Remark. By virtue of a result of Ehresmann (1947), local triviality of the Albanese map $J : M \to A(M)$ automatically follows from the fact that it has maximal rank $(=p)$ everywhere on M. Nevertheless, making use of the "transversal" Lie group action of P on M we preferred to give an explicit construction of local trivializations.

Corollary 4.38. *Let M be a compact oriented Riemannian manifold (with base point o) and assume that* $\operatorname{rank} J \le q = \operatorname{codim} K(M)$ *. Then there exists a finite covering projection*

$$\pi : \bar{M} \to M$$

such that $\bar{M}$ splits diffeomorphically as $T^q \times F$, where F is a compact manifold.

Proof. Preserving the notations above, by the proof of 4.36, the toroidal group P acts on M such that, for each $x \in M$, the isotropy subgroup $P_x = \{a \in P \mid a(x) = x\}$ is contained in the (finite) structure group H of the fibre bundle $J : M \to A(M)$ which is the largest subgroup of P preserving the fibres; in particular, $P_x = H_x (= \{a \in H \mid a(x) = x\})$. We first claim that there exists $x^* \in M$ such that P_{x^*} is contained in each isotropy subgroup of P on M. Indeed, as P is commutative, the isotropy subgroups along each orbit under P are the same and hence, fixing a fibre $J^{-1}(y)$, $y \in A(M)$, by $P(J^{-1}(y)) = M$, our task reduces to showing the existence of $x^* \in J^{-1}(y)$ such that

$H_{x^*} \subset H_x$, for all $x \in J^{-1}(y)$. The finite group H being considered to act on $J^{-1}(y)$, by continuity, for each $x \in J^{-1}(y)$, there is a neighbourhood $U \subset J^{-1}(y)$ of x such that $H_x \supset H_z$ for all $z \in U$. Now, a standard continuation argument establishes the existence of $x^* \in J^{-1}(y)$ with the required property.
By the proof of 4.36, the restriction

$$J|P(x^*) : P(x^*) \to A(M)$$

is a finite covering projection of the orbit $P(x^*)$ through x^* onto the Albanese torus $A(M)$. Consider the pull-back $J_o : \bar{M} \to P(x^*)$ of the fibre bundle $J : M \to A(M)$ along $J|P(x^*) : P(x^*) \to A(M)$. We then have the commutative diagram

$$\begin{array}{ccc} \bar{M} & \xrightarrow{\ \pi\ } & M \\ {\scriptstyle J_o}\downarrow & & \downarrow{\scriptstyle J} \\ P(x^*) & \xrightarrow{\ J|P(x^*)\ } & A(M) \end{array}$$

where $J|P(x^*) : P(x^*) \to A(M)$ and, hence, $\pi : \bar{M} \to M$ are finite covering projections. By the definition of the pull-back,

$$\bar{M} = \{(z,z') \in P(x^*) \times M \mid J(z) = J(z')\} .$$

For $(z,z') \in \bar{M}$, we set $\phi(z,z') = (z, a(z'))$, where $a \in P$ such that $a(z) = x^*$ holds. If $a' \in P$ is an other isometry with $a'(z) = x^*$ then $a'^{-1} \cdot a \in P_z = P_{x^*}$ and, P_{x^*} being contained in each isotropy subgroup of P on M , we get $a(z') = a'(z')$. It follows that $\phi(z,z')$ does only depend on the point $(z,z') \in M$ and, as

$$J(a(z')) = \hat{j}_o(a)(J(z')) = \hat{j}_o(a)(J(z)) = J(a(z)) = J(x^*),$$

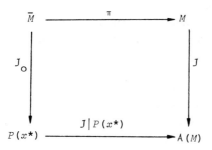

setting $J(x^*)=y$, the correspondence $(z,z') \to \phi(z,z')$, $(z,z')\in\bar{M}$, defines a map

$$\phi:\bar{M} \to P(x^*)\times J^{-1}(y) .$$

A routine calculation shows that ϕ is a diffeomorphism. The orbit $P(x^*)$, being a finite covering of the Albanese torus $A(M)$, is diffeomorphic with T^q and, setting $M_0=J^{-1}(y)$, the corollary follows. √

5. RIEMANNIAN MANIFOLDS WITH POSITIVE SEMIDEFINITE RICCI TENSOR FIELD

Following the ideas of Lichnerowicz, the methods developed in the previous section will now be applied to study the structure of compact oriented Riemannian manifolds with everywhere positive semidefinite Ricci tensor field. As a byproduct, we reduce the classification problem of such manifolds to those of zero first Betti number.

We begin with establishing some basic results concerning Killing vector fields and harmonic forms on an m-dimensional compact oriented Riemannian manifold M (with base point o and metric tensor $g=(,)$) which, throughout this section, will be fixed.

Proposition 5.1. *For any Killing vector field X on M,*

$$L_X\omega = 0 \tag{5.2}$$

holds for all harmonic r-forms ω on M, $r=0,\ldots,m$; in particular, for $\omega\in H^1(M)$,

$$\iota_X\omega = \text{constant} . \tag{5.3}$$

Proof. As the Laplacian Δ on M commutes with the isometries, by the very definition of the Lie differentiation, I.2,

$$\Delta(L_X(\omega)) = L_X(\Delta\omega) = 0 ,$$

in particular,

$$\partial(L_X\omega) = 0 .$$

On the other hand, by I.(3.6),

$$0 = \partial(L_X\omega) = (\partial\circ d\circ\iota_X)\omega + (\partial\circ\iota_X\circ d)\omega = (\partial\circ d)(\iota_X\omega)$$

since ω, being harmonic, is closed. Further,

$$0 = ((\ (\partial\circ d)(\iota_X\omega),\iota_X\omega)) = ((d(\iota_X\omega),d(\iota_X\omega)\ )) ,$$

or equivalently, $d(\iota_X\omega)=0$ (in particular, for $\omega\in H^1(M)$, the scalar $\iota_X\omega$ is constant). Again by I.(3.6), we have

$$L_X\omega = (d\circ\iota_X)\omega + (\iota_X\circ d)\omega = d(\iota_X\omega) = 0 . \checkmark$$

The definition of the ideal $K(M)\subset I(M)$ introduced in the last section can be reformulated via harmonic 1-forms as follows.

Proposition 5.4. *A Killing vector field* X *on* M *belongs to* $K(M)$ *if and only if*

$$\iota_X\omega = 0 \tag{5.5}$$

holds for any $\omega\in H^1(M)$ *. In particular, if* $\mathrm{Zero}(X)\neq\emptyset$ *then* $X\in K(M)$.

Proof. Let $(\phi_t)_{t\in\mathbf{R}}\subset i^o(M)$ denote the 1-parameter group of isometries induced by X. The vector field X can be lifted along the universal covering projection $\pi_M:\tilde{M}\to M$ yielding a Killing vector field $\tilde{X}$ on $\tilde{M}$ with induced 1-parameter isometry group $(\tilde{\phi}_t)_{t\in\mathbf{R}}\subset i^o(\tilde{M})$. The connecting equation

$$(\pi_M)_*(\tilde{X}) = X\circ\pi_M$$

implies

$$\pi_M\circ\tilde{\phi}_t = \phi_t\circ\pi_M , \quad t\in\mathbf{R} .$$

On the other hand, for $\omega \in H^1(M)$ fixed, let $u \in C^\infty(M)$ be a solution of the equation $du = \pi_M^*(\omega)$ (cf. I.3.14. and also the last section for the notations used here). Then, by the definition of the universal lift $\tilde{J}: \tilde{M} \to H^1(M)^*$ of the Albanese map, for $\tilde{x} \in \tilde{M}$ with $\pi_M(\tilde{x}) = x \in M$, we have

$$\frac{\partial}{\partial t}\{\tilde{J}(\tilde{\phi}_t(\tilde{x}))[\omega]\}_{t=0} = \frac{\partial}{\partial t}\{u(\tilde{\phi}_t(\tilde{x})) - u(\tilde{o})\}_{t=0} =$$

$$= (du)(\tilde{X}_{\tilde{x}}) = (\pi_M^*(\omega))(\tilde{X}_{\tilde{x}}) = \omega(X_x) = (\iota_X \omega)(x) .$$

It follows that $J_*(X)=0$, or equivalently, $\tilde{J}_*(\tilde{X})=0$ if and only if $\iota_X\omega = 0$, for all $\omega \in H^1(M)$, which completes the proof of the first assertion. The second follows directly from this and from (5.3). √

The interior product of harmonic 1-forms on M with respect to Killing vector fields defines, via (5.3), a bilinear pairing

$$\iota : I(M) \times H^1(M) \to \mathbb{R} ,$$

by setting

$$\iota(X,\omega) = \iota_X(\omega) , \quad X \in I(M) , \quad \omega \in H^1(M) ,$$

and, making use of 5.4, a standard result from linear algebra implies

$$\dim I(M) - \dim K(M) = \dim H^1(M) - \dim L^1(M) , \tag{5.6}$$

where $L^1(M) \subset H^1(M)$ is the linear subspace consisting of harmonic 1-forms ω on M which are annihilated by all interior products ι_X, $X \in I(M)$.

For each Killing vector field X on M the 1-form $\gamma(X) \in \mathcal{D}^1(M)$ which corresponds to X by duality is, by I.8.2, coclosed, i.e. the Hodge-de Rham decomposition of $\gamma(X)$ is easily seen to reduce to the form

$$\gamma(X) = \partial\lambda + H\gamma(X) , \tag{5.7}$$

where $H\gamma(X)$ denotes the harmonic part of $\gamma(X)$ (cf.I.(6.15)) and $\lambda \in \mathcal{D}^2(M)$. Define the linear map

$$\Phi : I(M) \to H^1(M)$$

by

$$\Phi(X) = H\gamma(X) \ , \qquad X \in I(M) \ .$$

We claim that

$$\ker \Phi = K(M) \ . \tag{5.8}$$

Indeed, by virtue of I.5.5 and (5.3), (5.7) above, for $X \in K(M)$, we have

$$((H\gamma(X), H\gamma(X))) = ((\gamma(X), H\gamma(X))) =$$

$$= \int_M (\gamma(X), H\gamma(X))\,\mathrm{vol}(M,g) = \tag{5.9}$$

$$= \int_M \iota_X(H\gamma(X))\,\mathrm{vol}(M,g) = \iota_X(H\gamma(X)) \int_M \mathrm{vol}(M,g) = 0$$

and hence $\Phi(X) = H\gamma(X) = 0$. Conversely, if $X \in \ker \Phi$ then, for any harmonic 1-form ω on M ,

$$(\iota_X \omega) \int_M \mathrm{vol}(M,g) = \int_M (\gamma(X), \omega)\,\mathrm{vol}(M,g) =$$

$$= \int_M (H\gamma(X), \omega)\,\mathrm{vol}(M,g) = 0 \ ,$$

i.e. we obtain $\iota_X \omega = 0$ and the claim follows.
In particular, if we set $q = \mathrm{codim}\, K(M) = \mathrm{codim}\, L^1(M)$, (5.8) implies that $\dim \mathrm{im}(\Phi) = q$.
A nonzero harmonic 1-form ω in $\mathrm{im}(\Phi)$ vanishes nowhere on M. Indeed, if $\omega_x = 0$ at some point $x \in M$ then, setting $\omega = H\gamma(X)$, $X \in I(M)$, the scalar $\iota_X H\gamma(X)$, by (5.3), being identically zero on M , the computation (5.9) leads to $\omega = H\gamma(X) = 0$. Applying this argument to arbitrary linear combinations of a base $\{\omega^1, \ldots, \omega^q\} \subset \mathrm{im}(\Phi)$ we obtain that the harmonic 1-forms $\omega^1, \ldots, \omega^q$ are linearly independent at any point of M .

Proposition 5.10. *Let M be a compact oriented Riemannian manifold (with base point o) such that the Ricci tensor field is everywhere positive semidefinite on M. Then*

$$\text{rank } J = \text{codim } K(M) = b_1(M) .$$

Proof. As J maps into the Albanese torus $A(M)$ with $\dim A(M)=b_1(M)$ and, by the proof of 4.36, $\text{codim } K(M) \leq \text{rank } J$, it is enough to show that

$$b_1(M) = q \ (=\text{codim } K(M)) ,$$

or equivalently, that the map

$$\Phi : I(M) \to H^1(M)$$

is surjective. Due to the curvature assumption, by I.6.7, each harmonic 1-form ω on M is parallel; i.e. setting $\omega=\gamma(X)$, $X \in V(M)$, by I.8.1(iv), we obtain that X is a Killing vector field on M. Hence $\Phi(X)=H\gamma(X)=H\omega=\omega$. √

Throughout the rest of this section, we take M as in 5.10 with $m=\dim M$ and $p=b_1(M)$. Theorem 4.36 can then be applied, in this special instance, showing that the Albanese map

$$J : M \to A(M)$$

is a harmonic (and hence, totally geodesic; 2.9) projection of a (locally trivial) fibre bundle over the Albanese torus with connected fibres and finite commutative structure group. Recall that the essential tool applied in obtaining the structure theorem 4.36 was to choose a proper direct summand P of the ideal $K(M) \subset I(M)$ in the centre $Z \subset I(M)$ such that the corresponding subgroup $P \subset i^{o}(M)$ is closed. The assumption here on the Ricci tensor field makes it possible to choose P is a more geometric way to gain more information about the geometric constituents of the Albanese map.

Proposition 5.11. *Let M be a compact oriented Riemannian* *manifold with everywhere positive semidefinite Ricci tensor field. Then the Lie algebra* $P_1(M)$ *of parallel vector fields on* M *is an ideal in* $I(M)$ *and we have the direct sum decomposition*

$$I(M) = P_1(M) \oplus K(M) . \tag{5.12}$$

Proof. Setting $X \in P_1(M) \cap K(M)$, as $\gamma(X)$ is parallel, in particular, harmonic, we have, by 5.4,

$$0 = \iota_X(\gamma(X)) = \| X \|^2$$

and hence $X=0$ showing that

$$P_1(M) \cap K(M) = \{0\} .$$

The restriction of the canonical isomorphism $\gamma : V(M) \to D^1(M)$ to $P_1(M) \subset V(M)$ establishes an isomorphism $P_1(M) \cong P^1(M)$. Due to the curvature assumption, by I.6.7, $P^1(M) = H^1(M)$, and hence $\dim P_1(M) = b_1(M)$. By 5.10, for reasons of dimensions, $P_1(M)$ and $K(M)$ actually span $I(M)$. Finally, we show that $P_1(M) \subset I(M)$ is an ideal. Setting $X \in I(M)$, for each $Y \in V(M)$, we have

$$L_X(\gamma(Y)) = \gamma([X,Y]) \tag{5.13}$$

(cf. Problem 8 for Chapter I). For $Y \in P_1(M)$, the associated 1-form $\gamma(X)$ is parallel, i.e. harmonic on M . The left hand side of (5.13) is also harmonic since $X \in I(M)$ and isometries preserve harmonicity. Thus $\gamma([X,Y]) \in H^1(M) = P^1(M)$ and, by the above, $[X,Y] \in P_1(M)$. √

To illustrate the use of the decomposition (5.12), we show that the fibres of the Albanese map $J : M \to A(M)$ are totally geodesic submanifolds of M with everywhere positive semidefinite Ricci tensor field. First, we claim that any parallel vector field X on M is everywhere orthogonal to the fibres of J . Indeed, for fixed $\tilde{x} \in \tilde{M}$ with $\pi_M(\tilde{x}) = x$ and $J(x) = y \in A(M)$, choose a curve $c : [0,1] \to J^{-1}(y)$ such that $c(0) = x$ and let $\tilde{c} : [0,1] \to \tilde{M}$ denote the universal lift of c along the projection $\pi_M : \tilde{M} \to M$ satisfying $\tilde{c}(0)=\tilde{x}$. As $\gamma(X)$ is harmonic, by fixing a scalar $u \in C^\infty(\tilde{M})$ with $du = \pi_M^*(\gamma(X))$, we have

$$(X_x, \dot{c}(0)) = \gamma(X)(\dot{c}(0)) = \gamma(X)((\pi_M)_* \dot{\tilde{c}}(0)) =$$

$$= (\pi_M^* \gamma(X))(\dot{\tilde{c}}(0)) = (du)(\dot{\tilde{c}}(0)) = \frac{\partial}{\partial t}\{u(\tilde{c}(t))\}_{t=0} =$$

$$= \frac{\partial}{\partial t}\{u(\tilde{c}(t)) - u(\tilde{o})\}_{t=0} = \frac{\partial}{\partial t} \tilde{J}(c(t))[\gamma(X)]_{t=0} = 0 ,$$

since, as

$$(p_M \circ \tilde{J})(\tilde{c}(t)) = (J \circ \pi_M)(\tilde{c}(t)) = J(c(t)) = y ,$$

(cf. (4.8)) the curve $\tilde{J} \circ \tilde{c}$ is constant.

It follows that the parallel vector fields on M span the horizontal distribution of the submersion $J : M \to A(M)$. If $Y, Z \in \in V(M)$ are everywhere tangent to the fibres of J then, for $X \in P_1(M)$,

$$(X, \nabla_Y Z) = Y(X,Z) - (\nabla_Y X, Z) = 0$$

and hence $\nabla_Y Z$ is also tangent to the fibres of J . This proves that $J^{-1}(y) \subset M$ is a totally geodesic submanifold of M , for all $y \in A(M)$. Now, for fixed $x \in M$ with $J(x) = y \in A(M)$, choose a local orthonormal moving frame $\{E^i\}_{i=1}^m$ around $x \in M$ such that $E^1, \ldots, E^p$, $p = b_1(M)$, are parallel and $E^{p+1}, \ldots, E^m$ are everywhere tangent to the fibres. Setting $F = J^{-1}(y)$, for $i, j = p+1, \ldots, m$, we have

$$\mathrm{Ric}^M(E^i, E^j) = \sum_{k=1}^{m} (R^M(E^k, E^i)E^j, E^k) =$$

$$= \sum_{k=1}^{m} (R^M(E^i, E^k)E^k, E^j) = \sum_{k=p+1}^{m} (R^M(E^i, E^k)E^k, E^j) =$$

$$= \sum_{k=p+1}^{m} (R^M(E^k, E^i)E^j, E^k) =$$

$$= \sum_{k=p+1}^{m} (R^F(E^k, E^i)E^j, E^k) = \mathrm{Ric}^F(E^i, E^j) .$$

Moreover, for $i=1,\ldots,m$ and $j=1,\ldots,p$,

$$\mathrm{Ric}^M(E^i,E^j) = \mathrm{Ric}^M(E^j,E^i) = \sum_{k=1}^{m} (R^M(E^k,E^i)E^j,E^k) = 0$$

and hence the Ricci tensor field Ric^F is positive semidefinite everywhere on F.

By the definition of the ideal $K(M) \subset I(M)$, each Killing vector field X in $K(M)$ is everywhere tangent to the fibres of $J:M \to A(M)$ since $J_*(X)=0$, and hence, setting $F=J^{-1}(y)$ for some $y \in A(M)$, we obtain that the restriction $X|F$ is a well-defined vector field on F. As the Riemannian metric on F is induced from that of M, $X|F$ is clearly a Killing vector field on F. We claim that

$$X|F \in K(F) .$$

Indeed, as $X \in K(M)$, by (5.7) and (5.8), the associated 1-form $\gamma(X) \in \mathcal{D}^1(M)$ has the Hodge-de Rham decomposition

$$\gamma(X) = \partial\lambda ,$$

for some $\lambda \in \mathcal{D}^2(M)$. Applying the Hodge-de Rham decomposition to λ again, we get

$$\gamma(X) = \partial\lambda = \partial(d\alpha) , \tag{5.14}$$

where $\alpha \in \mathcal{D}^1(M)$. Now, by 5.11, for $Y \in P_1(M)$, we have

$$[Y,X] \in P_1(M) \cap K(M) = \{0\} ,$$

and so, by (5.13), we obtain

$$L_Y(\gamma(X)) = 0 . \tag{5.15}$$

Substituting (5.14) into (5.15), we obtain

$$0 = L_Y(\gamma(X)) = L_X(\partial(d\alpha)) = \partial(d(L_Y\alpha)) ,$$

since d and ∂ commute with isometries. Hence,

$$0 = ((\partial(d(L_Y\alpha)), L_Y\alpha)) = ((d(L_Y\alpha), d(L_Y\alpha))) ,$$

or equivalently,

$$d(L_Y\alpha) = L_Y(d\alpha) = 0 .$$

On the other hand, as $Y \in P_1(M)$, for $Z, W \in V(M)$, we have

$$(L_Y(d\alpha))(Z,W) = Y((d\alpha)(Z,W)) - (d\alpha)([Y,Z],W) -$$

$$- (d\alpha)(Z,[Y,W]) = Y((d\alpha)(Z,W)) - (d\alpha)(\nabla_Y Z,W) -$$

$$- (d\alpha)(Z,\nabla_Y W) = (\nabla_Y(d\alpha))(Z,W) ,$$

i.e.

$$\nabla_Y(d\alpha) = 0$$

holds for all $Y \in P_1(M)$. Thus, for fixed $x \in F$, using the local orthonormal moving frame $\{E^i\}_{i=1}^m$ above,

$$\gamma(X) = \partial(d\alpha) = - \sum_{i=1}^{m} \iota_{E^i}\{\nabla_{E^i}(d\alpha)\} =$$

$$= - \sum_{i=p+1}^{m} \iota_{E^i}\{\nabla_{E^i}(d\alpha)\} ,$$

and so

$$\gamma(X)|F = \gamma(X|F) = \partial((d\alpha)|F) .$$

Again by (5.7) and (5.8), we obtain $X|F \in K(F)$.

We summarize the results obtained here in the following concise form (cf. also 4.36 and 4.38).

Theorem 5.16. *Let M be a compact oriented Riemannian manifold with base point o and first Betti number $p=b_1(M)$. Assume that the Ricci tensor field Ric^M is positive semidefinite everywhere on M. Then the Albanese map*

$$J : M \to A(M)$$

is a totally geodesic projection of a (locally trivial) fibre bundle over the Albanese torus with finite commutative structure group. Each fibre $F=J^{-1}(y)$, $y\in A(M)$, *is a connected totally geodesic submanifold of* M *with everywhere positive semi-definite Ricci tensor field and*

$$K(M)\,|\,F = K(F) \ . \tag{5.17}$$

Furthermore, there is a finite covering

$$\pi:\bar{M} \to M$$

such that $\bar{M}$ *splits diffeomorphically as*

$$\mathbf{T}^p \times F \ . \ \checkmark$$

Remark. As noted by Cheeger and Gromoll (1972) page 440, the diffeomorphism $\bar{M} \to \mathbf{T}^p\times F$ cannot, in general, be chosen to be isometric with respect to any product metric on $\mathbf{T}^p\times F$ with $\mathbf{T}^p$ flat. For example, take $\bar{M}=S^2\times\mathbf{R}$ with the canonical product metric and let $\mathbf{Z}$ act on $\tilde{M}$ by glide rotations

$$(x,t) \to (r^a(x), t+a) \ , \quad (x,t)\in\tilde{M} \ , \quad a\in\mathbf{Z} \ ,$$

where $r:S^2 \to S^2$ is a rotation through an angle ψ about some axis with $\frac{\psi}{2\pi}$ irrational. Then, the orbit space $M=\tilde{M}/\mathbf{Z}$ is diffeomorphic with $S^2\times S^1$ and satisfies the curvature condition of 5.16 but no covering of M is isometric with $S^2\times S^1$ endowed with any product metric.

On the other hand, assuming that M is *Ricci-flat* (i.e. $\mathrm{Ric}^M=0$), the splitting diffeomorphism can be chosen to be isometric. More precisely, Fischer and Wolf (1975) proved that a compact Ricci-flat Riemannian manifold M has a finite (normal) covering $\bar{M}=\Theta\times F$, with Θ a torus and F a compact simply connected manifold, such that the induced metric on $\bar{M}$ is the product of a flat metric on Θ and a Ricci-flat metric on F .

The Lichnerowicz reduction theorem which follows simplifies the classification problem of compact oriented Riemannian manifolds with everywhere positive semidefinite Ricci tensor field to those of zero first Betti number.

Theorem 5.18. *Let* M *be a compact oriented Riemannian manifold (with base point* o*) such that* Ric^M *is positive semidefinite everywhere on* M *. Let* $b \in \mathbb{Z}_+$ *denote the maximum of the first Betti numbers of finite covering manifolds of* M *. Then, there exists a finite covering projection*

$$\pi : \bar{M} \to M \ ,$$

with base point $\bar{o} \in \bar{M}$ *,* $\pi(\bar{o}) = o$ *, and* $b_1(\bar{M}) = b$ *such that the corresponding Albanese map*

$$\bar{J} = J_{\bar{M}} : \bar{M} \to A(\bar{M})$$

is a totally geodesic projection map of a (locally trivial) fibre bundle. Each fibre $\bar{F} = \bar{J}^{-1}(\bar{y})$ *,* $\bar{y} \in A(\bar{M})$*, is a connected totally geodesic submanifold of* $\bar{M}$ *with* $\mathrm{Ric}^{\bar{F}}$ *positive semidefinite everywhere on* $\bar{F}$ *and*

$$b_1(\bar{F}) = 0 \ .$$

Proof. We first note that the number b occurring in the theorem is finite; in fact, $b \leq m = \dim M$. (This follows here easily by taking into account that $H^1(\bar{M}) = P_1(\bar{M})$ and $\dim P_1(\bar{M}) \leq$ $\leq \dim \bar{M} = m$ hold for any finite (Riemannian) covering manifold $\bar{M}$ of M .)

Now, let

$$\pi : \bar{M} \to M$$

be a finite Riemannian covering projection with $b_1(\bar{M}) = b$ and fix a base point $\bar{o} \in \bar{M}$, $\pi(\bar{o}) = o$. Since the curvature properties of $\bar{M}$ are inherited from that of M , 5.16 applies showing that the Albanese map $\bar{J} : \bar{M} \to A(\bar{M})$ is a totally geodesic projection map of a (locally trivial) fibre bundle over the Albanese torus $A(\bar{M})$, with $\dim A(\bar{M}) = b_1(\bar{M}) = b$, such that each fibre $\bar{F} = \bar{J}^{-1}(\bar{y})$, $\bar{y} \in A(\bar{M})$, is connected. Applying the last statement of 5.16, we obtain a finite covering

$$\bar{\pi}:\bar{\bar{M}} \to \bar{M}$$

such that $\bar{\bar{M}}$ splits diffeomorphically as

$$\mathbf{T}^b \times \bar{F} \;;$$

in particular, by the Künneth rule (cf. Spanier (1966) page 247),

$$b_1(\bar{\bar{M}}) = b_0(\mathbf{T}^b)\cdot b_1(\bar{F}) + b_1(\mathbf{T}^b)\cdot b_0(\bar{F}) = b_1(\bar{F}) + b \quad .$$

By maximality of b, we obtain $b_1(\bar{F})=0$. √

Remark. For the sake of completeness, we mention here that, according to Cheeger and Gromoll (1972), a complete Riemannian manifold M with everywhere nonnegative sectional curvatures contains a compact totally geodesic and totally convex submanifold S (called the *soul* of M) such that M is diffeomorphic with the normal bundle of the inclusion map $S \to M$. By this result, the classification problem of complete nonnegatively curved Riemannian manifolds is reduced to that of compact ones.

PROBLEMS FOR CHAPTER II

1. Let $f:M\to N$ be an isometric immersion between Riemannian manifolds M and N . Show that, for each $X_x, Y_x \in T_x(M)$, $x\in M$,

$$(R^N(f_*(X_x),f_*(Y_x))f_*(Y_x),f_*(X_x)) = (R^M(X_x,Y_x)Y_x,X_x) -$$

$$- (\beta(f)(X_x,X_x),\beta(f)(Y_x,Y_x)) +$$

$$+ (\beta(f)(X_x,Y_x),\beta(f)(X_x,Y_x)) \quad .$$

(For hypersurfaces in $\mathbf{R}^3$ this reduces to Gauss' "Theorema Egregium".)

2. Let G be a compact Lie group endowed with a biinvariant Riemannian metric. Prove that if the multiplication map $\pi:G\times G\to G$

is totally geodesic then G is Abelian.

3. Show that the quaternionic Hopf-map $f:S^7 \to S^4$ is a Riemannian submersion.

4. Derive (3.10) from the classical Jacobi equation for geodesics. (Hint: Consider, for each curve $c:(-\varepsilon,\varepsilon) \to M$, $\varepsilon>0$, the variation of geodesics $t \to \exp(t \cdot v_{c(s)})$, $t\in\mathbf{R}$, $s\in(-\varepsilon,\varepsilon)$; cf. also Milnor (1963).)

5. Given a compact oriented Riemannian manifold M with *two* base points o and o', any curve $c:[0,1] \to M$, with $c(0)=o$ and $c(1)=o'$, gives rise to an isomorphism between the respective universal coverings. Using this as an identification, derive a formula relating the two Albanese maps corresponding to the base points $o,o'\in M$.

6. Given a compact oriented manifold M, the transpose of the de Rham map (4.4) is a linear isomorphism

$$ {}^t\!\int : H^r_{de\ R}(M;\mathbf{R}) \to H_r(M;\mathbf{R})^* , \qquad r\in\mathbf{Z} . $$

The lattice $H^r_{de\ R}(M;\mathbf{Z}) \subset H^r_{de\ R}(M;\mathbf{R})$ which corresponds (via ${}^t\!\int$) to the lattice of linear functionals in $H_r(M;\mathbf{R})^*$ which take integral values on $\operatorname{im} \rho_r \subset H_r(M;\mathbf{R})$ is said to be the *de Rham cohomology group with integral periods*. Show that the Q-module

$$ H^r_{de\ R}(M;\mathbf{Q}) = H^r_{de\ R}(M;\mathbf{Z}) \underset{\mathbf{Z}}{\otimes} \mathbf{Q} \subset H^r_{de\ R}(M;\mathbf{R}) , $$

of *de Rham cohomology classes with rational periods* is dense in $H^r_{de\ R}(M;\mathbf{R})$, $r\in\mathbf{Z}$.

7. Given a family $\{\alpha^i\}_{i=1}^q \subset \mathcal{D}^1(M)$ of pointwise linearly independent *closed* 1-forms on a compact oriented manifold M show that $\mathcal{D} = \bigcap_{i=1}^q \ker \alpha^i$ is a codimension q involutive distribution on M.

8. Let $\mathcal{D}$ be as in Problem 7. Prove that there exist closed 1-forms $\bar\alpha^1,\dots,\bar\alpha^q\in\mathcal{D}^1(M)$ such that, for $i=1,\dots,q$, $\bar\alpha^i$ is arbitrarily close to α^i (with respect to the C^∞-topology) and the involutive distribution $\bar{\mathcal{D}} = \bigcap_{i=1}^q \ker \bar\alpha^i$ does actually give rise to a fibre bundle

$$ f:M \to \mathbf{T}^q $$

(i.e. the space of leaves $M/\bar{\mathcal{D}}$ inherits a manifold structure from that of M such that $M/\bar{\mathcal{D}}=\mathbf{T}^q$ and the canonical projection f is the (smooth) projection map of a fibre bundle.) (Hint: With respect to a fixed Riemannian metric (,) on M, consider the Hodge-de Rham decomposition

$$\alpha^i = du^i + \omega^i ,$$

where $u^i \in C^\infty(M)$ and $\omega^i \in H^1(M)$, $i=1,\ldots,q$. For $\varepsilon>0$, use Problem 6 to find $\bar{\omega}^i \in H^1(M)$ with $\{\bar{\omega}^i\} \in H^1_{de\ R}(M;\mathbf{Q})$ such that

$$\| \omega^i - \bar{\omega}^i \| < \varepsilon , \quad i=1,\ldots,q .$$

Finally, setting $\bar{\alpha}^i = du^i + \bar{\omega}^i$, $i=1,\ldots,q$, imitate the construction of the Albanese map; cf. also Tischler (1970).)

9. Let G be a compact Lie group endowed with a biinvariant Riemannian metric. Show that the Albanese map $J_G : G \to A(G)$ (with respect to the base point $1 \in G$) is an epimorphism.

10. Using Problem 9, prove the following classical result of É. Cartan:

Any compact Lie group G can be written as

$$G = \frac{T \times S}{D} ,$$

where T is a toroidal group, S is a compact semisimple Lie group and $D \subset T \times S$ is a discrete central subgroup. (Hint: Describe the geometric constituents of the Albanese map J_G via the group structure on G.)

CHAPTER III

Harmonic Maps into Spheres

The next two chapters are built up in a parallel way reflecting the close relationship of harmonic and minimal maps between Riemannian manifolds. Each of them begins with variational calculus which, in the present chapter, gives a reinterpretation of harmonicity of maps via the energy functional. To investigate the qualitative behaviour of the energy near harmonic maps, we also derive here the second variation formula which, by a straightforward generalization of concepts in Morse theory, will naturally lead to the study of Jacobi fields along harmonic maps, Morse index, instability, etc. Our investigations, being mainly devoted to harmonic maps into space forms, are then specialized to maps into spherical codomains, e.g. we prove Leung's instability theorem for harmonic maps into spheres. The last four sections deal with the works of R.T. Smith who, by a series of brilliant ideas, has been able to develop an existence theory for harmonic maps between spheres which, for example, provides with harmonic representative each homotopy class of maps between spheres of dimension ≤ 7 . (Note that the (quick) proofs of Smith (1972a-b) and (1975a) are replaced here, following his own ideas, by very detailed arguments; in fact, it is our hope that this would lead to a better understanding of his results.)

1. FIRST AND SECOND VARIATION FORMULAS

Let M and N be Riemannian manifolds with metric tensors $g=(,)\in C^\infty(S^2(T^*(M)))$ and $h=(,)\in C^\infty(S^2(T^*(N)))$, respectively. Given a map $f:M\to N$, the scalar $e(f)\in C^\infty(M)$ defined (pointwise) by

$$e(f)(x) = \frac{1}{2}\,\mathrm{trace}\,\|f_*\|^2(x) = \frac{1}{2}\sum_{i=1}^{m}\|f_*(e^i)\|^2 \tag{1.1}$$

$$x \in M, \quad m = \dim M,$$

where $\{e^i\}_{i=1}^m \subset T_x(M)$ is an orthonormal base, is said to be the *energy density* of f. To simplify the matters, from here on, unless stated otherwise, we assume that M is compact and oriented. Then, the integral $E(f)$ of $e(f)$ with respect to the volume form $\mathrm{vol}(M,g) \in \mathcal{D}^m(M)$ on M is said to be the *energy of* f, i.e. setting

$$E(f) = \int_M e(f)\,\mathrm{vol}(M,g), \tag{1.2}$$

we obtain a functional E (called the *energy functional*) which associates to each map $f: M \to N$ its energy $E(f)$ via (1.2). As in the classical Morse theory, we are interested in the critical points of the energy functional, which are then maps $f: M \to N$ stable to first order with respect to variations. The *first variation formula* which follows reinterprets the harmonic maps, introduced in II.2 via an extension of the Hodge-de Rham theory, as critical points of the energy functional.

Theorem 1.3. *Let $f: M \to N$ be a map of a compact oriented Riemannian manifold M with metric tensor $g = (\,,)$ into a Riemannian manifold N with metric tensor $h = (\,,)$. Given a (smooth) variation*

$$t \to f_t, \quad t \in \mathbf{R}, \tag{1.4}$$

of f, we have

$$\left.\frac{\partial E(f_t)}{\partial t}\right|_{t=0} = -\int_M (\mathrm{trace}\,\beta(f), v)\,\mathrm{vol}(M,g), \tag{1.5}$$

where $\beta(f) \in C^\infty(S^2(T^*(M)) \otimes \tau^f)$ is the second fundamental form and $v = \left.\frac{\partial f_t}{\partial t}\right|_{t=0} \in C^\infty(\tau^f)$. In particular, f is a critical point of the energy functional E if and only if f is harmonic.

Proof. Using the notations introduced at the beginning of Section 3, Chapter II, the variation (1.4) gives rise to a 1-parameter family

$$t \to P(t) = \tau^t (f_t)_* \in \mathcal{D}^1(\tau^f) \ , \quad t \in \mathbf{R} \ ,$$

of 1-forms on M with values in the pull-back bundle τ^f. Then, by II.3.5, we have

$$\left.\frac{\partial E(f_t)}{\partial t}\right|_{t=o} = \frac{1}{2}\int_M \frac{\partial}{\partial t}\{\mathrm{trace}\,\|(f_t)_*\|^2\}_{t=o}\,\mathrm{vol}(M,g) =$$

$$= \frac{1}{2}\int_M \frac{\partial}{\partial t}\{\mathrm{trace}\,\|P(t)\|^2\}_{t=o}\,\mathrm{vol}(M,g) =$$

$$= \int_M \mathrm{trace}\left(f_* \ , \left.\frac{\partial P(t)}{\partial t}\right|_{t=o}\right)\mathrm{vol}(M,g) =$$

$$= \int_M \mathrm{trace}(f_* \ , \nabla v)\,\mathrm{vol}(M,g) = \int_M \mathrm{trace}\{\nabla(f_*,v)\}\mathrm{vol}(M,g) \ -$$

$$- \int_M (\mathrm{trace}\ \beta(f), v)\,\mathrm{vol}(M,g) \ .$$

To accomplish the proof, it is enough to show that the first integral of the right hand side is zero. Define a vector field X on M by

$$\gamma(X) = (f_*, v) \ .$$

Using an orthonormal base $\{e^i\}_{i=1}^m \subset T_x(M)$ at a point $x \in M$, we have

$$(\mathrm{div}\ X)(x) = \sum_{i=1}^m (\nabla_{e^i} X, e^i) = \sum_{i=1}^m \gamma(\nabla_{e^i} X)(e^i) =$$

$$= \sum_{i=1}^m (\nabla_{e^i}\gamma(X))(e^i) = \sum_{i=1}^m (\nabla_{e^i}(f_*, v))(e^i) =$$

$$= \mathrm{trace}\{\nabla(f_*,v)\}$$

and Green's theorem (Kobayashi and Nomizu (1963) pages 281-283), completes the proof. √

Remark. The vector field $\tau(f)=\text{trace}\,\beta(f)$ along f, occurring in the first variation formula (1.5), is said to be the *tension field of* f. It points to the direction at which the energy decreases most rapidly (at least infinitesimally). To prove the existence theorem for harmonic maps into compact nonpositively curved codomains, mentioned in II.3, Eells and Sampson considered the *heat equation*

$$\frac{\partial f_t}{\partial t} = \tau(f_t) \;, \quad f_0=f \;, \tag{1.6}$$

which, due to the curvature assumption, turned out to define a (smooth) variation of the given map $f:M\to N$ with harmonic (uniform) limit $f_\infty = \lim_{t\to\infty} f_t$ homotopic to f (cf. Eells and Sampson (1964) and Eells and Lemaire (1978)). Dropping the curvature assumption, the question of existence of a solution of (1.6) defined for all $t\in\mathbf{R}$ is still open and even if it exists with a (smooth) limit $f_\infty:M\to N$, it may well happen that f_∞ is not homotopic to f. (For further references of this problem, see Eells and Lemaire (1983), Lemaire (1977) and Sampson (1982).)

Continuing the analogue with classical Morse theory, the qualitative behaviour of the energy functional near a given harmonic map $f:M\to N$ can be studied via the *second variation formula* as follows.

Theorem 1.7. *Let* $f:M\to N$ *be a harmonic map of a compact oriented Riemannian manifold* M *with metric tensor* $g=(\,,)$ *into a Riemannian manifold* N *with metric tensor* $h=(\,,)$. *Given a (smooth) 2-parameter variation*

$$(s,t) \to f_{(s,t)} \;, \quad (s,t)\in\mathbf{R}^2 \;, \tag{1.8}$$

of f *(i.e.* $f_{(0,0)}=f$*), we have*

$$\left.\frac{\partial^2 E(f_{(s,t)})}{\partial s\,\partial t}\right|_{s=t=0} = \int_M \{(\nabla v,\nabla w) +$$

$$+ \operatorname{trace}(R(f_*,v)f_*,w)\}\operatorname{vol}(M,g) =$$

$$= \int_M (J_f(v),w)\operatorname{vol}(M,g) \ , \tag{1.9}$$

where

$$v = \left.\frac{\partial f_{(s,o)}}{\partial s}\right|_{s=o} \quad \textit{and} \quad w = \left.\frac{\partial f_{(o,t)}}{\partial t}\right|_{t=o} \ , \tag{1.10}$$

R is the Riemannian curvature tensor of N and $J_f \in \operatorname{Diff}_2(\tau^f,\tau^f)$ is given by

$$J_f = -\operatorname{trace} \nabla^2 + \operatorname{trace} R(f_*,.)f_* \tag{1.11}$$

(*cf.* II.(1.8)).

Proof. For fixed $t \in \mathbf{R}$, setting $f_t = f_{(o,t)}$ and $v^t = = \left.\frac{\partial f_{(s,t)}}{\partial s}\right|_{s=o} \in C^\infty(\tau^{f_t})$, $v^o = v$, by the proof of 1.3, we have

$$\left.\frac{\partial E(f_{(s,t)})}{\partial s}\right|_{s=o} = \int_M \operatorname{trace}((f_t)_*,\nabla v^t)\operatorname{vol}(M,g) \ . \tag{1.12}$$

Using the notations of Section 3, Chapter II, the variation

$$t \to f_t \ , \ t \in \mathbf{R} \ ,$$

defines a 1-parameter family

$$t \to P(t) = \tau^t (f_t)_* \in \mathcal{D}^1(\tau^f) \ , \quad t \in \mathbf{R} \ ,$$

and, differentiating (1.12) at $t=0$,

$$\left.\frac{\partial^2 E(f_{(s,t)})}{\partial s \partial t}\right|_{s=t=o} = \frac{\partial}{\partial t}\{\int_M \operatorname{trace}(P(t),\tau^t(\nabla v^t))\operatorname{vol}(M,g)\}_{t=o} =$$

$$= \int_M \operatorname{trace}(\left.\frac{\partial P(t)}{\partial t}\right|_{t=o}, \nabla v)\operatorname{vol}(M,g) + \tag{1.13}$$

$$+ \int_M \operatorname{trace}(f_*,\frac{\partial}{\partial t}\{\tau^t (\nabla v^t)\}_{t=o})\operatorname{vol}(M,g) \ .$$

The first integral on the right hand side of (1.13), by II.3.5 and Green's theorem, can be rewritten as

$$\int_M \operatorname{trace}\left(\left.\frac{\partial P(t)}{\partial t}\right|_{t=0}, \nabla v\right) \operatorname{vol}(M,g) =$$

$$= \int_M \operatorname{trace}(\nabla w, \nabla v) \operatorname{vol}(M,g) =$$

$$= \int_M \operatorname{trace}\{\nabla(\nabla v, w)\} \operatorname{vol}(M,g) -$$

$$- \int_M (\operatorname{trace} \nabla^2 v, w) \operatorname{vol}(M,g) =$$

$$= \int_M \operatorname{div}\{\gamma^{-1}(\nabla v, w)\} \operatorname{vol}(M,g) -$$

$$- \int_M (\operatorname{trace} \nabla^2 v, w) \operatorname{vol}(M,g) =$$

$$= -\int_M (\operatorname{trace} \nabla^2 v, w) \operatorname{vol}(M,g) ,$$

since, with respect to an orthonormal base $\{e^i\}_{i=1}^m \subset T_x(M)$, at $x \in M$, we have

$$\operatorname{trace}\{\nabla(\nabla v, w)\} = \sum_{i=1}^m (\nabla_{e^i}(\nabla v, w))(e^i) =$$

$$= \sum_{i=1}^m (\nabla_{e^i}\{\gamma^{-1}(\nabla v, w)\}, e^i) = \operatorname{div}\{\gamma^{-1}(\nabla v, w)\} .$$

To evaluate the second integral, we first compute

$$\frac{\partial}{\partial t}\{\tau^t(\nabla v^t)\}_{t=0} = \frac{\partial}{\partial t}\{(\tau^t \circ \nabla \circ \tau_t)(\tau^t(v^t))\}_{t=0} =$$

$$= \lim_{t \to 0} \frac{1}{t}\{(\tau^t \circ \nabla \circ \tau_t)(\tau^t(v^t)) - \nabla v\} = \tag{1.14}$$

$$= \lim_{t \to 0}(\tau^t \circ \nabla \circ \tau_t)\{\frac{1}{t}(\tau^t(v^t) - v)\} + \lim_{t \to 0} \frac{1}{t}\{(\tau^t \circ \nabla \circ \tau_t)v - \nabla v\} =$$

$$= \nabla\{\lim_{t\to 0} \frac{1}{t}(\tau^t(v^t)-v)\} + \lim_{t\to 0} \frac{1}{t}\{(\tau^t \circ \nabla \circ \tau_t)v - \nabla v\} \ .$$

In the first term of the right hand side of (1.14), $\lim_{t\to 0} \frac{1}{t}(\tau^t(v^t)-v)$ defines (pointwise) a vector field u along f, namely, for $x \in M$ fixed, u_x is the covariant differential of the section $t \to v_x^t$ along $t \to f_t(x)$, $t \in \mathbf{R}$, at $t=0$. Thus the first term equals ∇u. To compute the second, we fix $X_x \in T_x(M)$, at some point $x \in M$, and use I.4.11, by setting $C: (-\varepsilon,\varepsilon)\times(-\varepsilon,\varepsilon) \to N$, $C(s,t) = f_t(c(s))$, $s,t \in (-\varepsilon,\varepsilon)$, $\varepsilon > 0$, where $c: (-\varepsilon,\varepsilon) \to M$ is a curve with $\dot{c}(0) = X_x$, and $w_s = v_{c(s)} \in T_{C(s,0)}(N)$, $s \in (-\varepsilon,\varepsilon)$. Then, as

$$C_*(\frac{\partial}{\partial s}\Big|_{(0,0)}) = f_*(X_x) \quad \text{and}$$

$$C_*(\frac{\partial}{\partial t}\Big|_{(0,0)}) = w_x \ , \ C(0,0) = f(x) \ ,$$

where $\frac{\partial}{\partial s}$, $\frac{\partial}{\partial t} \in V((-\varepsilon,\varepsilon)\times(-\varepsilon,\varepsilon))$ denote the coordinate fields, respectively, using the notations of I.4.11, we have

$$\lim_{t\to 0} \frac{1}{t}\{(\tau^t \circ \nabla_{X_x} \circ \tau_t)v - \nabla_{X_x} v\} =$$

$$= \lim_{t\to 0} \frac{1}{t}\{(\tau_0^t(0) \circ \nabla_{\frac{\partial}{\partial s}\big|_{(0,t)}} \circ \tau_t^0(s))w_s - \tag{1.15}$$

$$- \nabla_{\frac{\partial}{\partial s}\big|_{(0,0)}} w_s\} = -R(f_*(X_x), w_x)v_x \ ,$$

or, in a more concise form,

$$\lim_{t\to 0} \frac{1}{t}\{(\tau^t \circ \nabla \circ \tau_t)v - \nabla v\} = -R(f_*, w)v \ .$$

Now, by making use of Green's theorem, the harmonicity of f and the curvature identity I.(4.29), we can evaluate the second integral of the right hand side of (1.13) as

$$\int_M \operatorname{trace}(f_*, \frac{\partial}{\partial t}\{\tau^t(\nabla v^t)\}_{t=0}) \operatorname{vol}(M,g) =$$

$$= \int_M \operatorname{trace}(f_*, \nabla u)\operatorname{vol}(M,g) +$$

$$+ \int_M \operatorname{trace}(f_*, -R(f_*, w)v)\operatorname{vol}(M,g) =$$

$$= \int_M \operatorname{trace} \nabla(f_*, u)\operatorname{vol}(M,g) -$$

$$- \int_M (\operatorname{trace} \nabla f_*, u)\operatorname{vol}(M,g) -$$

$$- \int_M \operatorname{trace}(R(f_*, w)v, f_*)\operatorname{vol}(M,g) =$$

$$= \int_M \operatorname{div} \gamma^{-1}(f_*, u)\operatorname{vol}(M,g) -$$

$$- \int_M \operatorname{trace}(R(v, f_*)f_*, w)\operatorname{vol}(M,g) =$$

$$= \int_M \operatorname{trace} (R(f_*, v)f_*, w)\operatorname{vol}(M,g)$$

and (1.9) follows. √

Remark 1. Keeping the analogue with the classical Morse theory, the *Hessian* of a harmonic map $f:M \to N$ is the symmetric bilinear pairing H_f on $C^\infty(\tau^f)$ defined by the right hand side of the second variation formula (1.9) as

$$H_f(v,w) = \int_M \{(\nabla v, \nabla w) + \operatorname{trace}(R(f_*, v)f_*, w)\}\operatorname{vol}(M,g) =$$

$$= \int_M (J_f(v), w)\operatorname{vol}(M,g) , \qquad v,w \in C^\infty(\tau^t) .$$

Remark 2. An immediate consequence of the second variation formula (1.9) is that if the sectional curvatures of N are nonpositive then every harmonic map $f:M \to N$ is a local minimum of the energy. In fact, for compact N, a much stronger assertion is true: any harmonic map $f:M \to N$ is an absolute minimum of the energy in its homotopy class. (Indeed, given a map $f_o:M \to N$ in the homotopy class of f, by the Eells-Sampson's existence theorem (cf. Eells and Sampson (1964)), the heat equation (1.6) deforms f_0 to a harmonic map f_∞ and, as the energy decreases along the trajectories of the heat equation, $E(f_\infty) \leq E(f_0)$. Moreover, as was remarked after II.3.16, the harmonic maps f and f_∞ are homotopic through harmonic maps; in particular, $E(f)=E(f_\infty)$. Thus, $E(f)$ is an absolute minimum. Note also that harmonic maps of minimum energy (e.g. holomorphic maps between Kähler manifolds) were extensively studied by Eells and Wood (1981), Lemaire (1978), Siu (1979), and Siu and Yau (1979).

Turning, for a moment, to the general situation, let $f:M \to N$ be a harmonic map between any Riemannian manifolds M and N. Then the differential operator $J_f \in \mathrm{Diff}_2(\tau^f, \tau^f)$ defined by

$$J_f(v) = -\mathrm{trace}\, \nabla^2 v + \mathrm{trace}\, R(f_*, v) f_*, \quad v \in C^\infty(\tau^f), \tag{1.16}$$

(cf. II.1.8) is said to be the *Jacobi operator* (*associated with the harmonic map* f). Elements of $J(f) = \ker J_f \subset C^\infty(\tau^f)$ are called *Jacobi fields along* f. By II.1.10, the Jacobi operator is elliptic. Define the *nullity* and the *Morse index* of the harmonic map $f:M \to N$ as

$$\dim J(f) = \dim \ker J_f \quad \text{and}$$

$$\mathrm{card}\{\text{negative eigenvalues of } J_f$$

$$\text{(counting multiplicities)}\}$$

respectively. Harmonic maps with zero Morse index are said to be *stable*.

Proposition 1.17. *If M is compact then the nullity and the Morse index of a harmonic map $f:M \to N$ are finite.*

Proof. First take M to be oriented. Then, the second variation formula (1.9) with I.(4.29) shows that J_f is self-adjoint, in particular, applying I.1.18, we obtain that the nullity of f is finite and $\mathrm{Spec}(J_f, M) \subset \mathbf{R}$ is discrete. Now, if $v \in C^\infty(\tau^f)$

is a (nonzero) eigenvector of J_f with eigenvalue $\lambda \in \mathrm{Spec}(J_f, M)$ then

$$\lambda\, ((v,v)) = ((J_f(v), v)) = \mathrm{trace}\, ((\nabla v, \nabla v)) +$$

$$+ ((\,\mathrm{trace}\, R(f_*, v) f_*, v)) \geq -\mathrm{trace}\, ((R(f_*, v) v, f_*)) \quad .$$

As M is compact,

$$\lambda \geq \frac{-\mathrm{trace}\,((\,R(f_*, v) v, f_*))}{((\,v, v))} \geq \text{const.}$$

and the proposition follows in this case.
If M is nonoriented, we consider its 2-sheeted oriented Riemannian covering $\bar{M}$. The natural projection $\pi : \bar{M} \to M$ being a local isometry, $f \circ \pi : \bar{M} \to N$ is harmonic (cf. II.2.31) and, for each vector field v along f , we have

$$J_{f\circ\pi}(v\circ\pi) = -\mathrm{trace}\, \nabla^2 (v\circ\pi) +$$

$$+ \mathrm{trace}\, R((f\circ\pi)_*, v\circ\pi)(f\circ\pi)_* =$$

$$= -\mathrm{trace}(\nabla^2 v)\circ\pi + (\mathrm{trace}(R(f_*, v) f_*)\circ\pi = (J_f(v))\circ\pi \quad .$$

Hence, precomposition with π carries J_f into $J_{f\circ\pi}$ and each eigenspace of J_f into an eigenspace of $J_{f\circ\pi}$ (corresponding to the same eigenvalue) and the first part of the proof applies.√

Example 1.18. Taking the codomain N to be flat (i.e. $R=0$) , a Jacobi field v along a harmonic map $f: M \to N$ satisfies

$$J_f(v) = -\mathrm{trace}\, \nabla^2 v = 0$$

and as

$$((\,\mathrm{trace}\, \nabla^2 v, v)) = -((\,\nabla v, \nabla v)) \quad ,$$

we obtain $v \in P^0(\tau^f)$. In particular, if $f = id_M$, where M is a compact (oriented) flat Riemannian manifold, then, by I.8.1(iv), I.8.2 and I.8.3, v is a Jacobi field along id_M if and only if v is a Killing vector field on M .

We close this section by the following.

Proposition 1.19. *Let* $f: M \to N$ *be a harmonic map between Riemannian manifolds* M *and* N *. If* $t \to f_t$ *,* $t \in \mathbf{R}$ *, is a variation of* f *through harmonic maps then*

$$v = \left.\frac{\partial f_t}{\partial t}\right|_{t=0} \in C^\infty(\tau^f)$$

is a Jacobi field along f *.*

Proof. As in II.3, the variation $t \to f_t$, $t \in \mathbf{R}$, gives rise to a 1-parameter family

$$t \to P(t) = \tau^t (f_t)_* \in \mathcal{D}^1(\tau^f) \ , \quad t \in \mathbf{R} \ ,$$

and, as $f_t: M \to N$ is harmonic for all $t \in \mathbf{R}$, we have

$$\operatorname{trace}\{\nabla (f_t)_*\} = \operatorname{trace}\{(\tau^t \circ \nabla \circ \tau_t) P(t)\} = 0 \ . \tag{1.20}$$

Differentiating (1.20) at $t=0$ and using II.3.5, a computation similar to the one in (1.15) gives

$$0 = \frac{\partial}{\partial t}\{\operatorname{trace}(\tau^t \circ \nabla \circ \tau_t) P(t)\}_{t=0} = \operatorname{trace}\{\nabla (\left.\frac{\partial P(t)}{\partial t}\right|_{t=0})\} +$$

$$+ \operatorname{trace} \frac{\partial}{\partial t}\{(\tau^t \circ \nabla \circ \tau_t) f_*\}_{t=0} = \operatorname{trace} \nabla^2 v -$$

$$- \operatorname{trace} R(f_*, v) f_* = -J_f(v) \ . \ \surd$$

2. INSTABILITY

Since the main purpose of this chapter is the study of harmonic maps into spheres, from now on we specialize the codomain N to be the Euclidean n-sphere S^n with metric tensor $(,)$ inherited from that of $\mathbf{R}^{n+1}$ via the inclusion $S^n \subset \mathbf{R}^{n+1}$. The Riemannian curvature tensor R of S^n is particularly simple, namely

$$R(X,Y)Z = (Y,Z)X - (X,Z)Y \,, \quad X,Y,Z\in V(S^n) \tag{2.1}$$

(cf. Problem 4 for Chapter I).

By the results of II.2.21, a map $f:M\to S^n$ of a Riemannian manifold M of dimension m into S^n is harmonic if and only if $\Delta\bar{f}$ is proportional to $\bar{f}$, where $\bar{f}:M\to \mathbf{R}^{n+1}$ denotes the vector function on M obtained from f via the inclusion $S^n\subset \subset\mathbf{R}^{n+1}$. On the other hand, for fixed $x\in M$, choosing a local orthonormal moving frame $\{E^i\}_{i=1}^m$ (adapted to an orthonormal base $\{e^i\}_{i=1}^m\subset T_x(M)$) over a normal coordinate neighbourhood centered at x, from $(\bar{f},\bar{f})=1$ we have

$$(\Delta\bar{f},\bar{f})(x) = -\sum_{i=1}^m (e^i(E^i(\bar{f})),\bar{f}(x)) =$$

$$= -\sum_{i=1}^m e^i(E^i(\bar{f}),\bar{f}) + \sum_{i=1}^m (e^i(\bar{f}),e^i(\bar{f})) =$$

$$= \sum_{i=1}^m (f_*(e^i),f_*(e^i)) = 2e(f)(x)$$

and, again by II.2.21, $f:M\to S^n$ is harmonic if and only if the equation

$$\Delta\bar{f} = 2e(f)\bar{f} \tag{2.2}$$

is satisfied.

Example 2.3. If $f:S^m\to S^n$ is a map such that all components of the induced vector function $\bar{f}:S^m\to\mathbf{R}^{n+1}$ are spherical harmonics of a given order k (cf.I.7) then, by (2.2), f is harmonic. We then call $f:S^m\to S^n$ a *harmonic map defined by spherical harmonics of order* k.

The main result of this section is the following instability theorem due to Leung (1981).

Theorem 2.4. *For* $n\geq 3$, *any nonconstant harmonic map* $f:M\to S^n$ *of a compact oriented Riemannian manifold* M *of dimension* m *into* S^n *has strictly positive Morse index.*

Proof. We shall use the notations introduced in II.2.12 and II.2.21, namely, $\check{}: T(\mathbf{R}^{n+1}) \to \mathbf{R}^{n+1}$ will denote the canonical identification obtained by translating tangent vectors on $\mathbf{R}^{n+1}$ to the origin of $\mathbf{R}^{n+1}$, and, for a vector field v along the inclusion map $S^n \to \mathbf{R}^{n+1}$, $v^\top$ and $v^\perp$ will stand for the tangential and orthogonal part of v.
Given a vector $\check{Z}\in\mathbf{R}^{n+1}$, the tangential part of the uniform extension $Z\in V(\mathbf{R}^{n+1})$ of $\check{Z}$ defines a vector field $Z^\top$ on S^n. For each $x\in M$, setting

$$q_x(\check{Z}) = \| \nabla(Z^\top\circ f) \|_x^2 + \mathrm{trace}(R(f_*, Z^\top\circ f)f_*, Z^\top\circ f)_x ,$$

$$\check{Z}\in\mathbf{R}^{n+1} ,$$

we obtain a quadratic form q_x on $\mathbf{R}^{n+1}$, and, by Remark 1 after the proof of 1.7,

$$H_f(Z^\top\circ f, Z^\top\circ f) = \int_M q(\check{Z})\,\mathrm{vol}(M,g) , \quad \check{Z}\in\mathbf{R}^{n+1} . \tag{2.5}$$

To compute $q_x(\check{Z})$ more explicitly, we fix an orthonormal base $\{e^i\}_{i=1}^m \subset T_x(M)$ and set $y=f(x)\in S^n$. Then, by (2.1), we have

$$q_x(\check{Z}) = \sum_{i=1}^m \| \nabla_{e^i}(Z^\top\circ f) \|^2 +$$

$$+ \sum_{i=1}^m (R(f_*(e^i), Z_y^\top)f_*(e^i), Z_y^\top) =$$

$$= \sum_{i=1}^m \| \nabla_{f_*(e^i)} Z^\top \|^2 + \sum_{i=1}^m (f_*(e^i), Z_y)^2 -$$

$$- 2\| Z_y^\top \|^2 e(f)(x) .$$

To evaluate the first sum on the right hand side, we claim that

$$\nabla_{Y_y} Z^\top = -(\check{Z}, y)\cdot Y_y \tag{2.6}$$

holds for any $Y_y\in T_y(S^n)$.

Indeed, by II.(2.17), as $Z \in V(\mathbf{R}^{n+1})$ is uniform,

$$\nabla_{Y_y} Z^{\top} = \nabla_{Y_y} (Z|S^n - Z^{\perp}) = (\nabla_{Y_y} (Z - Z^{\perp}))^{\top} = -(\nabla_{Y_y} Z^{\perp})^{\top}$$

and hence (2.6) reduces to

$$((\nabla_{Y_y} Z^{\perp})^{\top})^{\vee} = (\check{Z}, y)\check{Y}_y \ , \quad Y_y \in T_y(S^n) \ . \tag{2.7}$$

Assuming $\| Y_y \| = 1$, by I.(4.25), we have

$$(\nabla_{Y_y} Z^{\perp})^{\vee} = \frac{\partial}{\partial t}\{(Z^{\perp})^{\vee}(\cos t\cdot y + \sin t\cdot Y_y)\}_{t=0} =$$

$$= \frac{\partial}{\partial t}\{(\check{Z}, \cos t\cdot y + \sin t\cdot \check{Y}_y)(\cos t\cdot y + \sin t\cdot \check{Y}_y)\}_{t=0} =$$

$$= (\check{Z}, \check{Y}_y)y + (\check{Z}, y)\check{Y}_y$$

and, by tangential projection, (2.7) follows.
Thus, by making use of (2.6),

$$\sum_{i=1}^{m} \| \nabla_{f_*(e^i)} Z^{\perp} \|^2 = (\check{Z}, y)^2 \sum_{i=1}^{m} \| f_*(e^i) \|^2 =$$

$$= 2(\check{Z}, y)^2 \, e(f)(x) \ ,$$

and summarizing, we obtain

$$q_x(\check{Z}) = 2(\check{Z}, y)^2 e(f)(x) +$$

$$+ \sum_{i=1}^{m} (f_*(e^i), Z_y)^2 - 2\| Z_y^{\top} \|^2 e(f)(x) \ .$$

Now, choose an orthonormal base $\{\check{Z}^j\}_{j=0}^{n} \subset \mathbf{R}^{n+1}$ with $\check{Z}^0 = y$.
Then,

$$\text{trace } q_x = 2e(f)(x) \cdot \sum_{j=0}^{n} (\check{Z}^j, y)^2 +$$

$$+ \sum_{i=1}^{m} \sum_{j=0}^{n} (f_*(e^i), Z_y^j)^2 - 2e(f)(x) \cdot \sum_{j=0}^{n} \| (Z^j)_y^\top \|^2 =$$

$$= 2e(f)(x) + \sum_{i=1}^{m} \| f_*(e^i) \|^2 - 2ne(f)(x) =$$

$$= 2(2-n)e(f)(x) \ .$$

Assuming that $f:M \to N$ is a stable harmonic map, we have

$$H_f(Z^\top \circ f, Z^\top \circ f) = \int_M q(\check{Z})\,\mathrm{vol}(M,g) \geq 0$$

for all $\check{Z} \in \mathbf{R}^{n+1}$. Taking traces,

$$(2-n)e(f) \geq 0 \ ,$$

which means that the energy density of f vanishes everywhere on M, i.e. the harmonic map $f:M \to N$ is constant. ✓

Remark. For generalizations of 2.4 see also Leung (1981). Furthermore, note that, for *spherical domains*, Xin (1980) has obtained a similar instability theorem.

3. EXISTENCE OF HARMONIC MAPS BETWEEN SPHERES

In the rest of this chapter we describe a powerful method developed by R.T. Smith which produces a large number of harmonic maps between spheres and enables us to solve the existence problem posed in II.3 for homotopy classes of maps between spheres of dimension ≤ 7. Smith's method is essentially based on applying a homotopy operation (called the join) to pairs of (harmonic) maps defined as follows. Let $f:S^p \to S^q$ and $g:S^r \to S^s$ be maps between Euclidean spheres and define their *join*

$$f*g:S^{p+r+1} \to S^{q+s+1}$$

by

$$\overline{f*g}(x,y) = (\|x\| \cdot \bar{f}(\tfrac{x}{\|x\|}), \|y\| \cdot \bar{g}(\tfrac{y}{\|y\|})) \ , \qquad (3.1)$$

where we use polar coordinates on the domain (and on the codomain), i.e. the pair (x,y) with $x \in \mathbf{R}^{p+1}$, $y \in \mathbf{R}^{r+1}$, $\|x\|^2 + \|y\|^2 = 1$, is considered to be a point in S^{p+r+1} (similarly on S^{q+s+1}) and on the submanifolds defined by $x=0$ or $y=0$ we take the corresponding limits. In order to get a geometric view of the join, for fixed $u,v \in \mathbf{N}$, consider the scalar $\mu : S^{u+v+1} \to [-1,1]$ defined as the restriction (to S^{u+v+1}) of the polynomial $P:\mathbf{R}^{u+v+2} \to \mathbf{R}$ given by $P(x,y) = \|x\|^2 - \|y\|^2$, $x \in \mathbf{R}^{u+1}$, $y \in \mathbf{R}^{v+1}$. For $s \in (-1,1)$, the *level hypersurface* $M_s = \mu^{-1}(s) \subset S^{u+v+1}$ is isometric with the generalized torus $(\cos t \cdot S^u) \times (\sin t \cdot S^v)$, where $\cos(2t) = s$, $t \in (0,\frac{\pi}{2})$, and for $s=1$ and $s=-1$, the submanifolds $M_s = \mu^{-1}(s)$ (called the *focal varieties*) are isometric with S^u and S^v, respectively. (Thus apart from the focal varieties, the sphere S^{u+v+1} is foliated by generalized tori. Note also that this construction is an example of É. Cartan's isoparametric hypersurfaces.)

Now, setting $u=p$, $v=r$ and $u=q$, $v=s$, we consider on S^{p+r+1} and S^{q+s+1} the corresponding isoparametric families of hypersurfaces. Clearly, the join $f*g : S^{p+r+1} \to S^{q+s+1}$ restricts to direct products

$$(\cos t \cdot f) \times (\sin t \cdot g) : \cos t \cdot S^p \times \sin t \cdot S^r \to$$

$$\to \cos t \cdot S^q \times \sin t \cdot S^s \ , \quad t \in (0,\tfrac{\pi}{2}) \ ,$$

and, between the focal varieties, $f*g$ reduces to f and g, respectively. Thus the join $f*g$ can be viewed to be the glueing together of various homothetic products of f and g along É. Cartan's isoparametric hypersurfaces. It is also clear from the interpretation above that, in the special case $f = id_{S^p}$, the join $f*g$ reduces to the $(p+1)$-th *suspension* $S^{p+1}g$ of g. (For the elementary homotopy theory used here we refer to Hu (1959) and Spanier (1966).)

The main result of Smith, to be proved in Sections 3-5, is the following.

Theorem 3.2. *Let* $f : S^p \to S^q$ and $g : S^r \to S^s$ *be maps defined by spherical harmonics of order* k *and* ℓ, *respectively. Then* *the join* $f*g: S^{p+r+1} \to S^{q+s+1}$ *is homotopic to a harmonic map provided that the conditions*

$$k > \frac{1}{2}(\sqrt{2}-1)(p-1) \quad \text{and} \quad \ell > \frac{1}{2}(\sqrt{2}-1)(r-1) \tag{3.3}$$

are satisfied.

As a direct consequence, we obtain a (partial) solution of the existence problem for homotopy classes of maps between spheres.

Corollary 3.4. *If* $g: S^r \to S^s$ *is a map defined by spherical harmonics of order* $\ell > \frac{1}{2}(\sqrt{2}-1)(r-1)$ *then, for* $m \leq 7$, $S^{m-1}g$ *is homotopic to a harmonic map. In particular, any map* $h: S^m \to S^m$ *is homotopic to a harmonic map, provided that* $m \leq 7$.

Proof. For the first statement, set $f = id_{S^{m-2}}$ with $m \leq 7$. Then, as $k = 1 > \frac{1}{2}(\sqrt{2}-1)(m-3)$, 3.2 yields the result. For the second, we first note that, homotopy classes of maps $h: S^m \to S^m$ being parametrized by the Brouwer degree, it is enough to show the existence of harmonic maps $h: S^m \to S^m$, $m \leq 7$, with prescribed degree d. As the composition of a harmonic map $h: S^m \to S^m$ with a reflection through a hyperplane of $\mathbf{R}^{m+1}$ preserves harmonicity and changes the sign of the Brouwer degree, we may assume that $d \geq 0$. Now, define $g: S^1 \to S^1$ by $g(z) = z^d$, $z \in S^1$, where S^1 is considered to be the unit circle on the complex plane $\mathbf{C}$. Then, g is easily seen to be defined by spherical harmonics of order d and, by the first statement, $S^{m-1}g: S^m \to S^m$ is homotopic to a harmonic map $h: S^m \to S^m$, provided that $m \leq 7$. By elementary homotopy theory, $\deg(h) = \deg(S^{m-1}g) = \deg(g) = d$. ✓

Remark 1. Extending Smith's method, Baird has recently shown that, for any $m \in \mathbf{N}$, each map $f: S^m \to S^m$ is homotopic to a harmonic map of the Euclidean m-sphere S^m into S^m with a properly chosen Riemannian metric $g \in C^\infty(S^2(T^*(S^m)))$. (For details, see Baird (1983) and Karcher and Wood (1983).)

Remark 2. Note that, most recently, Eells and Lemaire (1984) have shown that any map $h: S^9 \to S^9$ is homotopic to a harmonic map.

The principal idea in proving Theorem 3.2 is to construct a harmonic map

$$h: S^{p+r+1} \to S^{q+s+1} \tag{3.5}$$

of the form

$$\bar{h}(x,y) = (\sin \alpha(t) \bar{f}(\frac{x}{\| x \|}), \cos \alpha(t) \bar{g}(\frac{y}{\| y \|})) ,$$

$$(0 \neq) x \in \mathbf{R}^{p+1} , \quad (0 \neq) y \in \mathbf{R}^{r+1} , \quad \| x \|^2 + \| y \|^2 = 1 , \tag{3.6}$$

with $t = \log(\frac{\| x \|}{\| y \|})$, where the function $\alpha : \mathbf{R} \to \mathbf{R}$ is to be determined. Depending on the asymptotic behaviour of α at $\pm\infty$ the map h may or may not be extended smoothly across the focal varieties defined by $x=0$ and $y=0$, resp., and our task is to choose an α such that h is well-defined, smooth and harmonic. (For $0 < t < \frac{\pi}{2}$ we have $\bar{h}(\cos t \cdot x, \sin t \cdot y) = (\sin \alpha(\log(\cot t)) \cdot \cdot \bar{f}(x), \cos \alpha(\log(\cot t)) \cdot \bar{g}(y))$, $x \in S^p$, $y \in S^r$, i.e. h maps the level hypersurfaces $\cos t \cdot S^p \times \sin S^r \subset S^{p+r+1}$ into level hypersurfaces of S^{q+s+1} .) To close this section we express the condition of harmonicity of h (apart from the focal varieties) in terms of the function α . To simplify the matters we set $p, r \geq 3$; the remaining cases can be treated differently but using the same ideas as below. We also assume that f and g are nonconstant, i.e. their component functions are eigenfunctions of the corresponding Laplacian with nonzero eigenvalue (cf.2.3).

To reformulate the condition of harmonicity of h , we first note that (3.6) defines $\bar{h}$ as a map $\bar{h}: \mathbf{R}^{p+1} \setminus \{0\} \times \mathbf{R}^{r+1} \setminus \{0\} \to \mathbf{R}^{q+s+2}$ which is radially constant. Thus, by I.(7.3), apart from the singular varieties,

$$\Delta^{S^{p+r+1}} (\bar{h} | S^{p+r+1}) = (\Delta^{\mathbf{R}^{p+r+2}} \bar{h}) | S^{p+r+1} . \tag{3.7}$$

Denoting by $\rho : \mathbf{R}^{p+1} \setminus \{0\} \to S^p$, $\bar{\rho}(x) = \frac{x}{\| x \|}$, $x \in \mathbf{R}^{p+1} \setminus \{0\}$, the canonical retraction and setting

$$\Delta = \Delta^{\mathbf{R}^{p+r+2}} = -\sum_{i=1}^{p+1}\left(\frac{\partial}{\partial x^i}\right)^2 - \sum_{j=1}^{r+1}\left(\frac{\partial}{\partial y^j}\right)^2 ,$$

where $\{\frac{\partial}{\partial x^i}\}_{i=1}^{p+1} \subset V(\mathbf{R}^{p+1})$ and $\{\frac{\partial}{\partial y^j}\}_{j=1}^{r+1} \subset V(\mathbf{R}^{r+1})$ are the canonical bases, we have

$$\Delta(\sin\alpha(t)\cdot(\bar{f}\circ\rho)(x)) = \bar{f}(\rho(x))\cdot\Delta(\sin\alpha(t)) - \tag{3.8}$$

$$- 2\sum_{i=1}^{p+1}\frac{\partial(\sin\alpha(t))}{\partial x^i}\cdot\frac{\partial(\bar{f}\circ\rho)}{\partial x^i}(x) + \sin\alpha(t)\cdot\Delta(\bar{f}\circ\rho)(x) .$$

To evaluate the right hand side of (3.8), by

$$\frac{\partial t}{\partial x^i} = \frac{x^i}{\| x\|^2} \quad\text{and}\quad \frac{\partial t}{\partial y^i} = -\frac{y^j}{\| y\|^2} , \tag{3.9}$$

$$i=1,\ldots,p+1 \; ; \quad j=1,\ldots,r+1 ,$$

we first compute

$$\Delta(\sin\alpha(t)) = -\sum_{i=1}^{p+1}\left(\frac{\partial}{\partial x^i}\right)^2(\sin\alpha(t)) -$$

$$-\sum_{j=1}^{r+1}\left(\frac{\partial}{\partial y^j}\right)^2(\sin\alpha(t)) =$$

$$= \frac{(\sin\alpha(t))\dot{\alpha}(t)^2 - (\cos\alpha(t))\ddot{\alpha}(t)}{\| x\|^2} -$$

$$- \frac{p-1}{\| x\|^2}(\cos\alpha(t))\cdot\dot{\alpha}(t) +$$

$$+ \frac{(\sin\alpha(t))\dot{\alpha}(t)^2 - (\cos\alpha(t))\ddot{\alpha}(t)}{\| y\|^2} +$$

$$+ \frac{r-1}{\| y\|^2}(\cos\alpha(t))\cdot\dot{\alpha}(t) .$$

Further, as $\bar{f}\circ\rho$ is radially constant, by (3.9), we have

$$\sum_{i=1}^{p+1} \frac{\partial(\sin\alpha(t))}{\partial x^i} \cdot \frac{\partial(\bar{f}\circ\rho)}{\partial x^i}(x) =$$

$$= (\cos\alpha(t))\cdot\dot{\alpha}(t)\,\frac{1}{\|x\|^2}\sum_{i=1}^{p+1} x^i\,\frac{\partial(\bar{f}\circ\rho)}{\partial x^i}(x) = 0\ ,$$

i.e. the second term on the right hand side of (3.8) vanishes. Finally, using again that $\bar{f}\circ\rho$ is radially constant and that the components of $\bar{f}$ are spherical harmonics of order k, by I.7.5, we have

$$(\Delta(\bar{f}\circ\rho))(x) = (\Delta^{\|x\|\cdot S^p}(\bar{f}\circ\rho))(x) =$$

$$= \frac{1}{\|x\|^2}(\Delta^{S^p}\bar{f})(\rho(x)) = \frac{1}{\|x\|^2}\lambda\cdot(\bar{f}\circ\rho)(x)\ ,$$

where $\lambda=k(k+p-1)$.

Summarizing, we obtain

$$\Delta(\sin\alpha\cdot\bar{f}\circ\rho)(x,y) = \{\frac{\lambda\cdot\sin\alpha}{\|x\|^2} + \frac{(\sin\alpha)\dot{\alpha}^2-(\cos\alpha)\ddot{\alpha}}{\|x\|^2\,\|y\|^2} +$$

$$+ (\cos\alpha)\dot{\alpha}(\frac{r-1}{\|y\|^2} - \frac{p-1}{\|x\|^2})\}(\bar{f}\circ\rho)(x)\ .$$

Similarly,

$$\Delta(\cos\alpha\cdot\bar{g}\circ\sigma)(x,y) = \{\frac{\nu\cdot\cos\alpha}{\|y\|^2} +$$

$$+ \frac{(\cos\alpha)\dot{\alpha}^2+(\sin\alpha)\ddot{\alpha}}{\|x\|^2\,\|y\|^2} - (\sin\alpha)\dot{\alpha}(\frac{r-1}{\|y\|^2} -$$

$$- \frac{p-1}{\|x\|^2})\}(\bar{g}\circ\sigma)(y)\ ,$$

where $\sigma:\mathbf{R}^{r+1}\backslash\{0\} \to S^r$ is the canonical retraction and

$\nu=\ell(\ell+r-1)$. By (2.2), the criterion of harmonicity of h is that $\Delta\bar{h}$ should be proportional to $\bar{h}$. By virtue of the identities $\|x\|^2=\frac{e^t}{e^t+e^{-t}}$ and $\|y\|^2=\frac{e^{-t}}{e^t+e^{-t}}$ this can be reformulated in terms of the second-order differential equation

$$\ddot{\alpha}(t)+\frac{1}{e^t+e^{-t}}\{((p-1)e^{-t}-(r-1)e^t)\dot{\alpha}(t)+ \\ +(\nu e^t-\lambda e^{-t})\sin\alpha(t)\cos\alpha(t)\}=0\,. \tag{3.10}$$

Finally, in terms of λ and ν , conditions (3.3) become

$$4\lambda>(p-1)^2 \quad\text{and}\quad 4\nu>(r-1)^2\,. \tag{3.11}$$

Remark. Physically, equation (3.10) describes the movement of a pendulum with variable damping acted on by a force of variable gravity. In order to prove 3.2 we have to choose a proper solution α of the pendulum equation (3.10) with the damping conditions (3.11).

4. STUDY OF THE PENDULUM EQUATION

The objective of this section is to prove the following.

Theorem 4.1. *There exists a solution* $\alpha:\mathbf{R}\to(0,\frac{\pi}{2})$ *of the pendulum equation* (3.10) *with the damping conditions* (3.11) *such that* α *is eventually strictly monotone and*

$$\lim_{t\to\infty}\alpha(t)=\frac{\pi}{2} \quad\textit{and}\quad \lim_{t\to-\infty}\alpha(t)=0\,. \tag{4.2}$$

Remark. Continuing the physical interpretation of the pendulum equation (3.10), notice first that the "gravity" in (3.10) is positive for large positive parameters and negative for large negative parameters. 4.1 says that there exists an "exceptional" trajectory along which the pendulum connects asymptotically the two equilibria $\alpha_\infty=\frac{\pi}{2}$ and $\alpha_{-\infty}=0$.

Taking α as in the theorem above we define $\bar{h}$ by (3.6).

Then, by (4.2), apart from the origin of R^{p+r+2}, the function $\bar{h}$ can be extended continuously across the linear subspaces of R^{p+r+2} defined by $x=0$ and $y=0$ and hence restriction to the unit sphere yields a continuous map $h:S^{p+r+1} \to S^{q+s+1}$. (By the construction above, apart from the focal varieties, h is smooth and satisfies the harmonicity condition (2.2).) Moreover, h is homotopic to the join $f*g$ by a homotopy H defined by

$$H(s,x,y) = \left(\frac{A(s,x,y)}{\sqrt{A(s,x,y)^2+B(s,x,y)^2}}\bar{f}\left(\frac{x}{\|x\|}\right),\right.$$

$$\left.\frac{B(s,x,y)}{\sqrt{A(s,x,y)^2+B(s,x,y)^2}}\bar{g}\left(\frac{y}{\|y\|}\right)\right), \quad 0\le s\le 1,$$

where

$$A(s,x,y) = s\|x\| + (1-s)\sin\alpha(t)$$

and

$$B(s,x,y) = s\|y\| + (1-s)\cos\alpha(t),$$

$t=\log\left(\frac{\|x\|}{\|y\|}\right)$, $x\in\mathbf{R}^{p+1}$, $y\in\mathbf{R}^{r+1}$, $\|x\|^2+\|y\|^2=1$, and, for $x=0$ or $y=0$, we take the corresponding limits.

The proof of 4.1 is broken up into several lemmas. First, we write (3.10) in the form

$$\ddot{\alpha}(t) = F(t,\alpha(t),\dot{\alpha}(t)), \tag{4.3}$$

where $F:\mathbf{R}^3 \to \mathbf{R}$ is a (smooth) function such that

$$F(t,0,0) = F(t,\tfrac{\pi}{2},0) = 0, \quad t\in\mathbf{R}, \tag{4.4}$$

and there exists $C_0>\frac{1}{2\pi}$ with

$$|F(t,\beta,\gamma)| \le C_0(1+|\gamma|), \quad t,\beta,\gamma\in\mathbf{R}. \tag{4.5}$$

Lemma 4.6. *For fixed $T>0$, there exists a solution α_T of (4.3) satisfying the boundary conditions $\alpha_T(-T)=0$, $\alpha_T(T)=\frac{\pi}{2}$ such that*

$$0 < \alpha(t) < \frac{\pi}{2}$$

holds for $|t|<T$.

Proof. We first show the existence of a solution α of (4.3) for which $\alpha(-T)=0$ and $\alpha(-T+\delta)=\frac{\pi}{2}$ hold for some $\delta\in(0,2T)$ and $\alpha|[-T,-T+\delta]$ is strictly increasing. Choose $M_0 > \max(4\pi C_0, \frac{\pi}{2T})$ and define

$$F = \{x\in C^2[-T,-T+\frac{\pi}{M_0}] \mid x(-T) = 0 , \quad \dot{x}(-T)=M_0 ,$$

$$|\dot{x}(t)-M_0| < 2\pi C_0 , \quad |\ddot{x}(t)| < C_0(1+2\pi C_0+M_0) ,$$

$$\text{for} \quad t\in[-T,-T+\frac{\pi}{M_0}]\}.$$

(Note that F consists of strictly increasing functions since $M_0>2\pi C_0$.) Then $F \subset C^2[-T,-T+\frac{\pi}{M_0}]$ is convex and closed (with respect to the C^2-topology). Define a map

$$\Psi : F \to C^2[-T,-T+\frac{\pi}{M_0}]$$

by

$$\Psi(x)(t) = y(t) = M_0(t+T) + \int_{-T}^{t}\int_{-T}^{s} F(u,x(u),\dot{x}(u))\,du\,ds ,$$

$$t\in[-T,-T+\frac{\pi}{M_0}] , \quad x\in F .$$

We claim that $\Psi(F)\subset F$. Indeed, the boundary conditions for $y=\Psi(x)$, $x\in F$, are clearly satisfied and, for $t\in[-T,-T+\frac{\pi}{M_0}]$, by (4.5), we have

$$|\dot{y}(t)-M_0| \le \int_{-T}^{t} |F(u,x(u),\dot{x}(u))|\,du \le$$

$$\le C_0 \int_{-T}^{t} (1+|\dot{x}(u)|)\,du < C_0(1+2\pi C_0+M_0)(t+T) \le$$

$$\le C_0(1+2\pi C_0+M_0)\frac{\pi}{M_0} \le C_0(4\pi C_0+M_0)\frac{\pi}{M_0} < 2\pi C_0$$

and

$$|\ddot{y}(t)| = |F(t,x(t),\dot{x}(t))| \le C_o(1+|\dot{x}(t)|) <$$

$$< C_o(1+2\pi C_o+M_o) ,$$

i.e. $y\in F$. It follows that Ψ restricts to a continuous (linear) operator $\Psi : F \to F$.

Next, we show that $\Psi(F)$ is *precompact*, i.e. for any sequence $(x_n)_{n\in\mathbf{N}} \subset F$ the image $(y_n)_{n\in\mathbf{N}} \subset F$, $y_n=\Psi(x_n)$, $n\in\mathbf{N}$, has an accumulation point in F . As $x_n\in F$, for $t_o,t_1\in$ $\in[-T,-T+\frac{\pi}{M_o}]$, we have

$$|x_n(t_o)-x_n(t_1)| < (2\pi C_o+M_o)|t_o-t_1| \quad \text{and}$$

$$|\dot{x}_n(t_o)-\dot{x}_n(t_1)| < C_o(1+2\pi C_o+M_o) ,$$

i.e. the sequences $(x_n)_{n\in\mathbf{N}}$ and $(\dot{x}_n)_{n\in\mathbf{N}}$ are uniformly bounded and equicontinuous. It follows easily that the same holds for the sequences $(y_n)_{n\in\mathbf{N}}$, $(\dot{y}_n)_{n\in\mathbf{N}}$, $(\ddot{y}_n)_{n\in\mathbf{N}}$, where $y_n=\Psi(x_n)$, $n\in\mathbf{N}$. Arzelà-Ascoli theorem will then imply that $(y_n)_{n\in\mathbf{N}}$ has an accumulation point, i.e. $\Psi(F)$ is precompact.

Now we are in the position to apply the Schauder fixed point theorem to Ψ (cf. Hartman (1964) page 405) yielding an element $\alpha^{M_o}\in F$ with $\Psi(\alpha^{M_o})=\alpha^{M_o}$. The last relation means that α^{M_o} satisfies (4.3) with the initial data $\alpha^{M_o}(-T)=0$, $\dot{\alpha}^{M_o}(-T)=M_o$ and, as $\alpha^{M_o}\in F$, we get by integration over $[-T,-T+\frac{\pi}{M_o}]$

$$(M_o-2\pi C_o)(T+t) < \alpha(t) , \quad t\in[-T,-T+\frac{\pi}{M_o}] ,$$

i.e. for $t=-T+\frac{\pi}{M_o}$

$$\frac{\pi}{2} < (M_o-2\pi C_o)\frac{\pi}{M_o} < \alpha(-T+\frac{\pi}{M_o}) .$$

As $\alpha|[-T,-T+\frac{\pi}{M_o}]$ is strictly increasing there is a unique $\delta\in$ $\in(0,2T)$ with $\alpha(-T+\delta)=\frac{\pi}{2}$ which accomplishes the first step.

Secondly, we consider the one-parameter family of solutions

α^M, $M\in[0,M_0]$ of the equation (4.3) given by the initial data $\alpha^M(-T)=0$, $\dot{\alpha}^M(-T)=M$. Then, α^0 being the trivial solution, by continuity, there exists $M_1\in(0,M_0)$ with $\alpha^{M_1}(T)=\frac{\pi}{2}$. Setting $\alpha_T=\alpha^{M_1}$, it remains only to prove that $0<\alpha_T(t)<\frac{\pi}{2}$ holds for $t\in(-T,T)$. Assuming that $\alpha_T|(-T,T)>0$ does not hold, denote by $t_0,t_1\in(-T,T)$ the first two zeros of α_T. Increasing the value of M from M_1 it follows the existence of a number $M_2\in(M_1,M_0)$ such that the solution α^{M_2} vanishes to first order somewhere on $[-T,T]$ which, by (4.4), contradicts the uniqueness. Similarly, $\alpha_T(t)<\frac{\pi}{2}$, $t\in(-T,T)$. √

By making use of (4.5), the next statement is a direct consequence of the Nagumo lemma (cf. Hartman (1964) page 428).

Lemma 4.7. *For fixed* $\tau_0>0$ *there exists a constant* $C>0$ *(depending only on* τ_0 *and* C_0*) such that for* $T>\tau_0$ *we have* $|\dot{\alpha}_T(t)|\leq C$, $t\in(-T,T)$. √

In order to construct a proper solution α of (3.10) consider a sequence $(\alpha_{T_n})_{n\in\mathbf{N}}$, where $\lim_{t\to\infty} T_n=\infty$, $T_n>\tau_0$, $n\in\mathbf{N}$. By (4.5) and 4.7, the sequences $(\alpha_{T_n})_{n\in\mathbf{N}}$, $(\dot{\alpha}_{T_n})_{n\in\mathbf{N}}$ and $(\ddot{\alpha}_{T_n})_{n\in\mathbf{N}}$ are uniformly bounded and equicontinuous and, again by the Arzelà-Ascoli theorem, there exists a subsequence $(\alpha_{T_n})_{n\in\mathbf{N}}$, denoted by the same symbol, which converges in the C^2-topology to a solution α of (3.10) uniformly on each compact subset of $\mathbf{R}$. To analyze the behaviour of α we write (3.10) as

$$\ddot{\alpha}(t) = v(t)\dot{\alpha}(t) - u(t)\sin\alpha(t)\cos\alpha(t)\,, \tag{4.8}$$

where $u,v:\mathbf{R}\to\mathbf{R}$ are functions defined by

$$u(t) = \frac{\nu e^t-\lambda e^{-t}}{e^t+e^{-t}} \quad\text{and}\quad v(t) = \frac{(r-1)e^t-(p-1)e^{-t}}{e^t+e^{-t}}\,,\quad t\in\mathbf{R}\,.$$

Lemma 4.9. $\alpha\neq 0$.

Proof. We distinguish two cases as follows.

(i) $r>1$. Define $\tau:\mathbf{R}_+\to\mathbf{R}$, with $\tau(0)=0$, by

$$\dot{\tau}(t) = \exp\{(r-1)t - \int_t^\infty [v(s)-(r-1)]ds\} , \quad t\geq 0 .$$

Then τ is strictly increasing and $\lim_{t\to\infty} \tau(t)=\infty$. Setting

$$y_T(\tau(t)) = \alpha_T(t) , \quad t\in[0,T] , \quad T>\tau_0 ,$$

(4.8) is transformed into

$$\ddot{y}_T(\tau) + (\frac{\tau}{\dot{\tau}})^2 u\cdot\frac{\sin y_T(\tau)}{y_T(\tau)} \cos y_T(\tau)\cdot\frac{y_T(\tau)}{\tau^2} = 0 . \quad (4.10)$$

Further, we have

$$\lim_{t\to\infty} \frac{\tau(t)}{\dot{\tau}(t)} = \lim_{t\to\infty} \frac{\dot{\tau}(t)}{\ddot{\tau}(t)} = \lim_{t\to\infty} \frac{1}{v(t)} = \frac{1}{r-1}$$

and hence

$$\lim_{\substack{t\to\infty \\ y\to 0_+}} (\frac{\tau(t)}{\dot{\tau}(t)})^2 u(t) \frac{\sin y}{y} \cos y = \frac{\nu}{(r-1)^2} .$$

By the damping conditions (3.11) we may choose $\delta>0$ such that

$$\frac{\nu}{(r-1)^2} > \frac{1}{4-\delta} .$$

Consider now the equation

$$\ddot{y}(\tau) + \frac{1}{4-\delta}\frac{y(\tau)}{\tau^2} = 0 \quad (4.11)$$

whose general solution is of the form

$$y(\tau) = c_1\sqrt{\tau} \sin(\frac{1}{2}\sqrt{\frac{\delta}{4-\delta}} \log \tau) +$$

$$+ c_2\sqrt{\tau} \cos(\frac{1}{2}\sqrt{\frac{\delta}{4-\delta}} \log \tau) , \quad c_1,c_2\in\mathbf{R} .$$

It follows that every solution of (4.11) has a zero on any

interval $[\tau_1,\tau_2]\subset \mathbf{R}$, $0<\tau_1<\tau_2$, provided that

$$\log\frac{\tau_2}{\tau_1}\geq\frac{2\pi}{\sqrt{\frac{\delta}{4-\delta}}} .$$

Assume, on the contrary, that $\alpha\equiv 0$, i.e. the sequence $(\alpha_{T_n})_{n\in\mathbf{N}}$ and hence $(y_{T_n})_{n\in\mathbf{N}}$ tends to zero in the C^2-topology on every compact subset of $\mathbf{R}$ and fix $\varepsilon>0$ and $R>0$ such that, for $y\in(0,\varepsilon)$ and $t\geq R$,

$$\left(\frac{\tau(t)}{\dot\tau(t)}\right)^2 u(t)\,\frac{\sin y}{y}\cos y > \frac{1}{4-\delta} .$$

Setting $\tau_1=\tau(R)$, choose $\tau_2(>\tau_1)$ such that $\log\frac{\tau_2}{\tau_1}\geq\frac{2\pi}{\sqrt{\frac{\delta}{4-\delta}}}$.

For $T_n>R$ denote by y_n the solution of (4.11) with initial data $y_n(\tau_1)=y_{T_n}(\tau_1)$ and $\dot y_n(\tau_1)=\dot y_{T_n}(\tau_1)$. Then as $n\to\infty$ we have $0<y_n(\tau_1)\to 0$ and $\dot y_n(\tau_1)\to 0$, i.e., for sufficiently large n , we have $0<y_n(\tau)<\varepsilon$ for $\tau\in[\tau_1,z_n)$, where z_n is the least zero of y_n in $(\tau_1,\tau_2]$. Hence, (4.10) is a Sturmian majorant for (4.11) and comparison of the solutions y_{T_n} and y_n of (4.10) and (4.11), resp., shows that y_{T_n} vanishes somewhere in $[\tau_1,z_n]$ which is a contradiction.

(ii) $r=1$. Define $\rho:\mathbf{R}_+\to\mathbf{R}$ by

$$\rho(t) = \exp\{\frac{p-1}{2}\int_t^\infty \frac{e^{-s}}{e^s+e^{-s}}\,ds\} \qquad t\geq 0 .$$

Then, setting $y_T=\rho\cdot\alpha_T$, (4.8) transforms into

$$\ddot y_\tau(t) + \frac{1}{\rho(t)}\{\ddot\rho(t)+\dot\rho(t)\,\frac{(p-1)e^{-t}}{e^t+e^{-t}}\}y_T(t) + \\ + u(t)\,\frac{\sin(\rho(t)y_T(t))}{y_T(t)}\cos(\rho(t)y_T(t)) = 0 . \tag{4.12}$$

Clearly, $\lim_{t\to\infty}\rho(t)=1$, $\lim_{t\to\infty}\dot\rho(t)=0$ and as a simple calculation shows

$$\lim_{t\to\infty} \{\ddot{\rho}(t)+\rho(t)\ \frac{(p-1)e^{-t}}{e^{t}+e^{-t}}\} = 0 \ .$$

Hence, setting $0<\delta<\nu$, the same comparison method as above yields $\alpha\not\equiv 0$. √

Remark. An analogous proof shows that $\alpha\not\equiv\frac{\pi}{2}$.

Lemma 4.13. *For sufficiently large* $R>0$ *the solution* α *is strictly increasing on* $[R,\infty)$.

Proof. Again, we distinguish two cases.
(i) $r>1$. Choose $R>0$ such that $u,v>0$ on $[R,\infty)$. If for some $t_o>R$ we have $\dot{\alpha}(t_o)\le 0$ then, by (4.8), $\ddot{\alpha}(t_o)<0$ and hence $\dot{\alpha}(t)<0$, $t>t_o$. Again by (4.8), $\ddot{\alpha}(t)<0$, $t>t_o$, and, as α remains in the strip $\mathbf{R}\times(0,\frac{\pi}{2})$, we have $\lim_{t\to\infty}\dot{\alpha}(t) = \lim_{t\to\infty}\ddot{\alpha}(t)=0$. Thus $\lim_{t\to\infty}\alpha(t)=0$ and, using the comparison method in the proof of 4.9(i), we obtain a contradiction.
(ii) $r=1$. Let R be as in the first case and assume that there exist $t_o,t_1\in\mathbf{R}$, $R<t_o<t_1$, with $\dot{\alpha}(t_o)=\dot{\alpha}(t_1)=0$. From (4.8) it follows that $\ddot{\alpha}(t_o)$, $\ddot{\alpha}(t_1)<0$. Choose $t^*\in(t_o,t_1)$ such that $\alpha|[t_o,t_1]$ attains its minimum at t^* . Then $\dot{\alpha}(t^*)=0$ and $\ddot{\alpha}(t^*)\ge 0$ which contradicts to (4.8). Thus, by a suitable choice of R , we may assume that $\dot{\alpha}(t)<0$, $t>R$, since otherwise the lemma is proved. Then α , being strictly decreasing on $[R,\infty)$, tends to a limit $\alpha_\infty\ge 0$. If $\alpha_\infty=0$ then we can use the comparison method above to get a contradiction. If $\alpha_\infty>0$ then, $\dot{\alpha}$ being bounded on $\mathbf{R}$, it follows from (4.8) that

$$\lim_{t\to\infty}\ddot{\alpha}(t) = -\nu\sin\alpha_\infty\cos\alpha_\infty < 0$$

which is again a contradiction. √

Remark. A similar argument shows that α is strictly increasing on $(-\infty,-R]$ for sufficiently large $R>0$. Thus the limits

$$\lim_{t\to\infty} \alpha(t) = \alpha_\infty \quad \text{and} \quad \lim_{t\to-\infty} \alpha(t) = \alpha_{-\infty}$$

exist with $0 \le \alpha_\infty , \alpha_{-\infty} < \frac{\pi}{2}$.

To complete the proof of 4.1, we need to show the following.

Lemma 4.14. $\alpha_\infty = \frac{\pi}{2}$ *and* $\alpha_{-\infty} = 0$.

Proof. We restrict ourselves to proving the first relation.
(i) $r > 1$. Assume, on the contrary, that $\alpha_\infty < \frac{\pi}{2}$. Choosing $R > 0$ as in the proof of 4.13, we set

$$L = \min_{t \ge R} \frac{u(t)}{v(t)} > 0$$

and define $Q = \{(x_1, x_2) \in \mathbf{R}^2 \mid 0 < x_2 < \frac{\pi}{2} , x_1 < L \sin x_2 \cdot \cos x_2\} \subset \mathbf{R}^2$. As $0 < \alpha < \frac{\pi}{2}$, there exists a sequence $(t_n)_{n\in\mathbf{N}}$ such that $\lim_{n\to\infty} t_n = \infty$ and $\lim_{n\to\infty} \dot{\alpha}(t_n) = 0$. Hence the curve $c : \mathbf{R} \to \mathbf{R}^2$, $c(t) = (\dot{\alpha}(t), \alpha(t))$, $t \in \mathbf{R}$, cuts Q at, say, $t_0 \in \mathbf{R}$. As $\ddot{\alpha}(t) < 0$ for $\alpha(t) \in Q$, by $(0, \alpha_\infty) \in Q$, the function $t \to \dot{\alpha}(t)$ is strictly decreasing and c never leaves Q for $t \ge t_0$. In particular, $\lim_{t\to\infty} \dot{\alpha}(t)$ exists and hence zero. Taking $t\to\infty$ in (4.8) we obtain

$$\lim_{t\to\infty} \ddot{\alpha}(t) = -\nu \sin \alpha_\infty \cos \alpha_\infty < 0$$

which is a contradiction.
(ii) $r = 1$. With $R > 0$ as above we have $\dot{\alpha}(t) > 0$, $t > R$, and hence

$$\ddot{\alpha}(t) = -\frac{p-1}{e^{2t}+1} \dot{\alpha}(t) - u(t) \sin \alpha(t) \cos \alpha(t) < 0 , \quad t > R ,$$

i.e. $\dot{\alpha}(t)$ is strictly decreasing and $\lim_{t\to\infty} \ddot{\alpha}(t) = -\nu \sin \alpha_\infty \cos \alpha_\infty$.
As $0 < \alpha < \frac{\pi}{2}$ we get $\sin \alpha_\infty \cos \alpha_\infty = 0$, i.e. $\alpha_\infty = \frac{\pi}{2}$. √

5. REGULARITY

To accomplish the proof of 3.2, we have to show that the continuous map $h:S^{p+r+1} \to S^{q+s+1}$ defined by (3.6) (homotopic to the join $f*g$ and (apart from the focal varieties satisfying the harmonicity condition) extends smoothly to S^{p+r+1}. This amounts to having additional information on the asymptotic behaviour of the solution α of the pendulum equation (3.10) obtained in the preceding section. Namely, we have the following.

Theorem 5.1. (i) *As* $t\to\infty$ *we have*

$$\ell - \frac{\dot\alpha(t)}{\sin\alpha(t)\cos\alpha(t)} \ll e^{-2t}, \quad \frac{\dot\alpha(t)}{\cos\alpha(t)} - \ell \ll e^{-2t}$$

and

$$e^{-\ell t} \ll \cos\alpha(t) \ll e^{-\ell t}.$$

(ii) *As* $t\to-\infty$ *we have*

$$k - \frac{\dot\alpha(t)}{\sin\alpha(t)\cos\alpha(t)} \ll e^{2t}, \quad \frac{\dot\alpha(t)}{\sin\alpha(t)} - k \ll e^{2t}$$

and

$$e^{kt} \ll \sin\alpha(t) \ll e^{kt}. \;\dagger$$

The proof of the theorem is postponed to the end of this section. Our present aim is to show how the smooth extension of h across the focal varieties can be derived from the asymptotic behaviour of α above.

Theorem 5.2. *The first and second partial derivatives of* $\bar h$ *can be continuously extended to* $\mathbf{R}^{p+r+2}\backslash\{0\}$.

Thus, the map $h:S^{p+r+1} \to S^{q+s+1}$ is of class C^2; in particular, its tension $\tau(h)$ is continuous. As $\tau(h)$ vanishes apart

† In what follows, we shall use standard notations of asymptotic integration theory, e.g., for given scalars $\mu,\mu'\in C^\infty(\mathbf{R})$,

$$\mu(t) \ll \mu'(t) \quad \text{as} \quad t\to\infty$$

means that there exists $c>0$ such that

$$\mu(t) < c\mu'(t)$$

holds for sufficiently large $t\in\mathbf{R}$.

from the focal varieties it is zero everywhere. By Remark before II.2.8, a C^2-map with vanishing tension field is of class C^∞ (in fact, analytic, since the spheres are analytic). Hence, $h: S^{p+r+1} \to S^{q+s+1}$ is harmonic and homotopic to the join $f*g$ which completes the proof of 3.2.

Proof of Theorem 5.2. We restrict ourselves to show the smooth extendability of the first component

$$h^1 : (\mathbf{R}^{p+1}\setminus\{0\}) \times (\mathbf{R}^{r+1}\setminus\{0\}) \to \mathbf{R}^{q+1}$$

defined by

$$h^1(x,y) = \sin \alpha(t) \bar{f}\left(\frac{x}{\|x\|}\right) , \quad (0\neq)\, x\in\mathbf{R}^{p+1} , \quad (0\neq)\, y\in\mathbf{R}^{r+1} ,$$

to the set $\{x=0, y\neq 0\} \subset \mathbf{R}^{p+r+2}$. We first claim that the function $\beta : (\mathbf{R}^{p+1}\setminus\{0\}) \times (\mathbf{R}^{r+1}\setminus\{0\}) \to \mathbf{R}$ given by

$$\beta(x,y) = \frac{\sin \alpha(t)}{\|x\|^k} , \quad (0\neq)\, x\in\mathbf{R}^{p+1} , \quad (0\neq)\, y\in\mathbf{R}^{r+1} ,$$

can be continuously extended over $\{x=0, y\neq 0\}$. For this, it is enough to show that the first derivatives $\frac{\partial\beta}{\partial x^i}(x,y)$, $i=1,\ldots$ $\ldots,p+1$, remain bounded when $x\to 0$ and y remains outside a neighbourhood of the origin. Now, we have

$$\frac{\partial\beta}{\partial x^i}(x,y) = \frac{1}{\|x\|^{2k}}\{\cos \alpha(t)\cdot\dot{\alpha}(t)\cdot$$

$$\cdot \frac{\partial t}{\partial x^i}\|x\|^k - k\cdot\sin\alpha(t)\cdot\|x\|^{k-1}\frac{\partial\|x\|}{\partial x^i}\} =$$

$$= \frac{1}{\|x\|^{2k}}\{\cos\alpha(t)\cdot\dot{\alpha}(t)\frac{x^i}{\|x\|^2}\|x\|^k - k\cdot \qquad (5.3)$$

$$\cdot \sin\alpha(t)\|x\|^{k-1}\frac{x^i}{\|x\|}\} =$$

$$= \frac{x^i}{\|x\|^{k+2}} \{\dot{\alpha}(t)\cos\alpha(t) - k\sin\alpha(t)\}$$

and, for the asymptotes of $\frac{\partial\beta}{\partial x^i}$, we show that

$$\dot{\alpha}(t) - k\sin\alpha(t) = O(\|x\|^{k+2}) . \qquad (5.4)$$

First, by 5.1(ii),

$$\sin\alpha(t) = O(\|x\|^k) \qquad (5.5)$$

and, as

$$\frac{1-\cos\alpha(t)}{\|x\|^{2k}} \leq \frac{(1-\cos\alpha(t))(1+\cos\alpha(t))}{\|x\|^{2k}} = \frac{\sin^2\alpha(t)}{\|x\|^{2k}},$$

we obtain

$$1 - \cos\alpha(t) = O(\|x\|^{2k}) . \qquad (5.6)$$

Secondly, by 5.1(ii) and (5.5),

$$\dot{\alpha}(t) - k\sin\alpha(t) \ll e^{2t}\sin\alpha(t) \ll \|x\|^{k+2} .$$

Furthermore, (5.5) and (5.6) imply

$$k\sin\alpha(t) - \dot{\alpha}(t) = k\sin\alpha(t)\cos\alpha(t) -$$

$$- \dot{\alpha}(t) + k\sin\alpha(t)(1-\cos\alpha(t)) =$$

$$= k\sin\alpha(t)\cos\alpha(t) - \dot{\alpha}(t) + O(\|x\|^{3k}) .$$

Finally, by 5.1(ii) and (5.5),

$$k\sin\alpha(t)\cos\alpha(t) - \dot{\alpha}(t) \ll$$

$$\ll e^{2t}\sin\alpha(t)\cos\alpha(t) \ll \|x\|^{k+2} ,$$

and hence

$$k\sin\alpha(t) - \dot{\alpha}(t) \ll \|x\|^{k+2} ,$$

i.e. (5.4) follows.

Thus,

$$\dot{\alpha}(t)\cos\alpha(t) - k\sin\alpha(t) =$$

$$= -\dot{\alpha}(t)(1-\cos\alpha(t)) + \dot{\alpha}(t) - k\sin\alpha(t) = \tag{5.7}$$

$$= \dot{\alpha}(t)O(\|x\|^{2k}) + O(\|x\|^{k+2}) = O(\|x\|^{k+2}) ,$$

since $\dot{\alpha}(t)=O(\|x\|^{k})$. It follows from (5.3) that $\frac{\partial\beta}{\partial x^i}(x,y) = \frac{x^i}{\|x\|^{k+2}}O(\|x\|^{k+2}) = O(\|x\|)$ which, in fact, tends to zero as $x\to 0$ and the claim follows.

For brevity, we shall be dealing with the continuous extendability of $(\frac{\partial}{\partial x^i})^2 h^1$ since the other cases can be treated analogously. For this, using (5.7), we evaluate $\ddot{\alpha}(t)$ as

$$\ddot{\alpha}(t) - k\dot{\alpha}(t) = (v(t)-k)\dot{\alpha}(t) - u(t)\sin\alpha(t)\cdot\cos\alpha(t) =$$

$$= (v(t)-k)k\sin\alpha(t) - u(t)\sin\alpha(t)\cos\alpha(t) +$$

$$+ O(\|x\|^{k+2}) = \{(v(t)-k)k-u(t)\}\cdot$$

$$\cdot O(\|x\|^{k}) + O(\|x\|^{k+2}) .$$

Furthermore,

$$(v(t)-k)k - u(t) = \frac{1}{e^{t}+e^{-t}}\cdot$$

$$\cdot\{e^{t}((r-k-1)k-\nu)+e^{-t}((-p-k+1)k+\lambda)\} =$$

$$= ((r-k-1)k-\nu)\frac{e^{t}}{e^{t}+e^{-t}} = O(e^{2t}) = O(\|x\|^{2}) ,$$

i.e. we obtain

$$\ddot{\alpha}(t) - k\dot{\alpha}(t) = O(\|x\|^{k+2}) . \tag{5.8}$$

Turning to the estimation of $(\frac{\partial}{\partial x^i})^2 h^1$, a straightforward computation (as in Section 3) yields

$$((\frac{\partial}{\partial x^i})^2 h^1)(x,y) = \cos\alpha(t)\cdot\dot{\alpha}(t)\cdot$$

$$\cdot\{\frac{(||x||^2 - 2(k+1)(x^i)^2)\bar{f}(x)}{||x||^{k+4}} + \frac{2x^i\bar{f}_i(x)}{||x||^{k+2}}\} +$$

$$+ \sin\alpha(t)\{\frac{||x||^2\bar{f}_{ii}(x) - 2kx^i\bar{f}_i(x)}{||x||^{k+2}} + \frac{k(k+2)(x^i)^2\bar{f}(x)}{||x||^{k+4}}\} +$$

$$+ \cos\alpha(t)\cdot\ddot{\alpha}(t)\frac{(x^i)^2\bar{f}(x)}{||x||^{k+4}} -$$

$$- \sin\alpha(t)\cdot\dot{\alpha}(t)^2\frac{(x^i)^2\bar{f}(x)}{||x||^{k+4}} - k\sin\alpha(t)\frac{\bar{f}(x)}{||x||^{k+2}},$$

where $\bar{f}_i(x) = \frac{\partial\bar{f}}{\partial x^i}(x)$ and $\bar{f}_{ii}(x) = \frac{\partial^2\bar{f}}{\partial(x^i)^2}(x)$. The last three terms can be estimated by making use of (5.7) to give

$$\cos\alpha(t)\cdot\ddot{\alpha}(t)\frac{(x^i)^2\bar{f}(x)}{||x||^{k+4}} = k\cos\alpha(t)\cdot\dot{\alpha}(t)\frac{(x^i)^2\bar{f}(x)}{||x||^{k+4}} +$$

$$+ O(||x||^k),$$

$$-\sin\alpha(t)\cdot\dot{\alpha}(t)^2\frac{(x^i)^2\bar{f}(x)}{||x||^{k+4}} =$$

$$= O(||x||^{4k})\frac{1}{||x||^{k+2}} = O(||x||^{3k-2})$$

and

$$-k\sin\alpha(t)\frac{\bar{f}(x)}{||x||^{k+2}} = -k\cdot\beta(x,y)O(||x||^{k-2}) =$$

$$= -k\cdot o(||x||)O(||x||^{k-2}) = o(||x||^{k-1}),$$

i.e. we have

$$((\frac{\partial}{\partial x^i})^2 h^1)(x,y) = \cos\alpha(t)\cdot\dot{\alpha}(t)\cdot$$

$$\cdot\ \{\frac{(\|x\|^2-(k+2)(x^i)^2)\bar{f}(x)}{\|x\|^{k+4}}+\frac{2x^i\bar{f}_i(x)}{\|x\|^{k+2}}\} +$$

$$+\sin\alpha(t)\{\frac{\|x\|^2\bar{f}_{ii}(x)-2kx^i\bar{f}_i(x)}{\|x\|^{k+2}} +$$

$$+\frac{k(k+2)(x^i)^2\bar{f}(x)}{\|x\|^{k+4}}\} + \varepsilon(x) ,$$

where $\lim_{x\to 0}\varepsilon(x)=0$. Finally, (5.7) reduces this to the form

$$((\frac{\partial}{\partial x^i})^2 h^1)(x,y) = \bar{f}_{ii}(x)\beta(x,y) + \varepsilon(x)$$

which, by the above, can be continuously extended over $\{x=0, y\neq 0\}$. ✓

To accomplish the proof of Smith's theorem 3.2 we have to show the validity of the asymptotic formulas given in 5.1 for the solution $\alpha\in C^\infty(\mathbf{R})$ obtained in Section 4. We restrict ourselves to proving (i), i.e. asymptotes as $t\to\infty$, since the second case can be treated similarly. The proof of 5.1(i) is broken up into four lemmas.

Lemma 5.9. *As* $t\to\infty$

$$\frac{\dot{\alpha}(t)}{\cos\alpha(t)} - \ell \ll e^{-2t} . \tag{5.10}$$

Proof. Setting

$$\gamma(t) = \frac{\dot{\alpha}(t)}{\cos\alpha(t)} ,\quad t\in\mathbf{R} , \tag{5.11}$$

we obtain a scalar $\gamma\in C^\infty(\mathbf{R})$ which satisfies the Riccati equation

$$\dot{\gamma}(t) = v(t)\gamma(t) - u(t)\sin\alpha(t) + \sin\alpha(t)\gamma(t)^2 , \quad t\in\mathbf{R} .$$

Let $\gamma^+ \in C^\infty(\mathbf{R})$ be the larger solution of the equation

$$v(t)\gamma(t) - \nu + \gamma(t)^2 = 0 , \quad t \in \mathbf{R} ,$$

i.e.

$$\gamma^+ = \frac{-v+\sqrt{v^2+4\nu}}{2} .$$

Then, $\gamma^+ \in C^\infty(\mathbf{R})$ is eventually strictly increasing and we have

$$\lim_{t\to\infty} \gamma^+(t) = \frac{-(r-1)+\sqrt{(r-1)^2+4\nu}}{2} = \ell .$$

Furthermore, $v(t)=r-1+(r+p-2)e^{-2t}+O(e^{-4t})$ and so, by an easy computation, we obtain

$$\gamma^+(t) = \ell + O(e^{-2t}). \tag{5.12}$$

For fixed (sufficiently large) $t \in \mathbf{R}$ consider the solution β of the initial value problem

$$\dot\beta(s) = \gamma^+(t)\cos\beta(s) , \quad t \le s , \tag{5.13}$$

$$\beta(t) = \alpha(t) . \tag{5.14}$$

Then, by integration,

$$\frac{1+\sin\beta(s)}{\cos\beta(s)} = \frac{1+\sin\beta(t)}{\cos\beta(t)}\, e^{\gamma^+(t)(s-t)} , \quad t \le s ,$$

and β is eventually strictly increasing with $\lim_{s\to\infty}\beta(s) = \frac{\pi}{2}$.
Furthermore, for $t \le s$, we have

$$\ddot\beta(s) = -\gamma^+(t)\sin\beta(s)\dot\beta(s) = -\gamma^+(t)^2\sin\beta(s)\cos\beta(s) =$$

$$= (v(t)\gamma^+(t)-\nu)\sin\beta(s)\cos\beta(s) = v(t)\sin\beta(s)\dot\beta(s) -$$

$$- \nu\sin\beta(s)\cos\beta(s) < v(s)\dot\beta(s) - \nu\sin\beta(s)\cos\beta(s) .$$

We claim that

$$\dot\alpha(t) < \dot\beta(t) . \tag{5.15}$$

Assuming $\dot{\alpha}(t) \geq \dot{\beta}(t)$, we have

$$\ddot{\beta}(t) < v(t)\dot{\beta}(t) - \nu \sin \beta(t) \cos \beta(t) \leq$$

$$\leq v(t)\dot{\alpha}(t) - \nu \sin \alpha(t) \cos \alpha(t) < \ddot{\alpha}(t) \ ,$$

i.e. there exists a maximal (possibly infinite) open interval $I=(t,t')$ such that

$$\dot{\beta} < \dot{\alpha} \tag{5.16}$$

holds on I. If t' is finite then $\dot{\alpha}(t')=\dot{\beta}(t')$ and we have

$$\ddot{\alpha}(t') - \ddot{\beta}(t') = v(t')\dot{\alpha}(t') - u(t')\sin \alpha(t')\cos \alpha(t') -$$

$$- v(t)\sin \beta(t')\dot{\beta}(t') + \nu \sin \beta(t')\cos \beta(t') =$$

$$= v(t')\{\dot{\alpha}(t')-\dot{\beta}(t')\} + v(t')\{1-\sin \beta(t')\}\dot{\beta}(t') +$$

$$+ \{v(t')-v(t)\}\sin \beta(t')\dot{\beta}(t') +$$

$$+ \{\nu-u(t')\}\sin \beta(t')\cos \beta(t') +$$

$$+ \{\sin \beta(t')\cos \beta(t')-\sin \alpha(t')\cos \alpha(t')\}u(t') > 0$$

since the last four terms in the brackets $\{\dots\}$ are positive, which is a contradiction. Thus, (5.16) holds for all $t \leq s$ and, integrating, we obtain

$$\alpha(s) - \beta(s) = \int_t^s (\dot{\alpha}-\dot{\beta})$$

which contradicts $\lim_{s\to\infty} \{\alpha(s)-\beta(s)\} = 0$ as the integrand is positive. Thus (5.15) follows and, by (5.13) and (5.14), we showed that

$$\gamma(t) < \gamma^{+}(t) \tag{5.17}$$

holds for sufficiently large $t \in \mathbf{R}$. Combining this with (5.12), we have

$$\gamma(t) - \ell < \gamma^+(t) - \ell \ll e^{-2t} \ . \ \checkmark$$

Lemma 5.18. $e^{-\ell t} \ll \cos\alpha(t)$ *as* $t \to \infty$.

Proof. Integrating (5.10), we obtain

$$\log \frac{1+\sin\alpha(t)}{1-\sin\alpha(t)} - 2\ell t \ll 1 \ .$$

Thus,

$$\frac{1}{\cos^2\alpha(t)} \le \frac{(1+\sin\alpha(t))^2}{\cos^2\alpha(t)} = \frac{1+\sin\alpha(t)}{1-\sin\alpha(t)} \ll e^{2\ell t} \ . \ \checkmark$$

Lemma 5.19. *As* $t \to \infty$

$$\ell - \frac{\dot\alpha(t)}{\sin\alpha(t)\cos\alpha(t)} \ll e^{-2t} \ . \tag{5.20}$$

Proof. Setting

$$\gamma(t) = \frac{\dot\alpha(t)}{\sin\alpha(t)\cos\alpha(t)} \ , \qquad t\in\mathbf{R} \ , \tag{5.21}$$

we obtain a scalar $\gamma\in C^\infty(\mathbf{R})$ which satisfies the Riccati equation

$$\dot\gamma(t) = v(t)\gamma(t) - u(t) - \cos(2\alpha(t))\gamma(t)^2 \ , \qquad t\in\mathbf{R} \ . \tag{5.22}$$

For large values of $t\in\mathbf{R}$, define

$$\gamma^+ = \frac{-v+\sqrt{v^2-4\cos 2\alpha\cdot u}}{-2\cos 2\alpha} \ ,$$

or equivalently, γ^+ is the larger solution of the equation

$$v\gamma - u - \cos(2\alpha)\gamma^2 = 0 \ .$$

Then, we have

$$\lim_{t\to\infty} \gamma^+(t) = \frac{(r-1)-\sqrt{(r-1)^2+4\nu}}{-2} = \frac{r-1-(2\ell+r-1)}{-2} = \ell \ ,$$

and so, for $\varepsilon>0$ fixed, there exists $t_0 \in \mathbf{R}$ such that $|\gamma^+(t)-\ell|<\varepsilon$ holds for all $t_0 \le t$. We claim that this relation is valid for γ as well; in particular $\lim_{t\to\infty} \gamma(t) = \ell$. If, for some $t_0<t_1$, the inequality $\gamma(t_1)<\ell-\varepsilon$ is valid then, by (5.22), γ is strictly decreasing for $t_1 \le t$ and, being nonnegative, $\lim_{t\to\infty} \gamma(t)=\gamma_\infty$ exists and $\gamma_\infty \ge 0$. On the other hand, as $\lim_{t\to\infty} \dot\alpha(t)=0$ we apply L'Hôpital's rule to get

$$\gamma_\infty = \lim_{t\to\infty} \gamma(t) = \lim_{t\to\infty} \frac{\dot\alpha(t)}{\cos\alpha(t)} = -\lim_{t\to\infty} \frac{\ddot\alpha(t)}{\dot\alpha(t)\sin\alpha(t)} =$$

$$= -\lim_{t\to\infty} \frac{v(t)\dot\alpha(t)-u(t)\sin\alpha(t)\cos\alpha(t)}{\dot\alpha(t)\sin\alpha(t)} =$$

$$= -\lim_{t\to\infty} \frac{v(t)}{\sin\alpha(t)} + \lim_{t\to\infty} \frac{u(t)}{\sin\alpha(t)} \cdot \frac{1}{\gamma(t)} =$$

$$= -(r-1) + \nu \lim_{t\to\infty} \frac{1}{\gamma(t)} .$$

Thus $\gamma_\infty>0$ and, by the above,

$$\gamma_\infty = -(r-1) + \frac{\nu}{\gamma_\infty} ,$$

i.e. $\gamma_\infty=\ell$ which is a contradiction.
If, for some $t_0<t_1$, the inequality $\gamma(t_1)>\ell+\varepsilon$ is valid then, by (5.22), γ is strictly increasing for $t_1 \le t$ and, by the same computation as above, $\lim_{t\to\infty} \gamma(t)=\infty$ which contradicts (5.10) since γ is bounded from above. Thus the claim follows.
Next, for large values of $t \in \mathbf{R}$, define

$$\tilde\gamma^+ = \frac{-v+\sqrt{v^2+4u}}{2} ,$$

i.e. $\tilde\gamma^+$ is the larger solution of the equation

$$v\gamma - u + \gamma^2 = 0 .$$

Then, $\tilde{\gamma}^+$ can be expanded into convergent series

$$\tilde{\gamma}^+ = \ell + a_1 e^{-2t} + a_2 e^{-4t} + a_3 e^{-6t} + \ldots \quad .$$

Let $a_n \in \mathbf{R}$ be the first nonvanishing coefficient, i.e. set

$$\tilde{\gamma}^+ = \ell + a_n e^{-2nt} + \ldots \quad . \tag{5.23}$$

If $a_n < 0$ then, by (5.23), $\tilde{\gamma}^+$ is strictly increasing, say, for $t_0 \le t$. We claim that

$$\tilde{\gamma}^+(t) < \gamma(t) \ , \quad t_0 \le t \ . \tag{5.24}$$

Indeed, assuming $\gamma^+(t_1) \ge \gamma(t_1)$ for some $t_0 \le t_1$, we have

$$\dot{\gamma}(t_1) = v(t_1)\gamma(t_1) - u(t_1) - \cos(2\alpha(t_1))\gamma(t_1)^2 <$$

$$< v(t_1)\gamma(t_1) - u(t_1) + \gamma(t_1)^2 <$$

$$< v(t_1)\tilde{\gamma}^+(t_1) - u(t_1) + \tilde{\gamma}^+(t_1)^2 = 0 \ ,$$

or equivalently,

$$\dot{\gamma}(t_1) < 0 < \dot{\tilde{\gamma}}^+(t_1) \ . \tag{5.25}$$

As $\tilde{\gamma}^+$ is strictly increasing for $t_0 \le t$, the argument above shows that relation (5.25) is preserved on $[t_1, \infty)$, which is a contradiction, since

$$\lim_{t\to\infty} \gamma(t) = \lim_{t\to\infty} \tilde{\gamma}^+(t) = \ell \ .$$

Thus, by (5.23) and (5.24), we have

$$\ell - \gamma(t) < \ell - \tilde{\gamma}^+(t) = (-a_n)e^{-2nt} - \ldots \ll e^{-2t} \ ,$$

and the lemma follows in this case.
If $a_n > 0$ then, by (5.23), $\tilde{\gamma}^+$ is strictly decreasing, say, for $t_0 \le t$. We claim that

$$\ell < \gamma(t) \ , \quad t_o \le t \ . \tag{5.26}$$

Indeed, assuming $\ell \ge \gamma(t_1)$ for some $t_o \le t_1$, we have

$$\dot{\gamma}(t_1) < v(t_1)\gamma(t_1) - u(t_1) + \gamma(t_1)^2 <$$

$$< v(t_1)\ell - u(t_1) + \ell^2 <$$

$$< v(t_1)\tilde{\gamma}^+(t_1) - u(t_1) + \tilde{\gamma}^+(t_1)^2 = 0$$

and hence γ is strictly decreasing for $t_1 \le t$ which is a contradiction.
Thus, by (5.26), $\ell - \gamma(t) \ll 0$ and the lemma follows in this case as well. √

Lemma 5.27. $\cos\alpha(t) \ll e^{-\ell t}$ *as* $t \to \infty$.

Proof. Integrating (5.20), we obtain

$$\ell t - \log\tan\ \alpha(t) \ll 1$$

and hence

$$e^{\ell t}\cos\alpha(t) < e^{\ell t}\cot\alpha(t) \ll 1 \ . \ \surd$$

6. EXAMPLES OF HARMONICALLY REPRESENTABLE HOMOTOPY CLASSES

Theorem 3.2 applies to various circumstances yielding harmonic maps in certain homotopy classes. In what follows, using homotopy theory (cf. Husemoller (1966) and Toda (1962)), we give some examples.

Example 6.1. The group $\pi_3(S^2)$ ($\cong$Z) being generated by (the homotopy class of) the Hopf-map $f{:}S^3 \to S^2$ (cf.II.2.27) the homotopy classes are parametrized by their Hopf-invariant. By 3.4, for $d \in Z$, there exists a harmonic map $h{:}S^2 \to S^2$ with degree d . As f is a Riemannian submersion, II.2.31 implies that $h \circ f{:}S^3 \to S^2$ is harmonic whose Hopf-invariant is known to be d^2 . By composition with the antipodal map $a=-id_{\mathbf{R}^3}|S^2$, we obtain harmonic representatives in each homotopy class in $\pi_3(S^2)$ with Hopf-invariant $\pm d^2$.

Example 6.2. As the group $\pi_{n+1}(S^n)\,(\cong\mathbf{Z}_2)$, $n\geq 3$, is generated by the iterated suspension $S^{n-2}f:S^{n+1}\to S^n$ of the Hopf-map f , by 3.4, we obtain a harmonic generator of the group $\pi_{n+1}(S^n)$ for $n=3,\ldots,8$. (Note that we cannot expect these harmonic maps to be defined by spherical harmonics, e.g. by R. Wood (1968), any map $h:S^4\to S^3$, whose components $h^i:S^4\to\mathbf{R}$, $i=1,2,3,4$, are restrictions (to S^4) of homogeneous polynomials on $\mathbf{R}^5$, is constant.)

Example 6.3. The group $\pi_7(S^5)\,(\cong\mathbf{Z}_2)$ is generated by $S^3f\circ S^4f\simeq f*f$, i.e., by 3.2, this generator is harmonically representable.

Example 6.4. The torsionfree part of the group $\pi_7(S^4)\,(\cong\mathbf{Z}+\mathbf{Z}_{12})$ is generated by the quaternionic Hopf-map $f:S^7\to S^4$. As in 6.1, composing f with harmonic maps $h:S^4\to S^4$ of arbitrary degree we obtain harmonic representatives of all homotopy classes of $\pi_7(S^4)$ with Hopf-invariant d^2 (cf. also Problem 3 for Chapter II).

Example 6.5. The group $\pi_{n+3}(S^n)\,(\cong\mathbf{Z}_{24})$ is generated by the iterated suspension $S^{n-4}f:S^{n+3}\to S^n$ of the quaternionic Hopf-map f and hence, by 3.2, $\pi_{n+3}(S^n)$ has harmonically represented generator for $n\leq 10$.

PROBLEMS FOR CHAPTER III

1. Given a map $f:M\to N$ between Riemannian manifolds M and N with metric tensors g and h , respectively, the *stress-energy tensor* of f is defined as

$$S(f) = e(f)\cdot g - f^*(h)\in C^\infty(S^2(T^*(M))) \;.$$

Assuming that M is compact oriented and N is complete, for a vector field $X \in V(M)$, we have the geodesic variation

$$t \to f_t = \exp \circ (t \cdot f_*(X)) , \quad t \in \mathbf{R} ,$$

of f and the deformation

$$t \to g_t = (\phi_t)^*(g) , \quad t \in \mathbf{R} ,$$

of the metric tensor g, where $(\phi_t)_{t \in \mathbf{R}}$ stands for the 1-parameter group of transformations induced by X. Let $E_{g_t}(f_t)$, $t \in \mathbf{R}$, denote the energy of $f_t : M \to N$ with respect to the Riemannian metric g_t on M. Show that

$$\left.\frac{\partial E_{g_t}(f_t)}{\partial t}\right|_{t=0} = \int_M \operatorname{trace}(f_*, \nabla(f_*(X)))\operatorname{vol}(M,g) +$$

$$+ \int_M (L_X g, S(f))\operatorname{vol}(M,g) ,$$

where, with respect to an orthonormal base $\{e^i\}_{i=1}^m \subset T_x(M)$, $m = \dim M$, at $x \in M$,

$$(L_X g, S(f))_x = \frac{1}{2} \sum_{i,j=1}^{m} (L_X g)(e^i, e^j) \cdot S(f)(e^i, e^j)$$

(cf. Baird and Eells (1981) and Eells and Lemaire (1983).)

2. Given a harmonic map $f : M \to N$ between Riemannian manifolds M and N show that the divergence of the stress-energy tensor $S(f)$ vanishes.

3. Let $f : M \to N$ be a submersion (almost everywhere) between Riemannian manifolds M and N. Prove that if $\operatorname{div} S(f) = 0$ then f is harmonic.

4. Let $f : S^1 \to S^n$, $n \geq 2$, denote the canonical inclusion and let $v \in C^\infty(\nu^f)$ be a parallel unit section. Show that, for the geodesic variation v, $\operatorname{div} S(f_t) = 0$, for all $t \in \mathbf{R}$, but $f_t : S^1 \to S^n$ is not harmonic unless $t \in \frac{\pi}{2}\mathbf{Z}$.

5. Using the second variation formula (1.9), prove Proposition 1.19 for compact oriented M.

CHAPTER IV

Minimal Submanifolds in Spheres

We start here with developing the (variational) theory of minimal maps in a parallel way as was done for harmonic maps before. The energy is then replaced by the volume functional whose critical points and second order behaviour (near critical points) are studied in Section 1. As we are dealing with a restricted class of maps, namely immersions, our earlier methods gain an additional power yielding much more precise statements as shown, for example, by the instability theorem for minimal isometric immersions into spheres to be proved in Section 2. Furthermore, the existence problem of harmonic maps treated in the second part of Chapter III is replaced here by the full (algebraic) classification of minimal isometric immersions between spheres, due to Do Carmo and Wallach, which is introduced by putting emphasis on its close relationship with the rigidity problem. Though the variational theories of harmonic and minimal maps are developed in a parallel way, the classification of minimal immersions requires new techniques based essentially on the representation theory of the special orthogonal group.

1. FIRST AND SECOND VARIATION FORMULAS

Let M be an m-dimensional compact oriented manifold and N a Riemannian manifold with metric tensor $h=(\,,\,)\in C^\infty(S^2(T^*(N)))$. Given an immersion $f:M\to N$, the integral of the volume form $\mathrm{vol}(M,f^*(h))\in\mathcal{D}^m(M)$ with respect to the pull-back metric $f^*(h)\in C^\infty(S^2(T^*(M)))$ is said to be the *volume* of f, i.e. setting

$$\mathrm{Vol}(f) = \int_M \mathrm{vol}(M,f^*(h)) \,, \tag{1.1}$$

we obtain a functional Vol (called the *volume functional*) which associates to each immersion $f:M\to N$ its volume via (1.1). If we endow M with the pull-back metric $g=(,)=f^*(h)$, the immersion $f:M\to N$ becomes isometric and so we can speak of the *volume of an isometric immersion* $f:M\to N$ between Riemannian manifolds M and N.

Given an isometric immersion $f:M\to N$ between any Riemannian manifolds M and N, recall from II.2.16 that the pull-back bundle τ^f splits into the orthogonal direct sum

$$\tau^f \cong T(M) \oplus \nu^f \tag{1.2}$$

of Riemannian-connected vector bundles, where ν^f is the normal bundle of f. Denoting by $\top$ and $\perp$ the tangential and orthogonal projections associated with the splitting (1.2), resp., the second fundamental form of f can be considered as a section $\beta(f)\in C^\infty(S^2(T^*(M))\otimes\nu^f)$ defined by

$$\begin{aligned}&\beta(f)(X_x,Y_x) = (\nabla_{X_x}(f_*(Y)))^\perp,\\ &X_x,Y_x\in T_x(M),\quad x\in M,\end{aligned} \tag{1.3}$$

where $Y\in V(M)$ is an extension of Y_x.

The following first variation formula for the volume functional gives an equivalent interpretation of minimality of isometric immersions introduced in II.2 via the second fundamental form.

Theorem 1.4. *Let* $f:M\to N$ *be an immersion of a compact oriented manifold of dimension* m *into a Riemannian manifold* N *with metric tensor* $h=(,)$. *Given a (smooth) variation*

$$t\to f_t\,,\quad t\in\mathbf{R} \tag{1.5}$$

of f *through immersions, we have*

$$\left.\frac{\partial \mathrm{Vol}(f_t)}{\partial t}\right|_{t=0} = -\int_M (\mathrm{trace}\,\beta(f),v)\,\mathrm{vol}(M,g)\,, \tag{1.6}$$

where $v = \left.\frac{\partial f_t}{\partial t}\right|_{t=0} \in C^\infty(\tau^f)$ *and* $g=f^*(h)$. *In particular,* $f:M\to N$ *is a critical point of the volume functional* Vol *if and only if* f *is minimal.*

Proof. Take $f:M\to N$ to be isometric by fixing on M , as a Riemannian structure, the pull-back metric $g=(,)=f^*(h)$. We first express $\mathrm{vol}(M,f_t^*(h))\in V^m(M)$, $t\in\mathbf{R}$, in terms of a local orthonormal moving frame $\{E^i\}_{i=1}^m$ (adapted to an orthonormal base $\{e^i\}_{i=1}^m \subset T_x(M)$ at a point $x\in M$) over a normal coordinate neighbourhood centered at x . Decomposing the tensor $g_t= =f_t^*(h)\in C^\infty(S^2(T^*(M)))$ with respect to the local base $\{\gamma(E^i)\otimes \otimes\gamma(E^j)\}_{i,j=1}^m$ as

$$g_t = \sum_{i,j=1}^{m} g_{ij}(t)\gamma(E^i) \otimes \gamma(E^j)$$

with

$$g_{ij}(t) = ((f_t)_*E^i,(f_t)_*E^j) , \quad g_{ij}(0) = \delta_{ij} , \quad i,j=1,\dots,m , \quad t\in\mathbf{R} , \tag{1.7}$$

we obtain

$$\mathrm{vol}(M,g_t) = \sqrt{g(t)}\,\gamma(E^1)\wedge\dots\wedge\gamma(E^m) = =\sqrt{g(t)}\,\mathrm{vol}(M,g) , \tag{1.8}$$

where $g(t)=\det(g_{ij}(t))_{i,j=1}^m$ (cf.I.1). By (1.7), (1.8) and the definition of the determinant, we have

$$\frac{\partial}{\partial t}\{\mathrm{vol}(M,g_t)_x\}_{t=0} = \left.\frac{\partial\sqrt{g(t)(x)}}{\partial t}\right|_{t=0} \mathrm{vol}(M,g) =$$

$$= \frac{1}{2}\left.\frac{\partial g(t)(x)}{\partial t}\right|_{t=0} \mathrm{vol}(M,g) = \tag{1.9}$$

$$= \frac{1}{2}\frac{\partial}{\partial t}\{\det(g_{ij}(t)(x))_{i,j=1}^m\}\ \mathrm{vol}(M,g) =$$

$$= \frac{1}{2}\,\mathrm{trace}\{\frac{\partial}{\partial t}(g_{ij}(t)(x))_{t=o}\}_{i,j=1}^{m}\mathrm{vol}(M,g) =$$

$$= \frac{1}{2}\sum_{i=1}^{m}\left.\frac{\partial g_{ii}(t)(x)}{\partial t}\right|_{t=o}\mathrm{vol}(M,g)\ .$$

Furthermore, using the notations of II.3, by virtue of II.3.5 and (1.7), for $i=1,\ldots,m$, we have

$$\frac{1}{2}\left.\frac{\partial g_{ii}(t)(x)}{\partial t}\right|_{t=o} = \frac{1}{2}\frac{\partial}{\partial t}(\ (f_t)_* e^i,(f_t)_* e^i)\,|_{t=o} =$$

$$= \frac{1}{2}\frac{\partial}{\partial t}(P(t)e^i,P(t)e^i)\,\Big|_{t=o} = (f_*(e^i),\ \left.\frac{\partial P(t)}{\partial t}\right|_{t=o}e^i) =$$

$$= (f_*(e^i),\nabla_{e^i}v) = e^i(f_*(E^i),v) - ((\nabla_{e^i}f_*)(e^i),v_x)\ .$$

Taking summation over $i=1,\ldots,m$,

$$\frac{1}{2}\sum_{i=1}^{m}\left.\frac{\partial g_{ii}(t)(x)}{\partial t}\right|_{t=o} = (\mathrm{div}\ \gamma^{-1}(f_*,v))(x) -$$

$$- (\mathrm{trace}\ \beta(f),v)_x$$

and thus, by (1.9) and Green's theorem, we have

$$\left.\frac{\partial \mathrm{Vol}(f_t)}{\partial t}\right|_{t=o} = \int_M \frac{\partial}{\partial t}\{\mathrm{vol}(M,g_t)\}_{t=o} =$$

$$= \int_M \mathrm{div}\ \gamma^{-1}(f_*,v)\,\mathrm{vol}(M,g) - \int_M (\mathrm{trace}\ \beta(f),v)\,\mathrm{vol}(M,g) =$$

$$= -\int_M (\mathrm{trace}\ \beta(f),v)\,\mathrm{vol}(M,g)\ .\ \checkmark$$

Remark 1. As the trace of the second fundamental form $\beta(f)$ of an immersion $f:M\to N$ is a section of the normal bundle ν^f , the tangential part $v^{\top}$ of v does not have any effect in the first variation formula (1.6).

Remark 2. Comparing the first variation formulas for the energy and volume functionals (cf. III.(1.5) and (1.6)), for compact and oriented M, we conclude, as in II.2.20, that an isometric immersion $f:M\to N$ is harmonic if and only if f is minimal.

Remark 3. Various generalizations of the energy and volume functionals can be obtained in the following way. Given a map $f:M\to N$ between Riemannian manifolds M and N with metric tensors $g=(,)$ and $h=(,)$, respectively, the tensor $f^*(h)\in C^\infty(S^2(T^*(M)))$ considered as a symmetric linear endomorphism on each of the tangent spaces $T_x(M)$, $x\in M$, has eigenvalues $\{\lambda_i(x)\}_{i=1}^m \subset \mathbf{R}$ with respect to g. For a fixed symmetric function $\sigma:\mathbf{R}^m\to\mathbf{R}$, setting

$$e_\sigma(f)(x) = \sigma(\lambda_1(x),\ldots,\lambda_m(x)), \quad x\in M,$$

we obtain a scalar $e_\sigma(f)\in C^\infty(M)$ called the *σ-energy density* of f.
For compact and oriented M, we then define the *σ-energy* $E_\sigma(f)$ of f, by setting

$$E_\sigma(f) = \int_M e_\sigma(f)\,\mathrm{vol}(M,g).$$

Clearly, the energy and volume functionals can be obtained as

$$E = E_{\frac{1}{2}\sigma_1} \quad \text{and} \quad \mathrm{Vol} = E_{\sqrt{\sigma_m}},$$

where σ_i, $i=1,\ldots,m$, denotes the i-th elementary symmetric function on m variables.

In order to get any further, we now digress to explain some concepts of multilinear algebra which will be used subsequently. Given an m-dimensional vector space V (over $\mathbf{R}$) with scalar product $(,)$, we identify the symmetric square $S^2(V)$ with the vector space of all symmetric linear endomorphisms of V. The identification is given by assigning to the symmetric product

$v \cdot w \in S^2(V)$, $v, w \in V$, the symmetric linear endomorphism $v \cdot w : V \to V$ defined by

$$(v \cdot w)u = \frac{1}{2}\{(v,u)w + (w,u)v\} , \quad u \in V . \tag{1.10}$$

Then $(,)$ can be naturally extended to a scalar product $(,)$ on the vector space of symmetric linear endomorphisms, namely, for two of its elements A and B , we set

$$(A,B) = \text{trace}\ {}^tB \circ A = \sum_{i=1}^{m} (A(e^i), B(e^i)) , \tag{1.11}$$

where $\{e^i\}_{i=1}^m \subset V$ is an orthonormal base and the dual V^* is, as usual, identified with V via $(,)$. By the isomorphism above, $S^2(V)$ inherits a scalar product $(,)$ such that, for $A \in S^2(V)$ and $v, w \in V$,

$$(A, v \cdot w) = (Av, w) . \tag{1.12}$$

Indeed, fixing an orthonormal base $\{e^i\}_{i=1}^m \subset V$, by (1.10), we have

$$(A, v \cdot w) = \sum_{i=1}^{m} (Ae^i, (v \cdot w)e^i) = \frac{1}{2} \sum_{i=1}^{m} (v, e^i)(Ae^i, w) +$$

$$+ \frac{1}{2} \sum_{i=1}^{m} (w, e^i)(Ae^i, v) = \frac{1}{2}(A(\sum_{i=1}^{m} (v, e^i)e^i), w) +$$

$$+ \frac{1}{2}(A(\sum_{i=1}^{m} (w, e^i)e^i), v) = \frac{1}{2}(Av, w) + \frac{1}{2}(Aw, v) = (Av, w) .$$

These considerations apply (pointwise) to the vector bundle $S^2(T^*(M)) \cong S^2(T(M))$, over a Riemannian manifold M with metric tensor $g = (,)$, which is then identified with the vector bundle of all symmetric linear endomorphisms of $T^*(M) \cong T(M)$. Given an isometric immersion $f : M \to N$, we first introduce the *shape-operator* $A(f)$ as a section of the vector bundle $\text{Hom}(\nu^f, S^2(T(M)))$ which corresponds to the second fundamental form $\beta(f)$ of f via the canonical isomorphism

$$S^2(T^*(M)) \otimes \nu^f \cong \text{Hom}(\nu^f, S^2(T(M))) .$$

For a section $v \in C^\infty(\nu^f)$, the value $A(f)^v$ of the shape-operator $A(f)$ on v can then be considered as a symmetric linear endomorphism on each of the tangent spaces $T_x(M)$, $x \in M$, given by the formula

$$(A(f)^{v_x}(X_x), Y_x) = (\beta(f)(X_x, Y_x), v_x) \,, \qquad v_x \in \nu_x^f \,, \quad X_x, Y_x \in T_x(M) \,. \tag{1.13}$$

Lemma 1.14. *For each section* $v \in C^\infty(\nu^f)$ *and* $X_x \in T_x(M)$, $x \in M$,

$$f_*(A(f)^{v_x}(X_x)) = -(\nabla_{X_x} v)^\top \,. \tag{1.15}$$

Proof. By virtue of (1.13), for $Y \in V(M)$, we have

$$(A(f)^{v_x}(X_x), Y_x) = (\beta(f)(X_x, Y_x), v_x) =$$

$$= ((\nabla_{X_x}(f_*(Y)))^\perp, v_x) = (\nabla_{X_x}(f_*(Y)), v_x) =$$

$$= X_x(f_*(Y), v) - (f_*(Y_x), \nabla_{X_x} v) =$$

$$= -(\nabla_{X_x} v, f_*(Y_x)) = -((\nabla_{X_x} v)^\top, f_*(Y_x)) \,. \quad \surd$$

With the notations above, setting

$$\tilde{A}(f) = {}^tA(f) \circ A(f) \in C^\infty(\mathrm{Hom}(\nu^f, \nu^f)) \,,$$

we obtain a symmetric positive semidefinite linear endomorphism of the normal bundle ν^f, since, for $v, w \in C^\infty(\nu^f)$,

$$(\tilde{A}(f)(v), w) = (A(f)^v, A(f)^w) \,. \tag{1.16}$$

We can now state the second variation formula for minimal isometric immersions.

Theorem 1.17. *Let* $f: M \to N$ *be a minimal isometric immersion of an* m*-dimensional compact oriented Riemannian manifold* M *with* *metric tensor* $g=(,)$ *into a Riemannian manifold* N *with metric tensor* $h=(,)$. *Given a 2-parameter variation*

$$(s,t) \to f_{(s,t)}\ ,\quad (s,t)\in\mathbf{R}^2\ , \tag{1.18}$$

of f *(i.e.* $f_{(0,0)}=f$*) through immersions, such that*

$$v = \left.\frac{\partial f_{(s,0)}}{\partial s}\right|_{s=0} \in C^\infty(\nu^f) \quad \text{and}$$

$$w = \left.\frac{\partial f_{(0,t)}}{\partial t}\right|_{t=0} \in C^\infty(\nu^f)\ , \tag{1.19}$$

we have

$$\left.\frac{\partial^2 \mathrm{Vol}(f_{(s,t)})}{\partial s\,\partial t}\right|_{s=t=0} = \int_M \{(\nabla^\perp v,\nabla^\perp w)+(\mathrm{trace}\ R(f_*,v)f_*,w)-$$

$$-(\tilde{A}(f)(v),w)\}\mathrm{vol}(M,g) = \int_M (J_f^\perp(v),w)\,\mathrm{vol}(M,g)\ , \tag{1.20}$$

where $J_f^\perp \in \mathrm{Diff}_2(\nu^f,\nu^f)$ *is given by*

$$J_f^\perp = -\mathrm{trace}(\nabla^\perp)^2 + (\mathrm{trace}\ R(f_*,.)f_*)^\perp - \tilde{A}(f) \tag{1.21}$$

(*cf.*III.1.11).

Proof. For fixed $t\in\mathbf{R}$, setting $f_t=f_{(0,t)}$, $v^t=\left.\frac{\partial f_{(s,t)}}{\partial s}\right|_{s=0}\in C^\infty(\tau^{f_t})$, $v^0=v$, and $g_t=f_t^*(h)$, by (1.6) ,

$$\left.\frac{\partial \mathrm{Vol}(f_{(s,t)})}{\partial s}\right|_{s=0} = -\int_M (\mathrm{trace}_{g_t}\beta(f_t),v^t)\,\mathrm{vol}(M,g_t)\ , \tag{1.22}$$

where trace_{g_t} indicates that the trace is taken with respect to the metric tensor g_t on M .
Differentiating (1.22) at $t=0$, we obtain

$$\left.\frac{\partial^2 \mathrm{Vol}(f_{(s,t)})}{\partial s\,\partial t}\right|_{s=t=0} =$$

$$= -\int_M \frac{\partial}{\partial t}(\mathrm{trace}_{g_t}\, \beta(f_t), v^t)\Big|_{t=0} \mathrm{vol}(M,g) -$$

$$- \int_M (\mathrm{trace}\, \beta(f), v)\, \frac{\partial}{\partial t}\{\mathrm{vol}(M, g_t)\}_{t=0} ,$$

and, as $f:M \to N$ is minimal, the second integral vanishes. To compute the first, choose a local orthonormal moving frame $\{E^i\}_{i=1}^m$ on M with respect to g (adapted to an orthonormal base $\{e^i\}_{i=1}^m \subset T_x(M)$ at $x \in M$) over a normal coordinate neighbourhood U centered at x . Furthermore, for each $t \in \mathbf{R}$, let $\{E_t^i\}_{i=1}^m \subset V(U)$ be a local orthonormal moving frame with respect to g_t such that, for all $i=1,\ldots,m$,

$$t \to E_t^i , \quad t \in \mathbf{R} ,$$

is smooth in t and $E_0^i = E^i$.
Similarly to the proof of 1.4, we decompose g_t and E_t^i , $t \in \mathbf{R}$, $i=1,\ldots,m$, as

$$g_t = \sum_{i,j=1}^m g_{ij}(t)\gamma(E^i) \otimes \gamma(E^j) \tag{1.23}$$

with

$$\begin{aligned} g_{ij}(t) &= ((f_t)_* E^i, (f_t)_* E^j) , \\ g_{ij}(0) &= \delta_{ij} , \quad i,j=1,\ldots,m , \end{aligned} \tag{1.24}$$

and

$$\begin{aligned} E_t^i &= \sum_{j=1}^m C_{ij}(t) E^j, \quad C_{ij}(0) = \delta_{ij} , \\ i,j&=1,\ldots,m , \end{aligned} \tag{1.25}$$

respectively. Using these decompositions, we have

$$\frac{\partial}{\partial t}(\mathrm{trace}_{g_t}\, \beta(f_t), v^t)_x \Big|_{t=0} =$$

$$= \frac{\partial}{\partial t} \sum_{i=1}^{m} (\beta(f_t)(E_t^i, E_t^i), v^t)_x \Big|_{t=0} =$$

$$= \sum_{i,j,k=1}^{m} \frac{\partial}{\partial t}\{C_{ij}(t)(x) C_{ik}(t)(x)(\beta(f_t)(e^j, e^k), v_x^t)\}_{t=0} =$$

$$= \sum_{i,j=1}^{m} \frac{\partial C_{ij}(t)(x)}{\partial t}\Big|_{t=0} (\beta(f)(e^j, e^i), v_x) +$$

$$+ \sum_{i,k=1}^{m} \frac{\partial C_{ik}(t)(x)}{\partial t}\Big|_{t=0} (\beta(f)(e^i, e^k), v_x) +$$

$$+ \sum_{i=1}^{m} \frac{\partial}{\partial t}(\beta(f_t)(e^i, e^i), v_x^t)\Big|_{t=0} = \sum_{i,j=1}^{m} \{\frac{\partial C_{ij}(t)(x)}{\partial t}\Big|_{t=0} +$$

$$+ \frac{\partial C_{ji}(t)(x)}{\partial t}\Big|_{t=0}\} (\beta(f)(e^i, e^j), v_x) +$$

$$+ \frac{\partial}{\partial t}(\text{trace } \beta(f_t), v^t)_x\Big|_{t=0} .$$

To evaluate the term in the bracket $\{\ldots\}$, we differentiate the identity

$$\delta_{ij} = g_t(E_t^i, E_t^j) = \sum_{k,\ell=1}^{m} C_{ik}(t) C_{j\ell}(t) \cdot g_{k\ell}(t) ,$$

$$i, j = 1, \ldots, m ,$$

at $t=0$, yielding

$$0 = \frac{\partial C_{ij}(t)(x)}{\partial t}\Big|_{t=0} + \frac{\partial C_{ji}(t)(x)}{\partial t}\Big|_{t=0} + \frac{\partial g_{ij}(t)(x)}{\partial t}\Big|_{t=0} ,$$

$$i, j = 1, \ldots, m .$$

On the other hand, using the notations of II.3, by II.3.5, we obtain from (1.15), (1.19) and (1.24),

$$\left.\frac{\partial g_{ij}(t)(x)}{\partial t}\right|_{t=o} = \frac{\partial}{\partial t}\left((f_t)_* e^i, (f_t)_* e^j\right)\Big|_{t=o} =$$

$$= \frac{\partial}{\partial t}(P(t)e^i, P(t)e^j)\Big|_{t=o} = \left(\left.\frac{\partial P(t)e^i}{\partial t}\right|_{t=o},\right.$$

$$\left. f_*(e^j)\right) + \left(f_*(e^i), \left.\frac{\partial P(t)e^j}{\partial t}\right|_{t=o}\right) = (\nabla_{e^i} w, f_*(e^j)) +$$

$$+ (f_*(e^i), \nabla_{e^j} w) = ((\nabla_{e^i} w)^{\top}, f_*(e^j)) + (f_*(e^i), (\nabla_{e^j} w)^{\top}) =$$

$$= -(A(f)^{w_x}(e^i), e^j) - (e^i, A(f)^{w_x}(e^j)) =$$

$$= -2(A(f)^{w_x}(e^i), e^j) ,$$

and so

$$\left\{\left.\frac{\partial C_{ij}(t)(x)}{\partial t}\right|_{t=o} + \left.\frac{\partial C_{ji}(t)(x)}{\partial t}\right|_{t=o}\right\} = 2(A(f)^{w_x}(e^i), e^j) .$$

Thus, by (1.11), (1.13) and (1.16), we have

$$\sum_{i,j=1}^{m} \left\{\left.\frac{\partial C_{ij}(t)(x)}{\partial t}\right|_{t=o} + \left.\frac{\partial C_{ji}(t)(x)}{\partial t}\right|_{t=o}\right\}\cdot$$

$$\cdot(\beta(f)(e^i, e^j), v_x) =$$

$$= 2\sum_{i,j=1}^{m} (A(f)^{w_x}(e^i), e^j)(A(f)^{v_x}(e^i), e^j) =$$

$$= 2\sum_{i=1}^{m} (A(f)^{v_x}(e^i), \sum_{j=1}^{m} (A(f)^{w_x}(e^i), e^j)e^j) =$$

$$= 2\sum_{i=1}^{m} (A(f)^{v_x}(e^i), A(f)^{w_x}(e^i)) =$$

$$= 2(A(f)^{v_x}, A(f)^{w_x}) = 2(\tilde{A}(f)(v_x), w_x) .$$

Summarizing the computations above,

$$\left.\frac{\partial^2 \mathrm{Vol}(f_{(s,t)})}{\partial s \partial t}\right|_{s=t=0} = -2\int_M (\tilde{A}(f)(v),w)\,\mathrm{vol}(M,g) -$$

$$-\frac{\partial}{\partial t}\{\int_M (\mathrm{trace}\ \beta(f_t), v^t)\,\mathrm{vol}(M,g)\}_{t=0} .$$

To evaluate the second integral, we first use Green's theorem to obtain

$$-\int_M (\mathrm{trace}\ \beta(f_t), v^t)\,\mathrm{vol}(M,g) =$$

$$= -\int_M (\mathrm{trace}\{\nabla(f_t)_*\}, v^t)\,\mathrm{vol}(M,g) =$$

$$= -\int_M \mathrm{trace}\{\nabla((f_t)_*, v^t)\}\,\mathrm{vol}(M,g) +$$

$$+ \int_M \mathrm{trace}(\ (f_t)_*, \nabla v^t)\,\mathrm{vol}(M,g) =$$

$$= \int_M \mathrm{trace}(\ (f_t)_*, \nabla v^t)\,\mathrm{vol}(M,g) ,$$

since $\mathrm{trace}\{\nabla(\ (f_t)_*, v^t)\} = \mathrm{div}\ \gamma^{-1}((f_t)_*, v^t)$.
Noticing that the last integral occurs in the proof of III.1.7 (cf.III.(1.12)), we have

$$\frac{\partial}{\partial t}\{\int_M \mathrm{trace}(\ (f_t)_*, \nabla v^t)\,\mathrm{vol}(M,g)\}_{t=0} =$$

$$= \int_M \mathrm{trace}(\nabla v, \nabla w)\,\mathrm{vol}(M,g) +$$

$$+ \int_M (\mathrm{trace}\ R(f_*, v) f_*, w)\,\mathrm{vol}(M,g) .$$

Finally, by II.(2.18), (1.15) and (1.16),

$$\mathrm{trace}(\nabla v, \nabla w)_x = \sum_{i=1}^{m} (\nabla_{e_i} v, \nabla_{e_i} w) =$$

$$= \sum_{i=1}^{m} (\ (\nabla_{e_i} v)^{\top}\ ,\ (\nabla_{e_i} w)^{\top}\ ) +$$

$$+ \sum_{i=1}^{m} ((\nabla_{e^i} v)^{\perp} , (\nabla_{e^i} w)^{\perp}) =$$

$$= \sum_{i=1}^{m} (A(f)^{v}{}_{x}(e^i), A(f)^{w}{}_{x}(e^i)) +$$

$$+ \sum_{i=1}^{m} (\nabla^{\perp}_{e^i} v, \nabla^{\perp}_{e^i} w) = (A(f)^{v}{}_{x}, A(f)^{w}{}_{x}) +$$

$$+ \operatorname{trace}(\nabla^{\perp} v, \nabla^{\perp} w)_x = (\tilde{A}(f)(v), w)_x + \operatorname{trace}(\nabla^{\perp} v, \nabla^{\perp} w)_x$$

and the first equality of (1.20) follows. As for the second, we have

$$\operatorname{trace}(\nabla^{\perp} v, \nabla^{\perp} w)_x = \sum_{i=1}^{m} (\nabla^{\perp}_{e^i} v, \nabla^{\perp}_{e^i} w) = \sum_{i=1}^{m} e^i (\nabla^{\perp}_{E^i} v, w) -$$

$$- \sum_{i=1}^{m} (\nabla^{\perp}_{e^i} (\nabla^{\perp}_{E^i} v), w_x) = \operatorname{div} \gamma^{-1} (\nabla^{\perp} v, w) -$$

$$- (\operatorname{trace} (\nabla^{\perp})^2 v, w)$$

and integration over M completes the proof. √

Remark. The second variation formula (1.20) defines a symmetric bilinear pairing $H^{\perp}_{f}$ on $C^{\infty}(\nu^f)$ (called the *normal Hessian* of the minimal isometric immersion f) by setting

$$H^{\perp}_{f}(v,w) = \int_M \{ (\nabla^{\perp} v, \nabla^{\perp} w) + (\operatorname{trace} R(f_*, v) f_*, w) -$$

$$- (\tilde{A}(f)(v), w) \} \operatorname{vol}(M,g) = \int_M (J^{\perp}_{f}(v), w) \operatorname{vol}(M,g) ,$$

$$v, w \in C^{\infty}(\nu^f) .$$

Turning to the general situation, let $f: M \to N$ be a minimal isometric immersion between any Riemannian manifolds. The differential operator $J^{\perp}_{f} \in \operatorname{Diff}_2(\nu^f, \nu^f)$ defined by

$$J^{\perp}_{f}(v) = -\operatorname{trace}(\nabla^{\perp})^2 v + (\operatorname{trace} R(f_*, v) f_*)^{\perp} -$$

$$- \tilde{A}(f)(v) , \quad v \in C^{\infty}(\nu^f) , \tag{1.26}$$

(occurring in (1.20)) is said to be the *normal Jacobi operator* (associated to the minimal isometric immersion $f: M \to N$). Sections belonging to the kernel $J^{\perp}(f)$ of $J^{\perp}_f$ are called *normal Jacobi fields along* f. By II.1.10, $J^{\perp}_f$ is an elliptic operator. *Normal nullity, Morse-index* and *stability* are defined analogously to the case of the energy functional treated in III.1. and, by similar arguments, we obtain that, for compact M, the normal nullity and normal Morse index of a minimal isometric immersion $f: M \to N$ are finite. Furthermore, for compact oriented M, the normal Jacobi operator is self-adjoint as shown by (1.20). (For more details on the structure of the normal Jacobi operator, see Simons (1968) and Nagura (1981) and (1982).)

2. INSTABILITY

From here on, being mainly concerned with minimal isometric immersions into spheres, we take the codomain N to be the Euclidean n-sphere. Given a minimal isometric immersion $f: M \to S^n$ of an m-dimensional Riemannian manifold M with metric tensor $g = (,)$ into S^n, the curvature term occurring in the normal Jacobi operator $J^{\perp}_f$ becomes (by Problem 4 for Chapter I)

$$\begin{aligned} (\text{trace } R(f_*, v) f_*)^{\perp} &= -\text{trace} \| f_* \|^2 v = \\ &= -m \cdot v , \quad v \in C^{\infty}(\nu^f) . \end{aligned} \tag{2.1}$$

To study the stability of minimal isometric immersions, we first compute the normal Morse index of the inclusion map

$$i : S^m \to S^n , \quad m \leq n ,$$

(induced by $\mathbb{R}^{m+1} \to \mathbb{R}^{n+1}$, $(x^1, \ldots, x^{m+1}) \to (x^1, \ldots, x^{m+1}, 0, \ldots, 0) \in \mathbb{R}^{n+1}$, $(x^1, \ldots, x^{m+1}) \in \mathbb{R}^{m+1}$).

Proposition 2.2. *The normal Morse index of the inclusion map* $i:S^m \to S^n$ *equals* $n-m$.

Proof. As i is totally geodesic, the second fundamental form $\beta(i)$ and hence the operator $\tilde{A}(i)$ vanish. By (2.1), the normal Jacobi operator reduces to

$$J_i^{\perp}(v) = -\mathrm{trace}(\nabla^{\perp})^2 v - m\cdot v \ , \qquad v\in C^{\infty}(\nu^i) \ .$$

Thus the eigenspaces of $J_i^{\perp}$ coincide with those of $-\mathrm{trace}(\nabla^{\perp})^2$, and if λ is an eigenvalue of $-\mathrm{trace}(\nabla^{\perp})^2$, then the corresponding eigenvalue of $J_i^{\perp}$ is $\lambda - m$. Now, let $\{\check{z}^j\}_{j=1}^{n-m} \subset \mathbf{R}^{n+1}$ be an orthonormal base in the orthogonal complement of $\mathbf{R}^{m+1} \subset \subset \mathbf{R}^{n+1}$. Then, as in the proof of III.2.4, for $j=1,\ldots,n-m$, the restriction $z^j \circ \bar{i}$ of the uniform extension $z^j \in V(\mathbf{R}^{n+1})$ of the vector $\check{z}^j \in \mathbf{R}^{n+1}$ defines a parallel section W^j of the normal bundle ν^i such that $\{W^j\}_{j=1}^{n-m} \subset C^{\infty}(\nu^i)$ is an orthonormal base on each of the fibres ν_x^i , $x\in S^m$. Thus, for any section $v\in C^{\infty}(\nu^i)$, we have

$$v = \sum_{j=1}^{n-m} v^i \cdot W^j \ , \quad v^j \in C^{\infty}(S^m),\ j=1,\ldots,n-m \ , \tag{2.3}$$

and, as

$$-\mathrm{trace}(\nabla^{\perp})^2 v = -\sum_{j=1}^{n-m} (\mathrm{trace}(\nabla^{\perp})^2)(v^j W^j) =$$

$$= -\sum_{j=1}^{n-m} (\mathrm{trace}\ \nabla^2 v^j)\cdot W^j = \sum_{j=1}^{n-m} (\Delta^{S^m} v^j)\cdot W^j \ ,$$

a normal section v belongs to the eigenspace of $-\mathrm{trace}(\nabla^{\perp})^2$ with eigenvalue λ if and only if, for each $j=1,\ldots,n-m$,

$$\Delta^{S^m} v^j = \lambda v^j \ .$$

As shown in I.7.5, $\mathrm{Spec}(S^m)=\{k(k+m-1)\,|\,k\in\mathbf{Z}_+\}$ and hence the only negative eigenvalue of $J_i^{\perp}$ is $(-m)$ which corresponds to $0\in \in\mathrm{Spec}(S^m)$. By I.6.8, the 0-eigenspace of Δ^{S^m} consists of the constant functions on S^m and is, therefore, 1-dimensional. Hence, the $(-m)$-eigenspace of $J_i^{\perp}$ is $(n-m)$-dimensional. ✓

The rest of this section is devoted to the proof of the following instability theorem due to Simons (1968).

Theorem 2.4. *Let $f:M\to S^n$ be a minimal isometric immersion of an m-dimensional compact oriented Riemannian manifold M with metric tensor $g=(\,,\,)$ into S^n. Then the normal Morse index of $f\geq n-m$ and equality holds if and only if $f:M\to S^n$ is totally geodesic. In particular, any normally stable minimal isometric immersion $f:M\to S^n$ with $m<n$ is constant.*

The verification of 2.4 is preceded by several lemmas and we shall use the notations introduced in the proof of III.2.4. Namely, the uniform extension of a vector $\check{Z}\in\mathbf{R}^{n+1}$ is denoted by $Z\in V(\mathbf{R}^{n+1})$; $^{\top}$ and $^{\perp}$ stand for the tangential and orthogonal projections with respect to the inclusion $S^n\subset\mathbf{R}^{n+1}$ and, for a given vector field v along the minimal isometric immersion $f:M\to S^n$ $v_f^{\top}\in V(M)$ and $v_f^{\perp}\in C^{\infty}(\nu^f)$ denote the tangential and orthogonal projections of v with respect to the orthogonal decomposition

$$\tau^f = T(M)\oplus\nu^f .$$

For the sake of brevity, we set

$$Y_f^{\top} = (Y\circ f)_f^{\top}, \quad Y_f^{\perp} = (Y\circ f)_f^{\perp}, \quad Y\in V(S^n),$$

and

$$W_f^{\top} = (W^{\top})_f^{\top}, \quad W_f^{\perp} = (W^{\perp})_f^{\perp}, \quad W\in V(\mathbf{R}^{n+1}) .$$

Finally, to simplify the notations, as far as the local computations are concerned, we identify M with $\mathrm{im}(f)$ and, consequently, a tangent vector $X_x\in T_x(M)$ with its image in $T_{f(x)}(S^n)$ under the differential f_* of f .

Lemma 2.5. *Let Z be a uniform vector field on $\mathbf{R}^{n+1}$. Then, for each $X_x\in T_x(M)$, $x\in M$,*

$$\nabla^{\perp}_{X_x}(Z_f^{\perp}) = -\beta(f)(X_x,Z_f^{\top}) \tag{2.6}$$

and

$$\nabla_{X_x}(Z_f^{\top}) = -(\check{Z},x)X_x + A(f)^{Z_f^{\perp}}(X_x) \ . \tag{2.7}$$

Proof. By III.(2.6), we have

$$\nabla^{\perp}_{X_x}(Z_f^{\perp}) = (\nabla_{X_x}(Z_f^{\perp}))_f^{\perp} = (\nabla_{X_x}(Z^{\top}) - \nabla_{X_x}(Z_f^{\top}))_f^{\perp} =$$

$$= (-(\check{Z},x)X_x - \nabla_{X_x}(Z_f^{\top}))_f^{\perp} = -(\nabla_{X_x}(Z_f^{\top}))_f^{\perp} = -\beta(f)(X_x, Z_f^{\top})$$

and (2.6) follows. Again by the same formula combined with II.(2.17) and (1.15),

$$\nabla_{X_x}(Z_f^{\top}) = (\nabla_{X_x}(Z_f^{\top}))_f^{\top} = (\nabla_{X_x} Z^{\top} - \nabla_{X_x}(Z_f^{\perp}))_f^{\top} =$$

$$= (-(\check{Z},x)X_x - \nabla_{X_x}(Z_f^{\perp}))_f^{\top} = -(\check{Z},x)X_x + A(f)^{Z_f^{\perp}}(X_x) \ . \checkmark$$

As we know from II.2.16, the second fundamental form of an isometric immersion is symmetric in its two arguments. Due to the fact that the codomain of the minimal isometric immersion $f: M \to S^n$ is a space of constant curvature, the covariant derivative $\nabla\beta(f)$ is also symmetric which can be expressed via Codazzi's equation as follows.

Lemma 2.8. $\nabla\beta(f) \in C^{\infty}(S^3(T^*(M)) \otimes \nu^f)$ *, or equivalently, for any tangent vectors* $X_x^1, X_x^2, Y_x \in T_x(M)$ *,* $x \in M$ *,*

$$(\nabla_{X_x^1}\beta(f))(X_x^2, Y_x) = (\nabla_{X_x^2}\beta(f))(X_x^1, Y_x) \ . \tag{2.9}$$

Proof. Extend X_x^1, X_x^2 and Y_x to vector fields X^1, X^2 and Y on M, respectively, such that all covariant derivatives of X^1, X^2 and Y with respect to each other vanish at $x \in M$. Then

$$(\nabla_{X_x^1}\beta(f))(X_x^2, Y_x) = \nabla^{\perp}_{X_x^1}(\beta(f)(X^2, Y)) = \nabla^{\perp}_{X_x^1}((\nabla^{S^n}_{X^2} Y)_f^{\perp}) =$$

$$= (\nabla^{S^n}_{X^1_x}(\nabla^{S^n}_{X^2} Y)^{\perp}_f)^{\perp}_f = (\nabla^{S^n}_{X^1_x}(\nabla^{S^n}_{X^2} Y))^{\perp}_f - (\nabla^{S^n}_{X^1_x}(\nabla^{S^n}_{X^2} Y)^{\top}_f)^{\perp}_f =$$

$$= (\nabla^{S^n}_{X^1_x}(\nabla^{S^n}_{X^2} Y))^{\perp}_f - \beta(f)(X^1_x, \nabla_{X^2_x} Y) = (\nabla^{S^n}_{X^1_x}(\nabla^{S^n}_{X^2} Y))^{\perp}_f .$$

Interchanging the role of X^1 and X^2, as $[X^1, X^2]_x = = \nabla_{X^1_x} X^2 - \nabla_{X^2_x} X^1 = 0$, we have

$$(\nabla_{X^1_x}\beta(f))(X^2_x, Y_x) - (\nabla_{X^2_x}\beta(f))(X^1_x, Y_x) = (R(X^1_x, X^2_x)Y_x)^{\perp}_f .$$

On the other hand, R being the curvature tensor of S^n,

$$(R(X^1_x, X^2_x)Y_x)^{\perp}_f = (\ (X^2_x, Y_x)X^1_x - (X^1_x, Y_x)X^2_x)^{\perp}_f = 0$$

and (2.9) follows. ✓

Lemma 2.10. *For each uniform vector field* Z *on* $\mathbf{R}^{n+1}$,

$$\operatorname{trace}(\nabla^{\perp})^2(Z^{\perp}_f) = -\tilde{A}(f)(Z^{\perp}_f) . \tag{2.11}$$

Proof. Let $x \in M$ be fixed and choose a local orthonormal moving frame $\{E^i\}^m_{i=1}$ (adapted to an orthonormal base $\{e^i\}^m_{i=1} \subset T_x(M)$) on a normal coordinate neighbourhood centered at x. Then, by making use of (2.6), (2.7) and (2.9), we have

$$\operatorname{trace}((\nabla^{\perp})^2(Z^{\perp}_f))_x = \sum_{i=1}^{m} \nabla^{\perp}_{e^i}(\nabla^{\perp}_{E^i}(Z^{\perp}_f)) =$$

$$= -\sum_{i=1}^{m} \nabla^{\perp}_{e^i}(\beta(f)(E^i, Z^{\top}_f)) = -\sum_{i=1}^{m} \{(\nabla_{e^i}\beta(f))(e^i, Z^{\top}_f) +$$

$$+ \beta(f)(\nabla_{e^i}E^i, Z^{\top}_f) + \beta(f)(e^i, \nabla_{e^i}(Z^{\top}_f))\} =$$

$$= -\sum_{i=1}^{m} (\nabla_{e^i}\beta(f))(e^i, Z^{\top}_f) - \sum_{i=1}^{m} \beta(f)(e^i, \nabla_{e^i}(Z^{\top}_f)) =$$

$$= -\sum_{i=1}^{m} (\nabla_{Z_f^{\top}} \beta(f))(e^i, e^i) - \sum_{i=1}^{m} \beta(f)(e^i, -(\check{Z}, x)e^i +$$

$$+ A(f)^{Z_f^{\perp}}(e^i)) = -\nabla_{Z_f^{\perp}}\{\text{trace } \beta(f)\} + (\check{Z}, x)\cdot \text{trace } \beta(f)_x -$$

$$- \sum_{i=1}^{m} \beta(f)(e^i, A(f)^{Z_f^{\perp}}(e^i)) = -(\tilde{A}(f)(Z_f^{\perp}))_x$$

since trace $\beta(f)=0$ and, for $v_x \in \nu_x^f$,

$$(\sum_{i=1}^{m} \beta(f)(e^i, A(f)^{Z_f^{\perp}}(e^i)), v_x) =$$

$$= \sum_{i=1}^{m} (A(f)^{v_x}(e^i), A(f)^{Z_f^{\perp}}(e^i)) =$$

$$= (A(f)(v_x), A(f)(Z_f^{\perp}))_x = (\tilde{A}(f)(Z_f^{\perp})_x, v_x) \; . \quad \checkmark$$

Proof of Theorem 2.4. For each uniform vector field Z on $\mathbf{R}^{n+1}$, by (2.1) and (2.11),

$$H_f^{\perp}(Z_f^{\perp}, Z_f^{\perp}) = \int_M (-\text{trace}(\nabla^{\perp})^2(Z_f^{\perp}) - mZ_f^{\perp} -$$

$$-\tilde{A}(f)(Z_f^{\perp}), Z_f^{\perp})\,\text{vol}(M,g) = -m \int_M \| Z_f^{\perp}\| \,\text{vol}(M,g) \; .$$

At each $x \in M$, the vectors $Z_x^{\top}$ span $T_x(S^n)$ and so $(Z_f^{\perp})_x$ span ν_x^f . It follows that the normal Hessian $H_f^{\perp}$ is negative definite on the $(n-m)$-dimensional linear subspace $\{Z_f^{\perp} \in C^{\infty}(\nu^f) \mid \check{Z} \in \mathbf{R}^{n+1}\}$ of $C^{\infty}(\nu^f)$ and hence the normal Morse index of $f \geq n-m$.
Assuming that equality holds, consider the vector spaces

$$L = \{Z^{\top} \in V(S^n) \mid \check{Z} \in \mathbf{R}^{n+1} \text{ and } Z_f^{\perp} = 0\}$$

and, for fixed $x \in M$,

$$L_x = \{Z^{\top} \in V(S^n) \mid \check{Z} \in \mathbf{R}^{n+1} \quad \text{and} \quad (Z_f)_x^{\perp} = 0\} .$$

Clearly, $L \subset L_x$. Furthermore,

$$\dim L_x = n+1 - (n-m) = m + 1$$

and, by hypothesis, $\dim\{Z_f^{\perp} \mid \check{Z} \in \mathbf{R}^{n+1}\} = n-m$. Hence

$$\dim L = n + 1 - (n-m) = m + 1 ,$$

and we obtain that $L = L_x$. This implies that, for any given tangent vector $Y_x \in T_x(M)$, there exists $\check{Z} \in \mathbf{R}^n$ such that

$$Z_x^{\top} = Y_x \quad \text{and} \quad Z_f^{\perp} = 0 .$$

Thus, by (2.6), for $X_x \in T_x(M)$, we have

$$\beta(f)(X_x, Y_x) = \beta(f)(X_x, Z_x^{\top}) =$$

$$= \beta(f)(X_x, (Z_f^{\top})_x) = -\nabla_{X_x}^{\perp}(Z_f^{\perp}) = 0 ,$$

i.e. the second fundamental form $\beta(f)$ vanishes and f is totally geodesic. Now, 2.2 accomplishes the proof. √

Remark. In Simons (1968) it is also shown that the normal nullity of a minimal isometric immersion $f: M \to S^n$ of an m-dimensional compact oriented Riemannian manifold M into S^n is always greater than or equal to $(m+1)(n-m)$ and equality holds if and only if f is totally geodesic (cf. also Problem 2 for this Chapter).

3. EXISTENCE OF MINIMAL IMMERSIONS OF A HOMOGENEOUS MANIFOLD INTO SPHERES

For the classification of minimal immersions into spheres toward which we are heading we specialize our study to equivariant maps with homogeneous domains. Given a compact Lie group G

acting on a manifold M (on the left) we say that a map $f:M\to S$ into the unit sphere S of a (finite dimensional) vector space V with scalar product (,) is *equivariant* if there exists a homomorphism $\rho:G\to O(V)$ such that

$$f\circ a = \rho(a)\circ f$$

holds for all $a\in G$. If $f:M\to S$ is *full* i.e. $\mathrm{im}(f)\subseteq S\subseteq V$ is not contained in any proper linear subspace of V then a homomorphism satisfying the equation above is unique since, for each $a\in G$, the orthogonal transformation $\rho(a)$ is completely determined by its values on a base of V contained in $\mathrm{im}(f)$. As noted above, we shall be dealing with maps from a homogeneous domain and, for convenience, we briefly summarize here some basic results concerning homogeneous manifolds to be used subsequently. (For more details, see Helgason (1978), Kobayashi and Nomizu (1969) and Lichnerowicz (1958).)

Given a compact (analytic) Lie group G , a (possibly disconnected) *closed* subgroup $K\subset G$ carries a unique analytic structure such that K is a Lie subgroup of G (cf. Helgason (1978) page 115). The space of left cosets G/K has also an analytic structure with the property that the canonical projection

$$\pi:G\to G/K$$

defined by

$$\pi(a) = aK\ ,\quad a\in G\ ,$$

is an analytic submersion. The analytic manifold G/K with base point $o=\{K\}\in G/K$ (called the *origin*) is said to be a (compact) *homogeneous manifold*. The correspondence which associates to each group element $a\in G$ the analytic diffeomorphism $a:G/K\to G/K$, $a(a'K)=(aa')K$, $a'\in G$, gives rise to the *canonical action* of G on G/K (cf. Helgason (1978) page 123). This action is *transitive*, i.e. each point of G/K is carried into any point of G/K by a suitably chosen group element $a\in G$. In what follows, we always assume that G/K is *effective*, i.e. the only group element which fixes each point of G/K is the identity $1 \in G$, or equivalently, K does not contain any proper closed subgroup normal in G. The inclusion $K^{o} \subset K$ of the identity component K^{o} into K induces a finite (analytic) covering projection

$$\bar{\pi}: G/K^{o} \to G/K$$

such that the diagram

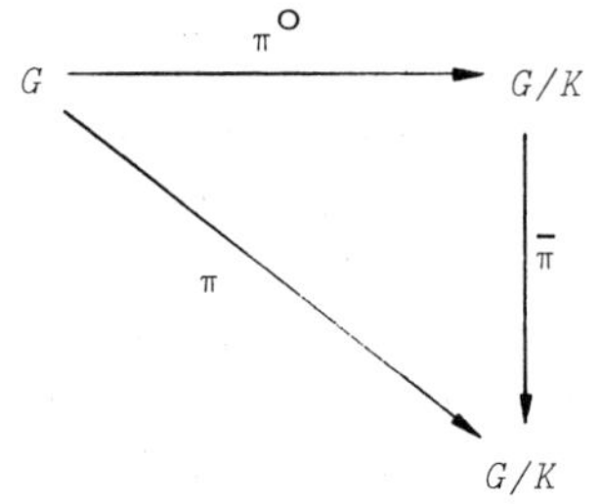

(3.1)

commutes, where $\pi^{o}: G \to G/K^{o}$ is the canonical projection. Let K and G denote the Lie algebras of K^{o} and G, respectively. Then K is a Lie subalgebra of G and the differential of $\pi: G \to G/K$ at $1 \in G$, having kernel K, factorizes through the canonical projection $\phi: G \to G/K$ yielding the commutative diagram

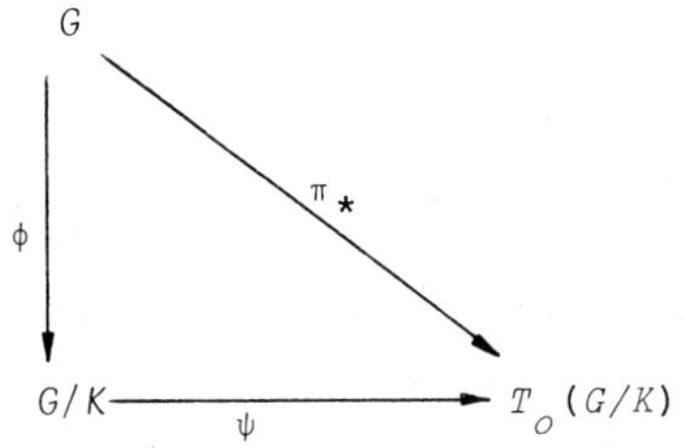

(3.2)

where $\psi: G/K \to T_{o}(G/K)$ is a linear isomorphism. The restriction $\mathrm{ad}_G|K$ of the adjoint representation ad_G of G on its Lie algebra G projects down via $\phi: G \to G/K$ inducing a linear representation (still denoted by) $\mathrm{ad}_G|K$ of the Lie group K on G/K. Furthermore, K acts on $T_{o}(G/K)$ via the isotropy representation and a simple computation in the use of (3.2) shows that the two representations are carried into each other by the linear isomorphism ψ . It also follows easily that there is a 1-1 correspondence between the *invariant Riemannian metrics* on G/K (i.e. for which G acts on G/K by isometries) and the $\mathrm{ad}_G|K$-invariant scalar products on G/K (cf. Kobayashi and Nomizu (1969) page 200).

Assume now that $M \subset G$ is a *reductive complement* of K in G, i.e. $M \subset G$ is an $\mathrm{ad}_G|K$-invariant linear subspace such that

$$G = K \oplus M$$

is a direct sum. Then M is identified with $T_o(G/K)$ via the linear isomorphism $\pi_*|M:M \to T_o(G/K)$. By the above, the $\mathrm{ad}_G|K$-invariant scalar products on M are in 1-1 correspondence with the invariant Riemannian metrics on G/K .

Conversely, let (,) be an $\mathrm{ad}_G|K$-invariant scalar product on G . Then, the orthogonal complement M of $K \subset G$ with respect to (,) is easily seen to be a reductive complement of K in G and hence the restriction of (,) to M (or rather, to $M \times M$) induces an invariant Riemannian metric on G/K (cf. Lichnerowicz (1958) page 50). If, moreover, (,) is an ad_G-invariant metric (which exists by compactness of G), the invariant Riemannian metric on G/K obtained in this way is said to be *naturally reductive*. In the special case $K=\{1\}$, a naturally reductive Riemannian metric on G is nothing but a *biinvariant metric* (i.e. a Riemannian metric which is invariant by both left and right translations; cf. Kobayashi and Nomizu (1969) page 203). Finally, if (,) is an ad_G-invariant scalar product on G then, by the above, it induces a biinvariant Riemannian metric on G . Moreover (,) being also $\mathrm{ad}_G|K$-invariant, an invariant Riemannian metric is induced on G/K with the property that the canonical projection

$$\pi : G \to G/K$$

is a Riemannian submersion (and, applying the same argument to K^o instead of K , (3.1) shows that $\bar{\pi}:G/K^o \to G/K$ is a

Riemannian covering).

The following existence theorem for minimal immersions into spheres is due to Hsiang (1966).

Theorem 3.3. *Let G be a compact Lie group and $K \subset G$ a (possibly disconnected) closed subgroup. Then, for sufficiently large $n \in \mathbf{N}$, there exists an equivariant minimal immersion*

$$f_o : G/K \to S^n$$

with respect to the induced (invariant) Riemannian metric on G/K, where G acts on G/K via the canonical action.

Proof. First note that the Mostow's embedding theorem (cf. Bredon (1972) page 111) guarantees the existence of an embedding

$$f : G/K \to S^n$$

which is equivariant with respect to a monomorphism

$$\rho : G \to SO(n+1) ,$$

i.e., for each $a \in G$, we have

$$f \circ a = \rho(a) \circ f . \tag{3.4}$$

(Note that as f is an embedding and G/K is effective, a homomorphism $\rho : G \to SO(n+1)$ satisfying (3.4) has necessarily trivial kernel.) The Lie group G then acts on S^n via ρ by isometries.

Let $\Omega \subset S^n$ denote the closed subset consisting of those points $y \in S^n$ for which K is contained in the isotropy subgroup $G_y = \{a \in G \mid \rho(a)(y) = y\}$ of G at y. Then, Ω is nonvoid since, by (3.4), $f(o) \in \Omega$. Furthermore, each point $y \in \Omega$ defines a map

$$f_y : G/K \to S^n$$

by setting

$$f_y(aK) = \rho(a)(y) \ , \quad a \in G \ ,$$

and f_y is clearly equivariant with respect to ρ . Next, define a function

$$c : \Omega \to \mathbf{R}$$

by

$$c(y) = \begin{cases} \mathrm{Vol}(\bar{\pi} \circ f_y) & \text{if } f_y \text{ is an immersion} \\ 0 & \text{otherwise,} \end{cases}$$

where $\bar{\pi} : G/K^{\mathrm{o}} \to G/K$ is the canonical projection. (Note that connectedness of K^{o} implies that G/K^{o} is oriented and hence $\mathrm{Vol}(\bar{\pi} \circ f_y)$ makes sense.)
We claim that c is continuous on Ω . Indeed, fix an ad_G-invariant scalar product $(,)$ on G and choose an orthonormal base $\{e^i\}_{i=1}^m$, $m = \dim G/K$, in the corresponding (reductive) complement M of K in G . Setting

$$\omega = \omega^1 \wedge \ldots \wedge \omega^m \ ,$$

where $\{\omega^i\}_{i=1}^m \subset M^*$ is the dual base, the m-form $\omega \in \Lambda^m(M^*)$ is invariant under $\mathrm{ad}_G | K^{\mathrm{o}}$. (Restricting ourselves to the identity component K^{o} is essential here since ω might change its sign under the action of an element of $\mathrm{ad}_G | K$.) Hence, the m-form $\bar{\omega}_o \in \Lambda^m(T_o^*(G/K^{\mathrm{o}}))$ which corresponds to ω by the isomorphism $M \tilde{=} T_o(G/K^{\mathrm{o}})$ given by $\bar{\pi}_*^{-1} \circ \pi_* | M$, is invariant under the canonical action of G on G/K^{o} . (Clearly, $\bar{\omega}$ is nothing but a volume form of G/K^{o} with respect to the invariant Riemannian metric on G/K^{o} induced by the ad_G-invariant scalar product $(,)$ on G ; cf. I.(1.10).) Let $y \in \Omega$ and assume that $f_y : G/K \to S^n$ is an immersion. Then,

$$(\bar{\pi}^* \circ f_y^*)(\mathrm{vol}(S^n)) = \bar{c}(y)\bar{\omega} \tag{3.5}$$

holds for some scalar $\bar{c}(y) \in C^\infty(G/K^{\mathrm{o}})$. Furthermore, by (3.5), for each $a \in G$, we have

$$a^*\{(\bar{\pi}^*\circ f_y^*)(\mathrm{vol}(S^n))\} = (f_y\circ\bar{\pi}\circ a)^*(\mathrm{vol}(S^n)) =$$

$$= (f_y\circ a\circ\bar{\pi})^*(\mathrm{vol}(S^n)) = (\rho(a)\circ f_y\circ\bar{\pi})^*(\mathrm{vol}(S^n)) =$$

$$= (f_y\circ\bar{\pi})^*\{\rho(a)^*(\mathrm{vol}(S^n))\} =$$

$$= \{(\bar{\pi}^*\circ f_y^*)(\mathrm{vol}(S^n)\circ\rho(a))\} = \{(\bar{\pi}^*\circ f_y^*)(\mathrm{vol}(S^n))\}\circ a$$

and hence $(\bar{\pi}^*\circ f_y^*)(\mathrm{vol}(S^n))$ is invariant under the canonical action of G on G/K°. As $\bar{\omega}$ is also invariant and the canonical action is transitive, we obtain from (3.5) that $\bar{c}(y)$ is constant (depending on $y\in\Omega$).

To compute $\bar{c}(y)$ in terms of ρ, we first note that the representation $\rho:G\to SO(n+1)$ on $\mathbf{R}^{n+1}$ induces a representation ρ of the Lie algebra G on $\mathbf{R}^{n+1}$ by setting

$$\rho(X)(y) = \frac{\partial}{\partial t}\{\rho(\exp(tX))(y)\}_{t=0}, \quad X\in G, \quad y\in\mathbf{R}^{n+1},$$

where $\exp:G\to G$ denotes the exponential map of the Lie group G.

Now, by the identification $M\cong T_o(G/K^{\circ})$, for $i=1,\ldots,m$, we have

$$(f_y\circ\bar{\pi})_*(e^i) = \frac{\partial}{\partial t}\{f_y(\bar{\pi}(\exp(te^i)K^{\circ}))\}_{t=0} =$$

$$= \frac{\partial}{\partial t}\{f_y(\exp(te^i)K)\}_{t=0} = \frac{\partial}{\partial t}\{\rho(\exp(te^i))(y)\}_{t=0} =$$

$$= \rho(e^i)(y)$$

and hence, by making use of (3.5), we obtain

$$\bar{c}(y) = \sqrt{\det(\rho(e^i)(y),\rho(e^j)(y))_{i,j=1}^m}\,.$$

Thus,

$$c(y) = \mathrm{Vol}(\bar{\pi}\circ f_y) = \int_{G/K^{\circ}} (\bar{\pi}^*\circ f_y^*)\mathrm{vol}(S^n) =$$

$$= \int_{G/K^{\circ}} \bar{c}(y)\bar{\omega} = \bar{c}(y)\int_{G/K}\bar{\omega}$$

and c is clearly continuous on the whole of Ω.
Now, as $\Omega \subset S^n$ is compact, the function c takes a maximum at some point $y_o \in \Omega$. Since $c(y_o) \neq 0$ the associated (equivariant) map

$$f_o = f_{y_o} : G/K \to S^n$$

is an immersion.
We claim that f_o is minimal with respect to the induced (invariant) Riemannian metric. First note that, $f_o : G/K \to S^n$ being equivariant with respect to ρ, the Lie group G acts naturally on the normal bundle ν^{f_o} via the restriction of the differentials $\rho(a)_*$, $a \in G$, to the fibres of ν^f (and this action induces the canonical action of G on the base G/K). Denoting by $\mu(f_o) \in C^\infty(\nu^{f_o})$ the mean curvature of f_o (cf. II. 2.16), for each $a \in G$, we have

$$a \cdot \mu(f_o) = \frac{1}{m}\, a \cdot \text{trace}\ \beta(f_o) = \frac{1}{m}\, \rho(a)_* \text{trace}\{\nabla(f_o)_*\} =$$

$$= \frac{1}{m}\, \text{trace}\{\nabla(\rho(a)_*(f_o)_*)\} = \frac{1}{m}\, \text{trace}\{\nabla(f_o)_*\} \circ a = \qquad (3.6)$$

$$= \mu(f_o) \circ a \quad .$$

Consider now the geodesic variation

$$t \to f_t = \exp \circ (t \cdot \mu(f_o)) \ , \quad t \in \mathbf{R} \ ,$$

of f_o. If $t \in \mathbf{R}$ is fixed then, by (3.6), for each $a \in G$, we have

$$f_t \circ a = \exp \circ (t \cdot \mu(f_o) \circ a) = \exp \circ (\rho(a)_* \mu(f_o)) =$$

$$= \rho(a) \circ \exp \circ (t \cdot \mu(f_o)) = \rho(a) \circ f_t \ ,$$

or equivalently, $f_t : G/K \to S^n$ is equivariant with respect to ρ; in particular, $y_t = f_t(o) \in \Omega$ since $K \subset G_{y_t}$, $t \in \mathbf{R}$. For sufficiently small $\varepsilon > 0$, the maps f_t are immersions for $t \in (-\varepsilon, \varepsilon)$. The mean curvature vector of $f_o \circ \bar{\pi} : G/K^o \to S^n$ being

$\mu(f_o)\circ\bar{\pi}$, as $\left.\frac{\partial(f_t\circ\bar{\pi})}{\partial t}\right|_{t=o} = \left.\frac{\partial f_t}{\partial t}\right|_{t=o}\circ\bar{\pi} = \mu(f_o)\circ\bar{\pi}$, applying the first variation formula (1.6), we obtain

$$\left.\frac{\partial \mathrm{Vol}(f_t\circ\bar{\pi})}{\partial t}\right|_{t=o} = \tag{3.7}$$

$$= -m\int_{G/K^o}(\mu(f_o)\circ\bar{\pi},\mu(f_o)\circ\bar{\pi})\,\mathrm{vol}(G/K^o,g_o)\ ,$$

where g_o is the metric tensor induced from that of S^n by pull-back via $f_o\circ\bar{\pi}:G/K^o\to S^n$. On the other hand, by the notations above, $f_t=f_{y_t}$, $y_t\in\Omega$, $t\in\mathbf{R}$, and, by the choice of f_o ,

$$\mathrm{Vol}(f_t\circ\bar{\pi}) \leq \mathrm{Vol}(f_o\circ\bar{\pi})$$

holds for all $t\in(-\varepsilon,\varepsilon)$. In particular,

$$\left.\frac{\partial \mathrm{Vol}(f_t\circ\bar{\pi})}{\partial t}\right|_{t=o} = 0$$

and, by (3.7), we obtain $\mu(f_o)=0$. √

Remark 1. Theorem 3.3 has first appeared in Hsiang (1966) without proof. The argument above follows the proof of Wallach (1972) (see also Hsiang and Lawson (1971)).

Remark 2. A complex analogue of 3.3 is given in Wallach (1973), page 152, namely a simply connected homogeneous Kähler manifold (=generalized flag manifold) can be holomorphically and equivariantly embedded into a complex projective space of sufficiently high dimension.

Hsiang's theorem establishes the existence of a minimal immersion of compact homogeneous manifolds into spheres of sufficiently large dimensions. In what follows, for a restricted class of domains, namely, for compact irreducible homogeneous manifolds, we give a fairly explicit method of building up minimal immersions into spheres known as standard minimal immersions, whose rich geometry will be studied subsequently.

First, recall that a compact homogeneous manifold G/K is said to be *irreducible* if the isotropy representation of the closed subgroup K on $T_o(G/K)$ is irreducible.

Proposition 3.8. *Let G/K be a compact irreducible homogeneous manifold with naturally reductive Riemannian metric $g\in C^\infty(S^2(T^*(G/K)))$. Given a tensor field $\bar{g}\in C^\infty(S^2(T^*(G/K)))$ which is invariant under the canonical action of G on G/K, we have*

$$\bar{g} = cg$$

for some $c\in\mathbf{R}$.

Proof. The restrictions $g_o,\bar{g}_o\in S^2(T_o^*(G/K))$ define a symmetric linear endomorphism

$$A_o\in S^2(T_o(G/K))$$

by

$$g_o(A_o(X_o),Y_o) = \bar{g}_o(X_o,Y_o)\ , \quad X_o,Y_o\in T_o(G/K)\ .$$

Invariance of g_o and $\bar{g}_o$ implies that the elements of the isotropy representation of K on $T_o(G/K)$ commute with A_o. As A_o is diagonalizable, irreducibility implies that $A_o=c\cdot I$ (=identity of $T_o(G/K)$), i.e. we get $\bar{g}_o=c\cdot g_o$. By invariance, $\bar{g}=cg$ holds on the whole of G/K. √

Let G/K be an m-dimensional compact oriented irreducible homogeneous manifold. Fix an invariant (and hence, naturally reductive) Riemannian metric $g=(,)\in C^\infty(S^2(T^*(G/K)))$ on G/K. By I.6, the Laplacian $\Delta\in \mathrm{Diff}_2(\varepsilon^1_{G/K}, \varepsilon^1_{G/K})$ is an elliptic self-adjoint differential operator whith respect to the global scalar product. By I.7, $\mathrm{Spec}(G/K)\subset\mathbf{R}$ is discrete and, for each $\lambda\in\mathrm{Spec}(G/K)$, the corresponding eigenspace V_λ is finite dimensional.

For fixed $(0\neq)\lambda\in\mathrm{Spec}(G/K)$, define a scalar product (,) on V_λ by

$$(\mu,\mu') = \frac{n(\lambda)+1}{\int_{G/K}\mathrm{vol}(G/K,g)}\int_{G/K}\mu\cdot\mu'\,\mathrm{vol}(G/K,g)\ , \qquad (3.9)$$

$$\mu, \mu' \in V_\lambda ,$$

where $\dim V_\lambda = n(\lambda)+1$. Choose an orthonormal base $\{f_\lambda^i\}_{i=0}^{n(\lambda)} \subset V_\lambda$ which, at the same time, identifies V_λ with $\mathbf{R}^{n(\lambda)+1}$ by a linear isometry sending the scalar f_λ^i to the i-th canonical base vector $e^i \in \mathbf{R}^{n(\lambda)+1}$, $i=0,\ldots,n(\lambda)$. Define a map

$$\bar{f}_\lambda : G/K \to V_\lambda \; (=\mathbf{R}^{n(\lambda)+1})$$

by

$$\bar{f}_\lambda(aK) = \sum_{i=0}^{n(\lambda)} f_\lambda^i(aK) f_\lambda^i = \sum_{i=0}^{n(\lambda)} f_\lambda^i(aK) e^i , \quad a \in G . \tag{3.10}$$

To analyze the geometric properties of $\bar{f}_\lambda$, we first note that the canonical action of G on G/K (by isometries with respect to g) gives rise to an action of G on $C^\infty(G/K)$ by setting

$$a \cdot \mu = \mu \circ a^{-1} , \quad a \in G . \tag{3.11}$$

Furthermore, as the Laplacian Δ commutes with isometries, this action leaves the eigenspace $V_\lambda \subset C^\infty(G/K)$ invariant and, hence, induces an action of G on V_λ which, by (3.9), preserves the scalar product $(,)$. With respect to the orthonormal base $\{f_\lambda^i\}_{i=0}^{n(\lambda)} \subset V_\lambda$, the action of each $a \in G$ is expressed by an orthogonal matrix $(A_{ij})_{i,j=0}^{n(\lambda)} \in SO(n(\lambda)+1)$ such that

$$a \cdot f_\lambda^i = \sum_{j=0}^{n(\lambda)} A_{ji} f_\lambda^j , \quad i=0,\ldots,n(\lambda) . \tag{3.12}$$

Then, setting

$$\bar{g} = \sum_{i=0}^{n(\lambda)} df_\lambda^i \otimes df_\lambda^i \in C^\infty(S^2(T^*(G/K))) ,$$

for each $a \in G$, we have

$$a^*(\bar{g}) = \sum_{i=0}^{n(\lambda)} d(f_\lambda^i \circ a) \otimes d(f_\lambda^i \circ a) =$$

$$= \sum_{i=0}^{n(\lambda)} d(a^{-1} \cdot f_\lambda^i) \otimes d(a^{-1} \cdot f_\lambda^i) =$$

$$= \sum_{i,j,k=0}^{n(\lambda)} A_{ij}A_{ik}df_\lambda^j \otimes df_\lambda^k = \sum_{j,k=0}^{n(\lambda)} \delta_{jk}df_\lambda^j \otimes df_\lambda^k = \bar{g} ,$$

i.e. the tensor $\bar{g}$ is invariant. Applying 3.8, we obtain the relation $\bar{g}=c\cdot g$ for some $c>0$; in particular, $\bar{f}_\lambda : G/K \to V_\lambda$ is an isometric immersion with respect to the naturally reductive Riemannian metric $\bar{g}=cg$ on G/K . Denoting by $\bar{\Delta}$ the Laplacian corresponding to $\bar{g}$, we have $\bar{\Delta}=\frac{1}{c}\Delta$ and, as $f_\lambda^i \in V_\lambda$, we obtain

$$\bar{\Delta} f_\lambda^i = \frac{\lambda}{c} f_\lambda^i , \quad i=0,\ldots,n(\lambda) . \tag{3.13}$$

We can now apply Takahashi's result, II.2.25, to get

$$\text{im } \bar{f}_\lambda \subset r\cdot S^{n(\lambda)} ,$$

where $r>0$ and $S^{n(\lambda)} \subset V_\lambda (=\mathbf{R}^{n(\lambda)+1})$ is the unit sphere. Furthermore, $\bar{f}_\lambda$ defines a minimal isometric immersion

$$f_\lambda : G/K \to r\cdot S^{n(\lambda)} .$$

To determine the actual value of r , we integrate the expression

$$(\bar{f}_\lambda(aK), \bar{f}_\lambda(aK)) = \sum_{i=0}^{n(\lambda)} (f_\lambda^i(aK))^2 = r^2 , \quad a\in G .$$

Then,

$$\sum_{i=0}^{n(\lambda)} \int_{G/K} (f_\lambda^i)^2 \text{vol}(G/K,g) =$$

$$= \frac{\int_{G/K} \text{vol}(G/K,g)}{n(\lambda)+1} \sum_{i=0}^{n(\lambda)} (f_\lambda^i, f_\lambda^i) =$$

$$= \int_{G/K} \text{vol}(G/K,g) = r^2 \int_{G/K} \text{vol}(G/K,g) ,$$

and hence $r=1$. Finally, again by II.2.25, equations (3.13) imply that $c=\frac{\lambda}{m}$.

Summarizing, we find that each nonzero eigenvalue λ of the Laplacian Δ on an m-dimensional compact oriented irreducible homogeneous manifold G/K with invariant Riemannian metric g gives rise to a minimal isometric immersion

$$f_\lambda : G/K \to S^{n(\lambda)} , \tag{3.14}$$

where $\dim V_\lambda = n(\lambda)+1$ and the Riemannian metric on G/K induced by f_λ equals $\frac{\lambda}{m} g$. The map f_λ is said to be the *standard minimal immersion* associated to the eigenvalue $\lambda \in \mathrm{Spec}(G/K)$. Clearly, $f_\lambda : G/K \to S^{n(\lambda)}$ is full and different choices of the orthonormal base in V_λ give rise to standard minimal immersions which differ by an isometry of the codomain. As noted above, the action of G on $C^\infty(G/K)$, in (3.11), restricts to an orthogonal action of G on V_λ . The identification $V_\lambda = \mathbf{R}^{n(\lambda)+1}$ defined by the orthonormal base $\{f_\lambda^i\}_{i=0}^{n(\lambda)} \subset V_\lambda$ transforms this action into an orthogonal representation

$$\rho : G \to SO(n(\lambda)+1)$$

which associates to each $a \in G$ the orthogonal matrix $(A_{ij})_{i,j=0}^{n(\lambda)} \in SO(n(\lambda)+1)$ via (3.12). We claim that $f_\lambda : G/K \to S^{n(\lambda)}$ is equivariant with respect to ρ . Indeed, for $a, a' \in G$, we have

$$(\rho(a) \circ \bar{f}_\lambda)(a'K) = \rho(a)\{\bar{f}_\lambda(a'K)\} =$$

$$= \rho(a) \{ \sum_{i=0}^{n(\lambda)} f_\lambda^i(a'K) f_\lambda^i \} = \sum_{i=0}^{n(\lambda)} f_\lambda^i(a'K) a \cdot f_\lambda^i =$$

$$= \sum_{i,j=0}^{n(\lambda)} f_\lambda^i(a'K) A_{ji} f_\lambda^j = \sum_{i,j=0}^{n(\lambda)} A_{ij} f_\lambda^j(a'K) f_\lambda^i =$$

$$= \sum_{i=0}^{n(\lambda)} (a^{-1} \cdot f_\lambda^i)(a'K) f_\lambda^i = \sum_{i=0}^{n(\lambda)} f_\lambda^i(aa'K) f_\lambda^i =$$

$$= \bar{f}_\lambda(aa'K) = (\bar{f}_\lambda \circ a)(a'K) ,$$

or equivalently,

$$f_\lambda \circ a = \rho(a) \circ f_\lambda \ , \quad a \in G \ , \tag{3.15}$$

and the claim follows.
Finally, setting $v^o = f_\lambda(o) \in V_\lambda$, by (3.15), the action $\rho|K$ of K on V_λ fixes v . The isotropy subgroup $H = G_{v^o} = \{a \in G \,|\, f_\lambda(a(o)) = f_\lambda(o)\}$ contains K and, by equivariance, $f_\lambda : G/K \to S^{n(\lambda)}$ factorizes through the canonical projection $\pi : G/K \to G/H$ (induced by the inclusion $K \subset H$) yielding an embedding $\hat{f}_\lambda : G/H \to S^{n(\lambda)}$ such that the diagram

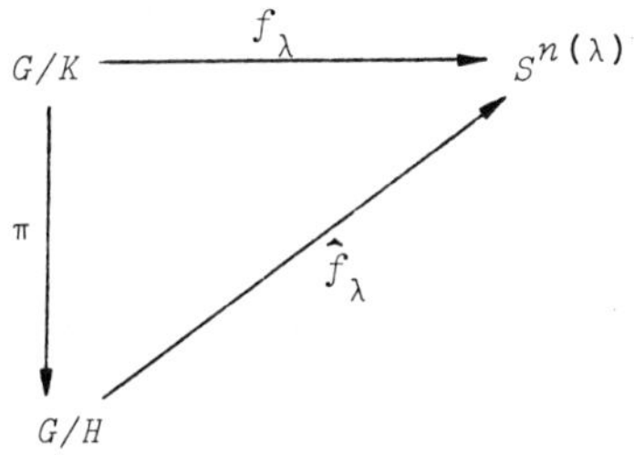

(3.16)

commutes. As f_λ is an immersion, $\pi : G/K \to G/H$ is a finite (Riemannian) covering and hence $\hat{f}_\lambda : G/H \to S^{n(\lambda)}$ is a minimal isometric embedding; in particular, $\mathrm{im}(f_\lambda) \subset S^{n(\lambda)}$ is a minimal submanifold. (In fact, this is true for any minimal isometric immersion; cf. Li (1981).)

For later purposes it is convenient to give a slightly more abstract setting. Given a (linear) representation ρ of G on a (real) vector space V , we say that V carries a *G-module structure*. *G-submodules* of a given G-module and *homomorphisms of G-modules* have the obvious meanings. For G-modules V and V' the collection of all G-module homomorphisms of V into V' is denoted by $\mathrm{Hom}_G(V, V')$. If a G-module V is endowed with a scalar product $(,)$ such that, for each $a \in G$, the linear automorphism $\rho(a)$ of V is orthogonal, then V is said to be *orthogonal*. Assuming, for the remainder of this section that V is finite dimensional, an orthogonal G-module structure on V is given by a homomorphism $\rho : G \to SO(V)$.

A class 1 *representation* of the pair (G,K) is an *irreducible* orthogonal G-module structure on V such that the closed subgroup K fixes a unit vector $v^o \in V$, i.e., for each $a \in K$, we have $\rho(a)v^o = v^o$.

Remark. If G/K is a compact oriented irreducible homogeneous manifold with invariant Riemannian metric g such that the action of G on the eigenspace $V_\lambda \subset C^\infty(G/K)$ is irreducible then the induced representation $\rho: G \to SO(V_\lambda)$, $V_\lambda = \mathbf{R}^{n(\lambda)+1}$, defined above, provides an example of a class 1 representation of the pair (G,K). These hypotheses are satisfied for the pair $(SO(m+1), SO(m))$ (cf. Problem 5 for this Chapter).
If we describe the construction of standard minimal immersions in the framework of class 1 representations we have:

Theorem 3.17. *Let G/K be a compact irreducible homogeneous manifold. Given a (nontrivial) class* 1 *representation* ρ *of* (G,K) *on* V *with K-fixed unit vector* $v^o \in V$, *the map*

$$f: G/K \to S \quad (S = \text{unit sphere in } V)$$

defined by

$$f(aK) = \rho(a)v^o, \quad a \in G,$$

is a minimal immersion which is equivariant with respect to $\rho: G \to SO(V)$.

To prove 3.17, we need some preliminary constructions. Fix an ad_G-invariant scalar product $(,)$ on the Lie algebra G of G. Then, it induces a biinvariant Riemannian metric $(,)$ on the Lie group G.

Lemma 3.18. *On* $C^\infty(G)$,

$$\Delta^G = -\sum_{i=1}^{\ell} (E^i)^2, \tag{3.19}$$

where $\{E^i\}_{i=1}^{\ell}$, $\ell = \dim G$, *is an orthonormal base of right invariant vector fields on* G .

Proof. Note first that biinvariance of the Riemannian metric $(,)$ on G implies

$$(\mathrm{Ad}_G(\exp(tZ))_* X\ ,\ \mathrm{Ad}_G(\exp(tZ))_* Y) = (X,Y) \tag{3.20}$$

for any right invariant vector fields X,Y,Z on G . Differentiating (3.20) at $t=0$,

$$([Z,X],Y) + (X,[Z,Y]) = 0 \tag{3.21}$$

follows. Now, as the scalar product of any two right invariant vector fields is constant, by the proof of I.4.20, we have

$$2(\nabla_X Y,Z) = X(Y,Z) + Y(X,Z) - Z(X,Y) +$$

$$+ ([X,Y],Z) - ([X,Z],Y) - ([Y,Z],X) =$$

$$= ([X,Y],Z) + ([Z,X],Y) + (X,[Z,Y]) =$$

$$= ([X,Y],Z)$$

and hence the Levi-Civita covariant differentiation ∇ on G can be characterized by the property that

$$\nabla_X Y = \frac{1}{2}[X,Y] \tag{3.22}$$

holds for any right invariant vector fields X,Y on G . In particular, $\nabla_{E^i} E^i = \frac{1}{2}[E^i,E^i] = 0$, $i=1,\ldots,\ell$, and so, for $\mu \in C^\infty(G)$,

$$\Delta^G \mu = \partial(d\mu) = -\sum_{i=1}^{\ell} \iota_{E^i}(\nabla_{E^i}(d\mu)) =$$

$$= -\sum_{i=1}^{\ell} E^i\{(d\mu)E^i\} = -\sum_{i=1}^{\ell} (E^i)^2 \mu \ . \checkmark$$

The ad_G-invariant scalar product $(,)$ on G induces an invariant Riemannian metric $g=(\,,\,)$ on the homogeneous manifold G/K such that the canonical projection $\pi : G \to G/K$ is a Riemannian submersion. Each fibre of π, being a left coset $aK \subset G$, $a \in G$, is minimal (in fact, totally geodesic; cf. Problem 6 for this chapter) and II.2.30 applies showing that π is a harmonic map. Thus, by II.2.34, for any scalar $\mu \in C^\infty(G/K)$,

$$\Delta^G(\mu \circ \pi) = (\Delta^{G/K}\mu) \circ \pi . \tag{3.23}$$

Our next purpose is, by making use of (3.23), to obtain an expression similar to (3.19) for the Laplacian of the naturally reductive homogeneous manifold G/K. Observe first that the Lie group G acts on G/K via the canonical action and so each element $X \in G$ induces a vector field X on G/K by setting

$$X_x = \frac{\partial}{\partial t}\{(\exp(tX))(x)\}_{t=0}, \quad x \in G/K .$$

Denoting also by X the right invariant extension of $X \in G$, for each $a \in G$, we have

$$X_{\pi(a)} = \frac{\partial}{\partial t}\{(\exp(tX))(\pi(a))\}_{t=0} =$$

$$= \frac{\partial}{\partial t}\{\pi((\exp(tX))\cdot a)\}_{t=0} =$$

$$= \frac{\partial}{\partial t}\{\pi(R_a(\exp(tX)))\}_{t=0} =$$

$$= (\pi_* \circ (R_a)_*)\frac{\partial}{\partial t}\{\exp(tX)\}_{t=0} = \pi_* \circ (R_a)_* X = \pi_*(X) ,$$

or equivalently, the right invariant vector field X on G and the induced vector field X on G/K are π-related.

Lemma 3.24. *With respect to the naturally reductive metric g on G/K,*

$$\Delta^{G/K} = -\sum_{i=1}^{\ell} (E^i)^2 \tag{3.25}$$

where $\{E^i\}_{i=1}^{\ell} \subset V(G/K)$ is induced by an orthonormal base of G.

Proof. By (3.19) and (3.23), for each $\mu \in C^{\infty}(G/K)$, we have

$$(\Delta^{G/K}\mu)\circ\pi = \Delta^{G}(\mu\circ\pi) = -\sum_{i=1}^{\ell} E^{i}\{E^{i}(\mu\circ\pi)\} =$$

$$= -\sum_{i=1}^{\ell} E^{i}\{(\pi_{*}E^{i})\mu\} = -\sum_{i=1}^{\ell} E^{i}\{(E^{i}(\mu))\circ\pi\} =$$

$$= -\sum_{i=1}^{\ell} \pi_{*}(E^{i})\{E^{i}(\mu)\} = -\{\sum_{i=1}^{\ell}(E^{i})^{2}\mu\}\circ\pi \ . \ \checkmark$$

For the next lemma, recall that, by (3.11), the Lie group G acts on $C^{\infty}(G/K)$.

Lemma 3.26. *(i) If* $V \subset C^{\infty}(G/K)$ *is a finite dimensional invariant linear subspace, then* $\Delta^{G/K}(V) \subset V$.
(ii) If, moreover, V *is irreducible then* $\Delta^{G/K}|V = \lambda\cdot I_V$ *(=identity of* V*) holds for some* $\lambda \in \mathbf{R}$, *i.e.* $V \subset V_\lambda$ *is an irreducible component of the eigenspace* V_λ *corresponding to the eigenvalue* $\lambda \in \mathrm{Spec}(G/K)$.

Proof. For each $X \in G$ and $\mu \in V$, we have

$$X(\mu) = \frac{\partial}{\partial t}\{\mu\circ(\exp(tX))\}_{t=0} =$$

$$= \frac{\partial}{\partial t}\{\mu\circ(\exp(-tX))^{-1}\}_{t=0} =$$

$$= \frac{\partial}{\partial t}\{\exp(-tX)\cdot\mu\}_{t=0} \in V$$

and so, by (3.25), (i) follows.
To show (ii), choose a scalar product $(\,,)$ on V which is invariant under the action of G on V. Then, for $X \in G$, we have

$$(\exp(tX)\cdot\mu, \exp(tX)\cdot\nu) = (\mu,\nu) \ , \quad \mu,\nu \in V \ ,$$

and differentiating at $t=0$,

$$(X(\mu),\nu) + (\mu, X(\nu)) = 0 \ ,$$

or equivalently, the vector field $X \in V(G/K)$ is skew-symmetric on V. Thus, by (3.25), $\Delta^{G/K}$ is a symmetric linear endomorphism on V, in particular diagonalizable. Furthermore, $\Delta^{G/K}$ commutes with the action of G on V and hence each eigenspace of $\Delta^{G/K}$ in V is invariant. By irreducibility, $\Delta^{G/K}$ has only one eigenvalue on V. √

Proof of Theorem 3.17. Let $\{e^i\}_{i=0}^n \subset V$, $\dim V=n+1$, $e^0=v^0$, be an orthonormal base and, for $i=0,\ldots,n$, define a scalar $f_\lambda^i \in C^\infty(G/K)$ by

$$f_\lambda^i(aK) = (\rho(a)e^0, e^i), \quad a \in G. \tag{3.27}$$

The linear map

$$\phi : V \to C^\infty(G/K)$$

given by

$$\phi(e)(aK) = (\rho(a)e^0, e), \quad e \in V, \quad a \in G, \tag{3.28}$$

is a G-module homomorphism with respect to the G-module structure of $C^\infty(G/K)$ in (3.11). This follows since, for $a, a' \in G$, we have

$$\phi(\rho(a)e)(a'K) = (\rho(a')e^0, \rho(a)e) = (\rho(a^{-1}a')e^0, e) =$$

$$= \phi(e)((a^{-1}a')K) = \phi(e)(a^{-1}(a'K)) = (a \cdot \phi(e))(a'K).$$

As V is irreducible, $\phi : V \to C^\infty(G/K)$ is a monomorphism. Setting $\bar{V}=\mathrm{im}(\phi)$ the linear map ϕ establishes a G-module isomorphism between V and $\bar{V}$. Applying 3.26(ii) to $\bar{V} \subset C^\infty(G/K)$, we find that

$$\Delta f_\lambda^i = \lambda f_\lambda^i, \quad i=0,\ldots,n, \tag{3.29}$$

holds for some $\lambda \in \mathbf{R}$.

Transporting the scalar product $(\,,\,)$ on V to that of $\bar{V}$ via ϕ, the isomorphism $V \cong \bar{V}$ becomes a linear isometry which, by $\phi(e^i)=f_\lambda^i$, $i=0,\ldots,n$, carries the orthonormal base $\{e^i\}_{i=0}^n$

to $\{f_\lambda^i\}_{i=0}^n$.
The Lie group G acts on $\bar{V}$ orthogonally and, with respect to the orthonormal base $\{f_\lambda^i\}_{i=0}^n \subset \bar{V}$, each $a \in G$ is expressed by an orthogonal matrix $(A_{ij})_{i,j=0}^n \in SO(n+1)$ such that

$$a \cdot f_\lambda^i = \sum_{j=0}^n A_{ji} f_\lambda^j , \quad i=0,\ldots,n .$$

As in the case of the standard minimal immersions, the tensor field

$$\bar{g} = \sum_{i=0}^n df_\lambda^i \otimes df_\lambda^i \in C^\infty(S^2(T^*(G/K)))$$

is invariant under the canonical action of G on G/K . By 3.8, $\bar{g}=cg$ and, again as in the case of standard minimal immersions, we find that the map

$$\bar{f}: G/K \to V$$

given by

$$\bar{f}(aK) = \rho(a)v^0 = \sum_{i=0}^n f^i(aK)e^i , \quad a \in G ,$$

defines a minimal isometric immersion

$$f: G/K \to S$$

with respect to the induced Riemannian metric $\bar{g}$. √

Example 3.30. Setting $G=SO(m+1)$, $K=SO(m)$ and $G/K=S^m$ with the Euclidean metric $g=(,)$ induced from that of $\mathbf{R}^{m+1}$ via the inclusion $S^m \subset \mathbf{R}^{m+1}$, by I.7.5 and I.7.9,

$$\mathrm{Spec}(S^m) = \{\lambda_k = k(k+m-1) \mid k \in \mathbf{Z}_+\}$$

and, for each $k \in \mathbf{Z}_+$, the eigenspace (corresponding to λ_k)

$$V_{\lambda_k} = \tilde{H}_m^k \tag{3.31}$$

is the vector space of spherical harmonics of order k on S^m with

$$\dim V_{\lambda_k} = \dim \tilde{H}^k_m = n(\lambda_k)+1 = \\ = (m+2k-1)\,\frac{(m+k-2)!}{k!\,(m-1)!}\;. \tag{3.32}$$

Denoting by S^m_κ the Euclidean m-sphere with constant sectional curvature κ, we find that the standard minimal immersion associated to the eigenvalue λ_k, $k\in\mathbf{N}$, takes the form

$$f_{\lambda_k}: S^m_{\kappa(k)} \to S^{n(k)}\;, \tag{3.33}$$

where

$$\kappa(k) = \frac{m}{\lambda_k} = \frac{m}{k(k+m-1)} \tag{3.34}$$

and

$$n(k) = n(\lambda_k) = (m+2k-1)\,\frac{(m+k-2)!}{k!\,(m-1)!} - 1\;. \tag{3.35}$$

For $k=1$, since $\kappa(1)=1$ and $n(1)=n(\lambda_1)=m$, it follows easily that

$$f_{\lambda_1}: S^m \to S^m$$

is nothing but an isometry.

For $k=2$,

$$\kappa(2) = \frac{m}{2(m+1)} \quad\text{and}\quad n(2) = \frac{m(m+3)}{2} - 1$$

and we give here an explicit description of the standard minimal immersion

$$f_{\lambda_2}: S^m_{\kappa(2)} \to S^{n(2)}\;.$$

Fixing an orthonormal base $\{f^i_\lambda\}_{i=0}^{n(2)} \subset \tilde{H}^2_m$, for each spherical harmonic $\chi\in\tilde{H}^2_m$, by (3.10), we have

$$(\bar{f}_{\lambda_2}(x),\chi) = \sum_{i=0}^{n(2)} f^i_\lambda(x)\,(f^i_\lambda,\chi) =$$

$$= \{\sum_{i=0}^{n(2)} (f^i_\lambda,\chi)\,f^i_\lambda\}(x) = \chi(x)\;, \qquad x\in S^m\;,$$

and hence the equation

$$(\bar{f}_{\lambda_2}(x),\chi) = \chi(x) \ , \quad \chi\in\tilde{H}^2_m \ , \quad x\in S^m \ , \tag{3.36}$$

can be thought of as defining the immersion

$$\bar{f}_{\lambda_2} : S^m \to \tilde{H}^2_m \ .$$

A homogeneous polynomial $p:\mathbf{R}^{m+1}\to\mathbf{R}$ of degree 2 can be written as

$$P = \sum_{k=1}^{m+1} a_k\phi_k + 2\sum_{i<j} b_{ij}\phi_{ij} \ , \tag{3.37}$$

where, for $k=1,\ldots,m+1$ and $1\leq i<j\leq m+1$, $a_k,b_{ij}\in\mathbf{R}$ and the homogeneous polynomials $\phi_k,\phi_{ij}:\mathbf{R}^{m+1}\to\mathbf{R}$ of degree 2 are defined by

$$\phi_k(x) = (x^k)^2 \quad \text{and} \quad \phi_{ij}(x) = x^i x^j \ ,$$

$$x = (x^1,\ldots,x^{m+1})\in\mathbf{R}^{m+1} \ ,$$

respectively. Furthermore, as

$$\Delta^{\mathbf{R}^{m+1}} P = -2\sum_{k=1}^{m+1} a_k \ ,$$

a spherical harmonic χ of order 2 on S^m is the restriction (to S^m) of a polynomial P in (3.37) satisfying the linear constraint

$$\sum_{k=1}^{m+1} a_k = 0 \ . \tag{3.38}$$

For brevity, put $\phi_k=\phi_k|S^m$ and $\phi_{ij}=\phi_{ij}|S^m$. Then, we have the orthogonality relations (with respect to the scalar product (3.9) on $\tilde{H}^2_m$)

$$(\phi_k,\phi_{ij}) = 0 \ , \quad k=1,\ldots,m+1 \ ; \quad 1\leq i<j\leq m+1 \ , \tag{3.39}$$

and

$$(\phi_{ij},\phi_{pq}) = 0 \ , \quad \text{if} \quad (i,j)\neq(p,q) \ ,$$

$$1 \le i < j \le m+1 \ ; \quad 1 \le p < q \le m+1 \ . \tag{3.40}$$

Furthermore, the constants $\| \phi_k \|^2$ and $\| \phi_{ij} \|^2$ do not depend on the indices $k=1,\ldots,m$ and $1 \le i < j \le m+1$ and, setting

$$I = \| \phi_k \|^2 \quad \text{and} \quad J = \| \phi_{ij} \|^2 \ , \tag{3.41}$$

we have $I > J$.

As $\operatorname{im} \bar{f}_{\lambda_2} \subset \tilde{H}^2_m$, we take

$$\bar{f}_{\lambda_2} = \sum_{r=1}^{m+1} A_r \phi_r + 2 \sum_{p<q} B_{pq} \phi_{pq} \ , \tag{3.42}$$

with

$$\sum_{r=1}^{m+1} A_r = 0 \ , \tag{3.43}$$

where the scalars $A_r, A_{pq} \in C^\infty(S^m)$, $r=1,\ldots,m+1$; $1 \le p < q \le m+1$, are to be determined from (3.36). Setting

$$P = \sum_{k=1}^{m+1} a_k \phi_k + 2 \sum_{i<j} b_{ij} \phi_{ij} \ , \quad \sum_{k=1}^{m+1} a_k = 0 \ , \tag{3.44}$$

for $\chi = P | S^m$, (3.36) gives

$$\Big(\sum_{r=1}^{m+1} A_r(x) \phi_r + 2 \sum_{p<q} B_{pq}(x) \phi_{pq} \ ,$$

$$\sum_{k=1}^{m+1} a_k \phi_k + 2 \sum_{i<j} b_{ij} \phi_{ij} \Big) = \sum_{k=1}^{m+1} a_k (x^k)^2 +$$

$$+ 2 \sum_{i<j} b_{ij} x^i x^j \ , \quad x = (x^1, \ldots, x^{m+1}) \in S^m \ .$$

By the orthogonality relations (3.39) and (3.40), this becomes

$$\sum_{k=1}^{m+1} \{ I A_k(x) + J \sum_{\substack{r=1 \\ r \ne k}}^{m+1} A_r(x) \} a_k + 2J \sum_{i<j} B_{ij}(x) b_{ij} =$$

$$\sum_{k=1}^{m+1} a_k (x^k)^2 + 2 \sum_{i<j} b_{ij} x^i x^j \ , \qquad x \in S^m \ .$$

Using (3.43), we obtain

$$\sum_{k=1}^{m+1} ((I-J)A_k(x) - (x^k)^2) a_k +$$

$$+ 2 \sum_{i<j} (JB_{ij}(x) - x^i x^j) b_{ij} = 0 \ .$$

As the constants b_{ij} can be chosen arbitrarily,

$$B_{ij}(x) = \frac{x^i x^j}{J} \ , \quad x \in S^m \ , \quad 1 \le i < j \le m+1$$

and hence

$$\sum_{k=1}^{m+1} ((I-J)A_k(x) - (x^k)^2) a_k = 0 \ , \quad x \in S^m \ .$$

Comparing this with $\sum_{k=1}^{m+1} a_k = 0$, it follows that the scalar

$$(I-J)A_k - \phi_k$$

does not depend on the index $k=1,\ldots,m+1$. Summing over $k=$ $=1,\ldots,m+1$, we have

$$(m+1)((I-J)A_k - \phi_k) = - \sum_{k=1}^{m} \phi_k^2 = -1$$

and so

$$A_k(x) = \frac{1}{I-J}\{(x^k)^2 - \frac{1}{m+1}\} \ , \quad x \in S^m \ , \quad k=1,\ldots,m+1 \ .$$

Summarizing, we obtain

$$\bar{f}_{\lambda_2}(x) = \frac{1}{I-J} \sum_{k=1}^{m+1} \{(x^k)^2 - \frac{1}{m+1}\}\phi_k +$$

$$+ \frac{2}{J} \sum_{i<j} x^i x^j \phi_{ij} \ , \quad x \in S^m \ . \qquad (3.45)$$

As the components of $\bar{f}_{\lambda_2}$ are homogeneous polynomials of degree 2, it follows that the standard minimal immersion $f_{\lambda_2}: S^m_{\kappa(2)} \to S^{n(2)}$ factors through the canonical projection $\pi: S^m_{\kappa(2)} \to \mathbf{R}P^m_{\kappa(2)}$ (defined by the antipodal map) yielding a minimal isometric immersion

$$\hat{f}_{\lambda_2}: \mathbf{R}P^m_{\kappa(2)} \to S^{n(2)} . \tag{3.46}$$

Furthermore, as (3.45) shows, the value $\bar{f}_{\lambda_2}(x)$ determines $x \in S^m$ up to sign which implies that $\hat{f}_{\lambda_2}$ is an embedding (called the *Veronese embedding*).

4. RIGIDITY

Throughout the rest of this chapter we study the classification problem of minimal isometric immersions of a fixed m-dimensional Riemannian manifold M with metric tensor $(,)$ into Euclidean spheres. As minimality of an isometric immersion $f: M \to S^n$ is preserved under isometries of the codomain, it is convenient to study the corresponding orbit space, i.e. we are led to study the equivalence classes of minimal isometric immersions of M into Euclidean spheres, where two maps $f, f': M \to S^n$ are said to be *equivalent* if there exists $U \in O(n+1)$ such that $f' = U \circ f$. In the simplest case of a unique equivalence class, any map $f: M \to S^n$ from this class is called rigid. Equivalently a *minimal* isometric immersion $f: M \to S^n$ is said to be *rigid*, if whenever $f': M \to S^n$ is another minimal isometric immersion then $f' = U \circ f$ holds for some $U \in O(n+1)$.

Remark. The concept of rigidity introduced here could, more precisely, be called *rigidity of the range* as harmonic maps which differ by precomposition with an isometry *on the domain* may belong to different equivalence classes.

Proposition 4.1. *If $f: M \to S^n$ is a rigid minimal isometric immersion of a compact Riemannian manifold M into S^n then f is equivariant with respect to a homomorphism*

$$\rho : i^{\circ}(M) \to SO(n+1) \ .$$

Proof. First note that the statement can easily be reduced to the case when $f:M \to S^n$ is full. Then, for each $a \in i^{\circ}(M)$, the map $f \circ a : M \to S^n$ being a minimal isometric immersion, rigidity of f implies the existence of an orthogonal matrix A such that

$$f \circ a = A \circ f \ . \tag{4.2}$$

As f is full, A is uniquely determined. Setting $\rho(a)=A$, we obtain a well-defined map

$$\rho : i^{\circ}(M) \to O(n+1) \ .$$

Then, a standard continuity argument in the use of (4.2) shows that ρ is a continuous homomorphism; in particular, $\mathrm{im}(\rho) \subset SO(n+1)$. ✓

For later purposes, we introduce a weaker notion which we call *linear rigidity*. An isometric immersion $f:M \to S^n$ is said to be linearly rigid if whenever $A \in M(n+1,n+1)$ such that

(i) $\mathrm{im}(A \circ \bar{f}) \subset S^n$, and

(ii) $A \circ \bar{f} : M \to \mathbf{R}^{n+1}$ is an isometric immersion, then $A \in O(n+1)$.

Clearly, for *minimal* isometric immersions, rigidity implies linear rigidity. Furthermore, by III.(2.2), any minimal isometric immersion $f:M \to S^n$ satisfies the equation

$$\Delta \bar{f} = m\bar{f} \ .$$

This follows because $e(f)=\frac{m}{2}$, and so, if $A \in M(n+1,n+1)$ such that $A \circ \bar{f} : M \to \mathbf{R}^{n+1}$ is an isometric immersion then, by linearity,

$$\Delta(A \circ \bar{f}) = m(A \circ \bar{f}) \ .$$

Then Takahashi's result, II.2.25, applies yielding that $\mathrm{im}(A \circ \bar{f}) \subset S^n$ and the corresponding map $A \circ f : M \to S^n$ is minimal. Thus, for minimal isometric immersions, (ii) implies (i).

Example 4.3. A standard minimal immersion $f_\lambda : G/K \to S^{n(\lambda)}$ is rigid if and only if f_λ is linearly rigid. In fact, if f_λ is linearly rigid then for any minimal isometric immersion $f: G/K \to S^{n(\lambda)}$ there exists a matrix $A \in M(n(\lambda)+1, n(\lambda)+1)$ such that $f = A \circ f_\lambda$ and, by linear rigidity of f_λ, $A \in O(n(\lambda)+1)$.

Proposition 4.4. *A minimal isometric immersion $f: M \to S^n$ is linearly rigid if and only if, whenever $A \in S^2(\mathbf{R}^{n+1})$ is a positive semidefinite endomorphism such that $A \circ \bar{f}: M \to \mathbf{R}^{n+1}$ is an isometric immersion, then $A = I_{n+1}$ (=identity of $\mathbf{R}^{n+1}$).*

Proof. Polar decomposition in the vector space $M(n+1, n+1)$ (of $(n+1)\times(n+1)$-matrices) says that any matrix $B \in M(n+1, n+1)$ can be *uniquely* written in the form

$$B = U \cdot A ,$$

where $U \in O(n+1)$ and $A \in S^2(\mathbf{R}^{n+1})$ is positive semidefinite. The rest is clear. √

We can now state the main classification theorem for minimal isometric immersions due to Do Carmo and Wallach.

Theorem 4.5. *Let $M = G/K$ be an m-dimensional compact oriented irreducible homogeneous manifold with invariant Riemannian metric $g = (,)$. Given $\lambda \in \mathrm{Spec}(G/K)$, the set of equivalence classes of full minimal isometric immersions $f: G/K \to S^n$, with induced Riemannian metric $\bar{g} = \frac{\lambda}{m} g$, can be (smoothly) parametrized by a compact convex body L lying in a (finite dimensional) vector space $W^\perp$. The interior points of L correspond to full minimal isometric immersions with maximal $n (= n(\lambda) = \dim V_\lambda - 1)$. In particular, any two such immersions can be deformed into each other by a smooth homotopy of the same type.*

Remark 1. Smoothness of the parametrization means here that the canonical map $\mathrm{int}\ L \times G/K \to S^{n(\lambda)}$ given by substitution is smooth.

Remark 2. The hypothesis on the choice of the constant $\frac{\lambda}{m}$,

$\lambda \in \mathrm{Spec}(G/K)$, is not really restrictive. In fact, given *any* minimal immersion $f:G/K \to S^n$ with induced Riemannian metric $\bar{g}=cg$, $c\in\mathbf{R}$, then

$$\bar{\Delta}\bar{f} = m\bar{f}$$

holds, where $\bar{\Delta}$ is the Laplacian with respect to $\bar{g}$. As $\bar{\Delta}=\frac{1}{c}\Delta$, this becomes

$$\Delta\bar{f} = \lambda\bar{f} ,$$

where $\lambda=mc$, i.e. the components $f^i \in C^\infty(G/K)$, $i=0,\ldots,n$, of $\bar{f}$ belong to V_λ.

To prove the classification theorem we start here with a slightly more general setting than necessary.
First, let $f:M\to S^n$ be a fixed minimal isometric immersion of an m-dimensional Riemannian manifold M with metric tensor $(,)$ into S^n. Then, $(\bar{f}_*(T_x(M)))\check{}\subset\mathbf{R}^{n+1}$ is a linear subspace for each point $x\in M$, and hence, its symmetric square $S^2((\bar{f}_*(T_x(M)))\check{})$ is contained in the vector space of all symmetric linear endomorphisms $S^2(\mathbf{R}^{n+1})$. Setting

$$W_f = \mathrm{span}_{\mathbf{R}}\{\bigcup_{x\in M} S^2((\bar{f}_*(T_x(M)))\check{})\}\subset S^2(\mathbf{R}^{n+1}) , \qquad (4.6)$$

denote by $W_f^\perp\subset S^2(\mathbf{R}^{n+1})$ the orthogonal complement of W_f in $S^2(\mathbf{R}^{n+1})$ with respect to the scalar product (1.11). Finally, set

$$L_f = \{C\in W_f^\perp \mid C+I_{n+1} \text{ is positive semidefinite}\}. \qquad (4.7)$$

Lemma 4.8. *Let $f:M\to S^n$ be a minimal isometric immersion of a Riemannian manifold M into S^n. Then,*

(i) For a positive semidefinite endomorphism $A\in S^2(\mathbf{R}^{n+1})$, the map $A\circ\bar{f}:M\to\mathbf{R}^{n+1}$ is an isometric immersion if and only if $A^2-I_{n+1}\in L_f$,

(ii) $W_f^0=\mathrm{span}_{\mathbf{R}}\{\bar{f}(x)^2\in S^2(\mathbf{R}^{n+1}) \mid x\in M\}\subset W_f$.

Proof. (i) For a positive semidefinite endomorphism $A\in S^2(\mathbf{R}^{n+1})$, the map $A\circ\bar{f}$ is isometric if and only if, for each tangent vector $X_x \in T_x(M)$, $x \in M$,

$$((A\circ\bar{f})_* X_x\,,\,(A\circ\bar{f})_* X_x) = (\bar{f}_* X_x, \bar{f}_* X_x)\,. \tag{4.9}$$

On the other hand, by (1.12), we have

$$((A\circ\bar{f})_* X_x\,,\,(A\circ\bar{f})_* X_x) = (\,((A\circ\bar{f})_* X_x)^{\vee}\,,$$

$$((A\circ\bar{f})_* X_x)^{\vee}) = (A((\bar{f}_* X_x)^{\vee})\,,\,A((\bar{f}_* X_x)^{\vee})\,) =$$

$$= (A^2((\bar{f}_* X_x)^{\vee})\,,\,(\bar{f}_* X_x)^{\vee}) = (A^2, ((\bar{f}_* X_x)^{\vee})^2)$$

and similarly,

$$(\bar{f}_* X_x, \bar{f}_* X_x) = (I_{n+1}, ((\bar{f}_* X_x)^{\vee})^2)\,.$$

Hence, by polarization, (4.9) holds if and only if

$$A^2 - I_{n+1} \in W_f^{\perp}\,.$$

(ii) Choose $C \in S^2(\mathbf{R}^{n+1})$ orthogonal to the linear subspace $W_f \subset S^2(\mathbf{R}^{n+1})$, i.e., for each $X_x \in T_x(M)$, $x \in M$,

$$(C, ((\bar{f}_* X_x)^{\vee})^2) = 0\,. \tag{4.10}$$

Fix $\varepsilon > 0$ such that the matrix $I_{n+1} + tC \in S^2(\mathbf{R}^{n+1})$ is positive semidefinite for $t \in (-\varepsilon, \varepsilon)$, and set

$$A = \sqrt{I_{n+1} + tC}\,, \qquad t \in (-\varepsilon, \varepsilon)\,.$$

Then, by (4.10), for $X_x \in T_x(M)$, $x \in M$, we have

$$((A\circ\bar{f})_* X_x\,,\,(A\circ\bar{f})_* X_x) = (A((\bar{f}_* X_x)^{\vee})\,,$$

$$A((\bar{f}_* X_x)^{\vee})\,) = (A^2, ((\bar{f}_* X_x)^{\vee})^2) =$$

$$= (I_{n+1}, ((\bar{f}_* X_x)^{\vee})^2) = (\bar{f}_* X_x, \bar{f}_* X_x)\,,$$

i.e. $A\circ\bar{f}:M\to\mathbf{R}^{n+1}$ is an isometric immersion. The argument before 4.3 shows that $\mathrm{im}(\bar{A}\circ f)\subset S^n$. Thus, for $x\in M$,

$$1 = (A(\bar{f}(x)),A(\bar{f}(x))) = (A^2(\bar{f}(x)),\bar{f}(x)) =$$

$$= (A^2,(\bar{f}(x))^2) = ((I_{n+1}+tC),(\bar{f}(x))^2) =$$

$$= (I_{n+1},(\bar{f}(x))^2) + t(C,(\bar{f}(x))^2) =$$

$$= (\bar{f}(x),\bar{f}(x)) + t(C,(\bar{f}(x))^2)$$

and hence C is orthogonal to W_f^o. Thus $W_f^o\subset W_f$. ✓

Lemma 4.11. *Let G/K be as in 4.5. Given a full minimal isometric immersion $f:G/K\to S^n$ with induced Riemannian metric $\bar{g}=\frac{\lambda}{m}g$, $\lambda\in\mathrm{Spec}(G/K)$, we have*

$$n+1 \le n(\lambda)+1 \ (=\dim V_\lambda) . \tag{4.12}$$

Also there exists a positive semidefinite endomorphism $A\in \in S^2(\mathbf{R}^{n(\lambda)+1})$ such that $\mathrm{im}(A\circ\bar{f}_\lambda)\subset S^{n(\lambda)}$ and the maps

$$A\circ f_\lambda\ ,\quad i\circ f:G/K\to S^{n(\lambda)} \tag{4.13}$$

are equivalent, where $f_\lambda:G/K\to S^{n(\lambda)}$ is a standard minimal immersion and $i:S^n\to S^{n(\lambda)}$ is the canonical inclusion map.

Proof. Fullness of f implies the linear independence of the system $\{f^i\}_{i=0}^n$ of its components contained in V_λ (cf. Remark 2 above); in particular, (4.12) follows. Fixing an orthonormal base $\{f_\lambda^i\}_{i=0}^{n(\lambda)}\subset V_\lambda$ with corresponding standard minimal immersion $f_\lambda:G/K\to S^{n(\lambda)}$, we have

$$\bar{f} = B_o\circ\bar{f}_\lambda$$

for some matrix $B_o\in M(n+1,n(\lambda)+1)$. Adding zero entries to B_o, if necessary, we obtain a matrix $B\in M(n(\lambda)+1,n(\lambda)+1)$ with the property

$$i\circ f = B\circ f_\lambda \ .$$

Now, polar decomposition of B completes the proof. √

Proof of Theorem 4.5. Let $f_\lambda : G/K \to S^n$ be a standard minimal immersion, with induced Riemannian metric $\bar{g}=\frac{\lambda}{m}\, g$, correspondin to an orthonormal base $\{f_\lambda^i\}_{i=o}^{n(\lambda)} \subset V_\lambda$.
Setting $W=W_{f_\lambda} \subset S^2(\mathbf{R}^{n(\lambda)+1})$ and $L=L_{f_\lambda}$ (cf. (4.6) and (4.7)), we claim that $L \subset W^\perp$ is an appropriate choice of the parameter space.
Given $C\in L(\subset S^2(\mathbf{R}^{n(\lambda)+1}))$, the map

$$\bar{f}_o : G/K \to \mathbf{R}^{n(\lambda)+1}$$

defined by

$$\bar{f}_o = \sqrt{C+I_{n(\lambda)+1}}\circ \bar{f}_\lambda$$

is, by 4.8(i), an isometric immersion with respect to $\bar{g}$ and hence, by Takahashi's result, II.2.25, restricts to a minimal isometric immersion

$$f_o : G/K \to S^{n(\lambda)} \ .$$

Clearly, there exists a full minimal isometric immersion

$$f : G/K \to S^n$$

such that $i\circ f : G/K \to S^{n(\lambda)}$ is equivalent to f_o , where $i : S^n \to S^{n(\lambda)}$ is the canonical inclusion map, and the equivalence class of f is uniquely determined by f_o . The parametrization in question is then given by associating to $C\in L$ the equivalence class of f .
To construct the inverse, let $f : G/K \to S^n$ be a full minimal isc metric immersion with induced metric $\bar{g}=\frac{\lambda}{m}\, g$. Then, by 4.11, $n\leq n(\lambda)$, and there exists a positive semidefinite endomorphism

$A\in S^2(\mathbf{R}^{n(\lambda)+1})$ such that $\mathrm{im}(A\circ\bar{f}_\lambda)\subset S^{n(\lambda)}$ and the maps $A\circ f_\lambda$ and $i\circ f$ are equivalent. An easy computation, using the fullness of f_λ and uniqueness of the polar decomposition, shows that A is uniquely determined by the equivalence class of f. Associate then to f the matrix

$$C = A^2 - I_{n(\lambda)+1} .$$

Again by 4.8(i), $C\in L$ and the correspondence $f\to C$ is clearly the inverse of the one given above.
So far we have shown that $L\subset W^\perp$ could serve as a parameter space. By definition

$$L = \{C\in W^\perp \mid C+I_{n(\lambda)+1} \text{ is positive semidefinite}\}$$

is a closed convex body in $W^\perp$. To see that L is actually compact, let $A\in S^2(\mathbf{R}^{n(\lambda)+1})$ be positive semidefinite such that $A^2-I_{n(\lambda)+1}\in L$. Then, as $\mathrm{im}(A\circ\bar{f}_\lambda)\subset S^{n(\lambda)}$, for each $x\in G/K$, we have

$$(A(f_\lambda(x)),A(f_\lambda(x))) = 1 ,$$

i.e., by (3.10),

$$\sum_{i,j=0}^{n(\lambda)} f_\lambda^i(x) f_\lambda^j(x)(Ae^i,Ae^j) = 1 .$$

Integrating over G/K with respect to $\mathrm{vol}(G/K,g)$, we obtain

$$\sum_{i,j=0}^{n(\lambda)} (A^2e^i,e^j)\int_{G/K} f_\lambda^i f_\lambda^j \,\mathrm{vol}(G/K,g) = \int_{G/K} \mathrm{vol}(G/K,g)$$

and, as $\{f_\lambda^i\}_{i=0}^{n(\lambda)}\subset V_\lambda$ is an orthonormal base, (3.9) implies that

$$\mathrm{trace}\, A^2 = \sum_{i=0}^{n(\lambda)} (A^2e^i,e^i) = n(\lambda) + 1 .$$

Thus $\mathrm{trace}(A^2-I_{n(\lambda)+1})=0$, and hence $\mathrm{trace}\, L=0$. It follows that the eigenvalues of elements in L are bounded and L is compact.

Finally, int $L=\{C\in W^{\perp} \mid C+I_{n(\lambda)+1}$ is positive definite$\}$ and this corresponds to *full* minimal isometric immersions into $S^{n(\lambda)}$. Smoothness of the parametrization is clear. √

Remark. The case of rigid standard minimal immersions corresponds to $W^{\perp}=\{0\}$ which, as we shall see in Section 6, may well happen. In what follows, in case of spherical domains, by refining the analysis developed here we shall give a lower estimate for $\dim W^{\perp}$.

5. HIGHER FUNDAMENTAL FORMS

The key to obtain a more precise description of the parameter space L in 4.5 is to study the higher order behaviour of minimal immersions of a homogeneous manifold into spheres. Let $f: G/K \to S^n$ be an isometric immersion of an m-dimensional compact homogeneous manifold G/K with invariant Riemannian metric $(,)$ into the Euclidean n-sphere S^n and, to simplify the matters, assume that f is equivariant with respect to a homomorphism $\rho: G \to SO(n+1)$, where G acts on G/K via the canonical action. (For a more general treatment, see Wallach (1972).) As noted in II. 2.16, the pull-back vector bundle $\tau^f = f^*(T(S^n))$ splits (isomorphically) into the orthogonal direct sum of Riemannian-connected vector bundles

$$\tau^f \cong T(G/K) \oplus \nu^f , \tag{5.1}$$

where ν^f stands for the normal bundle of f and the differential f_* of f embeds $T(G/K)$ into τ^f as a direct summand (denoted by the same symbol). Furthermore, due to the fact that f is equivariant, each vector bundle occurring in (5.1) is actually a G-bundle and (5.1) is a splitting of G-bundles.

We define the *first fundamental form* β_1 of f by setting

$$\beta_1 = f_*$$

and the *first osculating bundle* O^1 of f by

$$O^1 = \mathrm{im}\, \beta_1 (\tilde{=} T(G/K)) .$$

Then, at each point $x \in G/K$, β_1 is a linear map

$$\beta_1 : T_x(G/K) \to O_x^1$$

onto the linear subspace $O_x^1 = f_*(T_x(G/K)) \subset T_{f(x)}(S^n)$ called the *first osculating space* of f at $x \in G/K$.
Let $\perp_1 = (=\perp)$ denote the orthogonal projection onto the orthogonal complement $\nu^1 (=\nu^1)$ of $O^1 \subset \tau^f$. As in II.2.16, the *second fundamental form* $\beta_2 (=\beta(f))$ of f is a section of the vector bundle $T_2^0(G/K) \otimes \nu^1$ defined (pointwise) by

$$\beta_2(X_x^1, X_x^2) = ((\nabla_{X_x^1} \beta_1)(X_x^2))^{\perp_1} , \quad X_x^1, X_x^2 \in T_x(G/K) .$$

Furthermore, β_2 is symmetric, i.e., $\beta_2 \in C^\infty(S^2(T^*(G/K)) \otimes \nu^1)$ and, at each point $x \in G/K$, can then be viewed as a linear map

$$\beta_2 : S^2(T_x(G/K)) \to \nu_x^1 ,$$

with image

$$O_x^2 = \mathrm{im}\, \beta_2 \subset \nu_x^1 .$$

By equivariance of f ,

$$O^2 = \bigcup_{x \in G/K} O_x^2$$

inherits a G-bundle structure (over G/K) from that of τ^f .
The direct sum

$$O^1 \oplus O^2 \subset \tau^f$$

is called the *second osculating bundle* of f whose fibre $O_x^1 \oplus O_x^2 \subset T_{f(x)}(S^n)$ over $x \in G/K$ is the *second osculating space* of f at x .
Let $\perp_2$ denote the orthogonal projection onto the orthogonal

complement ν^2 of $O^1 \oplus O^2 \subset \tau^f$. The *third fundamental form* β_3 of f is a section of the vector bundle $T_3^0(G/K) \otimes \nu^2$ defined (pointwise) by

$$\beta_3(X_x^1, X_x^2, X_x^3) = ((\nabla_{X_x^1}\beta_2)(X_x^2, X_x^3))^{\perp_2} ,$$

$$X_x^1, X_x^2, X_x^3 \in T_x(G/K) , \quad x \in G/K .$$

Codazzi's equation (2.9) says that β_3 is symmetric in all its arguments, i.e., $\beta_3 \in C^\infty(S^3(T^*(G/K)) \otimes \nu^2)$, and, at each point $x \in G/K$, can then be viewed as a linear map

$$\beta_3 : S^3(T_x(G/K)) \to \nu_x^2 ,$$

with image

$$O_x^3 = \mathrm{im}\, \beta_3 \subset \nu_x^2 .$$

By equivariance of f,

$$O^3 = \bigcup_{x \in G/K} O_x^3$$

inherits a G-bundle structure from that of τ^f. The direct sum

$$O^1 \oplus O^2 \oplus O^3 \subset \tau^f$$

is called the *third osculating bundle* of f whose fibre $O_x^1 \oplus O_x^2 \oplus O_x^3 \subset T_{f(x)}(S^n)$ over $x \in G/K$ is the *third osculating space* of f at x.

Now, we proceed inductively, assuming that the $(r-1)$*-th fundamental form* $\beta_{r-1} \in C^\infty(S^{r-1}(T^*(G/K)) \otimes \nu^{r-2})$ is defined with $(r-1)$*-th osculating bundle*

$$O^1 \oplus \ldots \oplus O^{r-1} \subset \tau^f$$

and orthogonal projection $\perp_{r-1}$ onto ν^{r-1} with kernel $O^1 \oplus \ldots \oplus O^{r-1}$. The *r-th fundamental form* β_r of f is then a section of the vector bundle $T_r^0(G/K)\otimes\nu^{r-1}$ defined (pointwise) by

$$\beta_r(X_x^1,\ldots,X_x^r) = ((\nabla_{X_x^1}\beta_{r-1})(X_x^2,\ldots,X_x^r))^{\perp r-1} ,$$

$$X_x^1,\ldots,X_x^r \in T_x(G/K) , \; x\in G/K .$$

By an easy generalization of the proof of 2.8, we obtain that β_r is symmetric in all its arguments, i.e., $\beta_r \in C^\infty(S^r(T^*(G/K))\otimes \nu^{r-1})$, and at each point $x\in G/K$, can then be viewed as a linear map

$$\beta_r : S^r(T_x(G/K)) \to \nu_x^{r-1} , \quad x\in G/K ,$$

with image

$$O_x^r = \operatorname{im}\beta_r \subset \nu_x^{r-1} .$$

Again by equivariance of f ,

$$O^r = \bigcup_{x\in G/K} O_x^r$$

inherits a G-bundle structure from that of τ^f . The direct sum

$$O^1\oplus\ldots\oplus O^r \subset \tau^f$$

is called the *r-th osculating bundle* of f whose fibre over $x\in G/K$ is the *r-th osculating space* of f at x .
The procedure above must eventually stabilize, i.e. (for non-constant f) there is a unique $k\in\mathbf{N}$ such that $\beta_k\neq 0$ and $\beta_\ell = 0$ for $\ell > k$. We have then the orthogonal decomposition

$$\tau^f = O^1\oplus\ldots\oplus O^k\oplus\nu^k , \quad O^1 \cong T(G/K) , \tag{5.2}$$

of τ^f into G-bundles. We call k the *degree* of the isometric immersion $f: G/K \to S^n$. To unify the notations, set

$$S(T(G/K)) = \bigoplus_{r=0}^{\infty} S^r(T(G/K)) , \quad S^0(T(G/K)) = \varepsilon^1_{G/K} ,$$

and

$$S^+(T(G/K)) = \bigoplus_{r=1}^{\infty} S^r(T(G/K))$$

whose fibres over $x \in G/K$ are the symmetric algebra $S(T_x(G/K))$ (over the vector space $T_x(G/K)$) and its subalgebra $S^+(T_x(G/K))$, respectively. Then, by the inclusions $\nu^r \subset \tau^f$, $r=1,\ldots,k$,

$$\beta = \beta_1 + \ldots + \beta_k$$

gives rise to a bundle map

$$\beta : S(T(G/K)) \to \tau^f$$

which is *equivariant* (with respect to the G-bundle structure of $S(T(G/K))$ induced from that of $T(G/K)$) in the sense that, for each $a \in G$, β commutes with the respective actions of a on $S(T(G/K))$ and τ^f.
In particular, restricting all group actions to the closed subgroup $K \subset G$, (5.2) (over the origin $o \in G/K$) reduces to the decomposition

$$T_{f(o)}(S^n) = O_o^1 \oplus \ldots \oplus O_o^k \oplus \nu_o^k \,, \quad O_o^1 \cong T_o(G/K) \,, \tag{5.3}$$

into K-submodules. Moreover, for each $r=1,\ldots,k$,

$$\beta_r : S^r(T_o(G/K)) \to O_o^r \tag{5.4}$$

is a K-module epimorphism, where the K-module structure on the symmetric power $S^r(T_o(G/K))$ is given by the natural extension of the isotropy representation of K on $T_o(G/K)$.

Proposition 5.5. *Let $f : G/K \to S^n$ be a minimal equivariant isometric immersion of an m-dimensional compact homogeneous manifold G/K with invariant Riemannian metric $(,)$ into S^n. Setting $R^2 \in C^\infty(S^2(T(G/K)))$ with*

$$R_x^2 = \sum_{i=1}^{m} e^i \cdot e^i \,, \quad x \in G/K \,, \tag{5.6}$$

where $\{e^i\}_{i=1}^m \subset T_x(G/K)$ *is an orthonormal base, we have*

$$R^2 \cdot S(T(G/K)) \subset \ker \beta \ . \tag{5.7}$$

Proof. The condition of minimality being equivalent to $\beta(R^2)=$ $=0$, we proceed inductively assuming that $\beta(R_x^2 \cdot \sigma_x)=0$ holds for all $\sigma_x \in S^{r-2}(T_x(G/K))$, $x \in G/K$. If $\tau_x \in S^{r-1}(T_x(G/K))$ then

$$\tau_x = \sum_{j=1}^{p} \sigma_x^j \cdot X_x^j \ , \ X_x^j \in T_x(G/K) \ ,$$

$$\sigma_x^j \in S^{r-2}(T_x(G/K)) \ , \quad j=1,\ldots,p \ ,$$

and

$$\beta(R_x^2 \cdot \tau_x) = \sum_{j=1}^{p} \beta(R_x^2 \cdot \sigma_x^j \cdot X_x^j) \ . \tag{5.8}$$

For each $j=1,\ldots,p$, let $\sigma^j \in C^\infty(S^{r-2}(T(G/K)))$ be an extension of σ_x^j. Then, by (5.8), we have

$$\beta(R_x^2 \cdot \tau_x) = \sum_{j=1}^{p} (\nabla_{X_x^j}(\beta(R^2 \cdot \sigma^j)))^{\perp_{r-1}} = 0 \ ,$$

since, by hypothesis, $\beta(R^2 \cdot \sigma^j)=0$. ✓

In what follows, we consider a full minimal isometric immersion $f:G/K \to S^n$ of an m-dimensional compact homogeneous manifold G/K with invariant Riemannian metric (,) into S^n and assume that f is equivariant with respect to a homomorphism $\rho:G \to SO(n+1)$. By II.2.16, f is a harmonic map between the analytic Riemannian manifolds G/K and S^n and hence the remark before II.2.8 applies showing that f is analytic. Using the canonical identification $\check{}:T(\mathbf{R}^{n+1}) \to \mathbf{R}^{n+1}$, (5.3) becomes

$$\mathbf{R}^{n+1} = \mathbf{R}\cdot\bar{f}(o) \oplus \check{O}_o^1 \oplus \ldots \oplus \check{O}_o^k \oplus \check{\nu}_o^k \ .$$

Assuming $\nu^k \neq 0$, each vector $(0\neq)\, v_o \in \check{\nu}_o^k \subset \mathbf{R}^{n+1}$ gives rise to a linear map

$$\mu_o : \mathbf{R}^{n+1} \to \mathbf{R}$$

defined by

$$\mu_o(y) = (v_o, y) , \quad y \in \mathbf{R}^{n+1} .$$

By construction of the osculating spaces, the composition

$$\mu_o \circ \bar{f} : G/K \to \mathbf{R}$$

vanishes in infinite order at $o \in G/K$. On the other hand, as noted above, $\mu_o \circ \bar{f}$ is analytic, and hence identically zero on G/K . This implies that

$$\mathrm{im}\, \bar{f} \subset \ker \mu_o \cap S^n$$

which contradicts to the fullness of f . Hence the vector bundle ν^k is zero and (5.2) reduces to the form

$$\tau^f = O^1 \oplus \ldots \oplus O^k . \tag{5.9}$$

Now, we specialize our study to the pair $(G,K) = (SO(m+1), SO(m))$ by considering a full equivariant minimal isometric immersion

$$f : S^m_\kappa \to S^n ,$$

where $\kappa > 0$.
Fixing an orthonormal base $\{e^i\}_{i=1}^m \subset T_o(S^m_\kappa)$ establishes an isomorphism $T_o(S^m_\kappa) \cong \mathbf{R}^m$ by means of which the isotropy representation of $SO(m)$ on $T_o(S^m_\kappa)$ corresponds to the canonical $SO(m)$-module structure on $\mathbf{R}^m$ (given by matrix multiplication). Hence, for each $r = 1, \ldots, k$ (=degree of f) , the $SO(m)$-module $S^r(T_o(S^m_\kappa))$ is carried into the $SO(m)$-module $S^r(\mathbf{R}^m)$ of homogeneous polynomials of degree r on $\mathbf{R}^m$. Furthermore, by I.7.6,

$$S^r(\mathbf{R}^m) = H^r_{m-1} \oplus r_o^2 \cdot S^{r-2}(\mathbf{R}^m) , \tag{5.10}$$

where H^r_{m-1} is the $SO(m)$-module of homogeneous harmonic polynomials of degree r on $\mathbf{R}^m$ (viewed also as the $SO(m)$-module $\tilde{H}^r_{m-1}$ of spherical harmonics of order r on S^{m-1}) and $r_o^2 \in S^2(\mathbf{R}^m)$ is given by

$$r_o^2(x) = \sum_{i=1}^{m} (x^i)^2 , \quad x=(x^1,\ldots,x^m)\in\mathbf{R}^m .$$

Using the isomorphism $S^r(T_o(S_\kappa^m))\cong S^r(\mathbf{R}^m)$, $r=1,\ldots,k$, which, for $r=2$, associates r_o^2 to R_o^2 introduced in (5.6), the decomposition (5.10) is carried over to

$$S^r(T_o(S_\kappa^m)) = \tilde{H}_{m-1}^r \oplus R_o^2\cdot S^{r-2}(T_o(S_\kappa^m)) ,$$

where $\tilde{H}_{m-1}^r$ stands for the $SO(m)$-module of spherical harmonics of order r on the unit sphere of $T_o(S_\kappa^m)\cong\mathbf{R}^m$. By (5.7), for each $r=1,\ldots,k$, the $SO(m)$-module epimorphism $\beta_r:S^r(T_o(S_\kappa^m))\to O_o^r$ factorizes through the projection $S^r(T_o(S_\kappa^m))\to\tilde{H}_{m-1}^r$ yielding an epimorphism

$$\bar{\beta}_r:\tilde{H}_{m-1}^r \to O_o^r$$

of $SO(m)$-modules. Finally, $\tilde{H}_{m-1}^r$ is an irreducible $SO(m)$-module (cf. Problem 5 for this Chapter) and hence $\bar{\beta}_r$ is an isomorphism. Hence we obtain the following.

Theorem 5.11. *Given a full equivariant minimal isometric immersion*

$$f:S_\kappa^m \to S^n$$

the $SO(m+1)$-bundle τ^f admits the decomposition

$$\tau^f = O^1\oplus\ldots\oplus O^k \quad (k=\textit{degree of } f)$$

into $SO(m+1)$-bundles such that, for each $r=1,\ldots,k$, $\bigoplus_{j=1}^{r} O^j$ is the r-th osculating bundle of f and, over the origin $o\in S_\kappa^m$, O_o^r is equivalent to the $SO(m)$-module $\tilde{H}_{m-1}^r$ of spherical harmonics of order r on S^{m-1} . ✓

Remark. 5.11 specializes to the standard minimal immersion

$$f:S_{\kappa(k_o)}^m \to S^{n(k_o)}$$

corresponding to the eigenvalue $\lambda_{k_o}=k_o(k_o+m-1)\in\mathrm{Spec}(S^m)$. By making use of (5.9), we then have

$$n(k_o) = \sum_{r=1}^{k} \dim O_o^r = \sum_{r=1}^{k} \dim \tilde{H}^r_{m-1} =$$

$$= \sum_{r=1}^{k} (m+2r-2)\frac{(m+r-3)!}{r!(m-2)!}$$

which, by an easy computation, yields $k_o=k$ (=degree of f).

6. RIGIDITY OF MINIMAL IMMERSIONS BETWEEN SPHERES

By means of the concept of degree introduced in the previous section, we are now able to carry out our program for describing the rigid (and hence equivariant; cf. 4.1) minimal isometric immersions between spheres. We start with the following general result.

Theorem 6.1. *Let $f:G/K\to S^n$ be a full equivariant minimal isometric immersion of a compact homogeneous manifold G/K with invariant Riemannian metric (,) into S^n . If the degree of $f\leq 3$ then f is linearly rigid.*

Proof. The argument after the proof of 5.5 (in the use of analyticity of f) shows that

$$\tau^f = O^1\oplus O^2\oplus O^3$$

which, at a point $x\in G/K$, restricts to

$$\mathbf{R}^{n+1} = \mathbf{R}\cdot\bar{f}(x)\oplus\check{O}^1_x\oplus\check{O}^2_x\oplus\check{O}^3_x \ , \qquad (6.2)$$

where $O^1_x=f_*(T_x(G/K))$, $O^2_x=\beta_2(S^2(T_x(G/K)))$ and $O^3_x=$ $=\beta_3(S^3(T_x(G/K)))$.
Let $c:(-\varepsilon,\varepsilon)\to G/K$, $\varepsilon>0$, be a geodesic segment with $c(0)=x$. We claim that

$$\bar{f}(x)^2 \ , \ \bar{f}(x)\cdot(\bar{f}\circ c)^{\cdot}(0) \ , \ \bar{f}(x)\cdot(\bar{f}\circ c)^{\cdot\cdot}(0) \ ,$$

$$\bar{f}(x)\cdot(\bar{f}\circ c)^{\cdots}(0)\in W_f\ , \tag{6.3}$$

where $W_f \subset S^2(\mathbf{R}^{n+1})$ is given in (4.6). Indeed, by 4.8(ii), $\bar{f}(c(t))^2 \in W_f^o \subset W_f$ holds for all $t\in(-\varepsilon,\varepsilon)$; in particular, $\bar{f}(x)^2 \in W_f$. Differentiating, we have $\frac{\partial}{\partial t}\{(\bar{f}(c(t)))^2\} = 2\bar{f}(c(t))\cdot$ $\cdot(\bar{f}\circ c)^{\cdot}(t)\in W_f$; in particular, $\bar{f}(x)\cdot(\bar{f}\circ c)^{\cdot}(0)\in W_f$. Differentiating again, it follows that $\bar{f}(c(t))\cdot(\bar{f}\circ c)^{\cdot\cdot}(t)\in W_f$, since $((\bar{f}\circ c)^{\cdot}(t))^2 = \{(\bar{f}_*(\dot{c}(t)))^{\vee}\}^2 \in W_f$, by definition. In particular, $\bar{f}(x)\cdot(\bar{f}\circ c)^{\cdot\cdot}(0)\in W_f$. Finally, differentiating again, we get

$$(\bar{f}\circ c)^{\cdot}(0)\cdot(\bar{f}\circ c)^{\cdot\cdot}(0) + \bar{f}(x)\cdot(\bar{f}\circ c)^{\cdots}(0)\in W_f$$

and, as

$$\frac{\partial}{\partial t}\{((\bar{f}\circ c)^{\cdot}(t))^2\}_{t=0} = 2(\bar{f}\circ c)^{\cdot}(0)\cdot(\bar{f}\circ c)^{\cdot\cdot}(0)\in W_f\ ,$$

the last relation $\bar{f}(x)\cdot(\bar{f}\circ c)^{\cdots}(0)\in W_f$ follows.

To prove the theorem, by 4.4 and 4.8(i), it is enough to show that $W_f^{\perp}=\{0\}$. Let $C\in W_f^{\perp}$ be fixed. Then, by (6.3),

$$(C,\bar{f}(x)^2) = (C,\bar{f}(x)\cdot(\bar{f}\circ c)^{\cdot}(0)) =$$

$$= (C,\bar{f}(x)\cdot(\bar{f}\circ c)^{\cdot\cdot}(0)) =$$

$$= (C,\bar{f}(x)\cdot(\bar{f}\circ c)^{\cdots}(0)) = 0\ .$$

By making use of (1.12), these relations become

$$(C\cdot\bar{f}(x),\bar{f}(x)) = (C\cdot\bar{f}(x),(\bar{f}\circ c)^{\cdot}(0)) =$$

$$= (C\cdot\bar{f}(x),(\bar{f}\circ c)^{\cdot\cdot}(0)) =$$

$$= (C\cdot\bar{f}(x),(\bar{f}\circ c)^{\cdots}(0)) = 0\ .$$

Now, if $c:(-\varepsilon,\varepsilon)\to G/K$ runs through all geodesic segments with $c(0)=x$ then, as a straightforward computation shows, the vectors $(\bar{f}\circ c)^{\cdot}(0)$, $(\bar{f}\circ c)^{\cdot\cdot}(0)$ and $(\bar{f}\circ c)^{\cdots}(0)$ span

$\check{O}_x^1 \oplus \check{O}_x^2 \oplus \check{O}_x^3$. Thus, by making use of the decomposition (6.2), it follows that $C \cdot \bar{f}(x) = 0$. As $f: G/K \to S^n$ is full, we obtain $C=0$.√

Remark. For a more general statement, see Wallach (1972) page 32.

Corollary 6.4. *If* $f: S^m_\kappa(k) \to S^n$, $k=1,2,3,$ *is a full equivariant minimal isometric immersion then* $n=n(k)$ *and* f *is a rigid standard minimal immersion.*

Proof. A standard minimal immersion

$$f_{\lambda_k}: S^m_\kappa(k) \to S^{n(k)}$$

has, by 5.11, degree k . As $k \le 3$, 6.1 applies implying linear rigidity of f_{λ_k} . Hence, by 4.3, f_{λ_k} is rigid. Now, as f is full, by 4.11, $n \le n(k)$, and, denoting by $i: S^n \to S^{n(k)}$ the canonical inclusion map, $i \circ f$ and f_{λ_k} are equivalent since the latter is rigid. In particular, $n=n(k)$ and f , being equivalent to f_{λ_k} , is rigid and standard. √

Next, specializing to spherical domains, we state the following rigidity theorem (involving no restrictions on the degree) due to Calabi (1967) (see also Do Carmo and Wallach (1970)).

Theorem 6.5. *Any full minimal isometric immersion* $f: S^2 \to S^n$ *is rigid.*

Remark. For a local version, see Wallach (1970).

The proof of 6.5 is broken up into a few lemmas.

Lemma 6.6. *Let* ρ *be a class* 1 *representation of the pair* $(SO(m+1), SO(m))$ *on* V *with* $SO(m)$*-fixed unit vector* $v^0 \in V$. *Denoting also by* ρ *the induced representation on the symmetric square* $S^2(V)$, *the* $SO(m+1)$*-submodule*

$$W^0 = \mathrm{span}_{\mathbf{R}}\{\rho(a)((v^0)^2) \mid a \in SO(m+1)\} \subset S^2(V)$$

contains all subrepresentations of $S^2(V)$ *which are class* 1 *for* $(SO(m+1), SO(m))$.

Proof. By 3.17, the map

$$f: SO(m+1)/SO(m) \to S \quad (S=\text{unit sphere in } V)$$

defined by

$$f(a\cdot SO(m)) = \rho(a)v^o, \quad a\in SO(m+1),$$

is a minimal immersion. The Riemannian metric (,) on $SO(m+1)/$ $/SO(m)$ induced by f is invariant and, by irreducibility (cf. 3.8), makes $SO(m+1)/SO(m)$ a space of constant curvature $\kappa>0$, i.e. the map f can be thought of as a minimal isometric immersion

$$f: S^m_\kappa \to S .$$

By definition, f is equivariant with respect to $\rho: SO(m+1) \to$ $\to SO(V)$ and, hence, irreducibility of V (as an $SO(m+1)$-module) implies that f is full. The $SO(m+1)$-bundle τ^f, by 5.11, decomposes as

$$\tau^f = O^1\oplus\ldots\oplus O^k, \quad (k=\text{degree of } f), \tag{6.7}$$

where, for each $r=1,\ldots,k$, $\overset{r}{\underset{j=1}{\oplus}} O^j$ is the r-th osculating bundle of f and O^r_o is equivalent to the $SO(m)$-module $\tilde{H}^r_{m-1}$. If we set $V^o=\mathbf{R}\cdot\bar{f}(o)$ and $V^r=\check{O}^r_o$, $r=1,\ldots,k$, over the origin $o\in S^m_\kappa$, (6.7) implies

$$V = V^o\oplus V^1\oplus\ldots\oplus V^k . \tag{6.8}$$

This is an equality of $SO(m)$-modules, where each $SO(m)$-submodule V^r, $r=0,\ldots,k$, is irreducible (cf. Problem 5 for this chapter). Furthermore, the $SO(m)$-module structure of V in (6.8) is given by restricting the irreducible representation $\rho: SO(m+1) \to SO(V)$ to the subgroup $SO(m) \subset SO(m+1)$. Indeed, each group element $a\in SO(m)$ acts on the fibre $T_{v^o}(S^n)(=\tau_o^f)$ by the differential $\rho(a)_*$. As $\rho(a)$ is linear and fixes v^o, this linear automorphism on $T_{v^o}(S^n)$ is carried into $\rho(a)$ by the isomorphism $V\cong\mathbf{R}\cdot v^o\oplus(T_{v^o}(S^n))^{\check{}}$. (Note that (6.8) is a typical example of a branching phenomenon to be treated in the next section. In fact, given the irreducible $SO(m+1)$-module V, by restricting the module structure to $SO(m)\subset SO(m+1)$ the irreducibility of V is lost, and V splits into a direct sum (6.8) of irreducible $SO(m)$-modules.)

We first claim that if A is a symmetric positive semidefinite $SO(m)$-module endomorphism of V with the property $A(\rho(G)(v^o))\subset S$ then $A=I_V$(=identity of V).

Indeed, the hypothesis that A preserves the $SO(m)$-module structure of V implies that A commutes with $\rho(a)$, for each $a\in SO(m)$; in particular, $Av^o\in V$ is left fixed by $SO(m)$. As A is positive semidefinite, we get $Av^o=v^o$, i.e. A acts on $\mathbf{R}v^o$ as the identity. For $r,s=0,\ldots,k$, $r\neq s$, the irreducible $SO(m)$-submodules V^r and V^s are inequivalent and hence $\mathrm{Hom}_{SO(m)}(V^r,V^s)=0$. By Schur's lemma, for each $r=0,\ldots,k$, $A(V^r)\subset V^r$ and

$$A|V^r = \lambda_r\cdot I_{V^r}\,, \tag{6.9}$$

where $\lambda_r\in\mathbf{R}$ is nonnegative. To prove the claim, we need to show that $\lambda_r=1$ for all $r=0,\ldots,k$. By the above, $\lambda_o=1$ and so we may proceed by induction assuming $\lambda_o=\ldots=\lambda_{s-1}=1$, $s\leq k$.

Let $c:(-\varepsilon,\varepsilon)\to G/K$, $\varepsilon>0$, be a geodesic segment with $c(0)=o$. We set

$$\tilde{c}^{(r)} = \frac{\partial^r\tilde{c}}{\partial t^r}\,,\quad r=0,\ldots,k\,,$$

where $\tilde{c}=\bar{f}\circ c:(-\varepsilon,\varepsilon)\to S^n$. By definition of the higher fundamental forms, we have

$$\tilde{c}^{(r)}(0)\in V^o\oplus V^1\oplus\ldots\oplus V^r\,,\quad r=0,\ldots,k\,,$$

and the V^r-component of $\tilde{c}^{(r)}(0)$ is nonzero. By (6.9), it is enough to show that

$$(A\cdot\tilde{c}^{(s)}(0),\ A\cdot\tilde{c}^{(s)}(0)\ ) = (\tilde{c}^{(s)}(0),\ \tilde{c}^{(s)}(0)\ )\ .$$

As $(A\circ\bar{f},\bar{f})=1$, we have

$$0 = \frac{d^{2s}}{dt^{2s}}(A\circ\tilde{c},A\circ\tilde{c}) = \sum_{r=0}^{2s}\binom{2s}{r}(A\circ\tilde{c}^{(r)},A\circ\tilde{c}^{(2s-r)})\ ;$$

in particular,

$$(A\cdot\tilde{c}^{(s)}(0),A\cdot\tilde{c}^{(s)}(0)\ ) =$$

$$= \sum_{r=0}^{s-1} a_r^s(A\cdot\tilde{c}^{(r)}(0),A\cdot\tilde{c}^{(2s-r)}(0)\ )\ , \qquad (6.10)$$

where $a_r^s=-2\binom{2s}{r}\binom{2s}{s}^{-1}$, $r=0,\ldots,s-1$. For fixed $r=0,\ldots,s-1$, with $q=2s-r$, we determine

$$(A\cdot\tilde{c}^{(r)}(0),A\cdot\tilde{c}^{(q)}(0)\ )\ .$$

We set $\tilde{c}^{(q)}(0)=w_0+w_1+\ldots+w_q$, $w_p\in V^p$ $p=0,\ldots,q$. Schur's lemma entails $(\tilde{c}^{(r)}(0),A\cdot w_p)=0$ for all $p=s,\ldots,q$. Thus, by the induction hypothesis

$$(A\cdot\tilde{c}^{(r)}(0),A\cdot\tilde{c}^{(q)}(0)\ ) = (\tilde{c}^{(r)}(0),A\cdot\tilde{c}^{(q)}(0)\ ) =$$

$$= \sum_{p=0}^{q}(\tilde{c}^{(r)}(0),A\cdot w_p) = \sum_{p=0}^{s-1}(\tilde{c}^{(r)}(0),A\cdot w_p) =$$

$$= \sum_{p=0}^{s-1}(\tilde{c}^{(r)}(0),w_p) = \sum_{p=0}^{q}(\tilde{c}^{(r)}(0),w_p) =$$

$$= (\tilde{c}^{(r)}(0),\tilde{c}^{(q)}(0)\ )\ .$$

Hence, (6.10) becomes

$$(A\cdot\tilde{c}^{(s)}(0),A\cdot\tilde{c}^{(s)}(0)\ ) =$$

$$= \sum_{r=0}^{s-1} a_r^s(\tilde{c}^{(r)}(0), \tilde{c}^{(2s-r)}(0)) = (\tilde{c}^{(s)}(0), \tilde{c}^{(s)}(0)) ,$$

where the last equality follows from the fact that (6.10) holds, in particular, for $A=I_V$ since the coefficients a_r^s , $r=0,\ldots$ $\ldots,s-1$, do not depend on A . We obtain $\lambda_s=1$, i.e. the induction is complete and the claim follows. Turning to the proof of the lemma, assume that $\bar{W}$ is a (nontrivial) class 1 subrepresentation (for $(SO(m+1),SO(m))$) of $S^2(V)$ not contained in W^o . By orthogonal projection, we may assume that $(\bar{W},W^o)=0$. As $\bar{W}$ is class 1 for $(SO(m+1),SO(m))$, we may select $(0\neq)\,C\in W$ such that C is left fixed by $SO(m)$. This implies that C is a symmetric $SO(m)$-module endomorphism of V . Choose $t>0$ with the property that $tC+I_V$ is positive semidefinite. As $\bar{W}$ is orthogonal to W^o , for each $a\in SO(m+1)$, we have $(tC+I_V ,$ $\rho(a)((v^o)^2)) = (\rho(a)v^o , \rho(a)v^o) = 1$, or equivalently,

$$(A(\rho(a)v^o), A(\rho(a)v^o)) = 1 , \quad a\in SO(m+1) ,$$

where $A=\sqrt{tC+I_V}$. Thus, A is a symmetric positive semidefinite $SO(m)$-module endomorphism of V such that $A(\rho(G)(v^o))\subset S$. By the above, $A=I_V$ which implies $C=0$, a contradiction. √

Lemma 6.11. *Let $\{\chi^j\}_{j=0}^n \subset \tilde{H}_2^k$ be a linearly independent system of spherical harmonics of order k on S^2 such that*

$$\sum_{j=0}^{n} (\chi^j)^2 = 1 . \qquad (6.12)$$

Then, $n=n(k)=2k$ and $\{\chi^j\}_{j=0}^n \subset \tilde{H}_2^k$ is an orthonormal base.

Proof. Fixing an orthonormal base $\{f_{\lambda_k}^j\}_{j=0}^{2k} \subset \tilde{H}_2^k$ gives rise to a standard minimal immersion

$$f_{\lambda_k}: S^2_{\kappa(k)} \to S^{2k} , \quad \kappa(k) = \frac{2}{k(k+1)} ,$$

with

$$\bar{f}_{\lambda_k} = \sum_{j=0}^{2k} f_{\lambda_k}^j \cdot e^j$$

(cf.(3.10)).
Furthermore, by Section 3, f_{λ_k} is equivariant with respect to a homomorphism

$$\rho : SO(3) \to \mathbf{R}^{2k+1}$$

which, at the same time, endows $\mathbf{R}^{2k+1}$ $(=\tilde{H}_2^k)$ with an irreducible $SO(3)$-module structure. As $v^{\circ}=f_{\lambda_k}(o)$ is left fixed by $\rho(SO(2))$, the homomorphism ρ actually gives rise to a class 1 representation on $\mathbf{R}^{2k+1}$ for the pair $(SO(3),SO(2))$.
The relation (6.12) defines a map

$$f : S^2_{\kappa}(k) \to S^n$$

with

$$\bar{f} = \sum_{j=o}^{n} \chi^j e^j .$$

By the linear independence of the system $\{\chi^j\}_{j=o}^n$, we have $n \le 2k$ and there exists a matrix $A \in M(2k+1,2k+1)$ such that $\mathrm{im}(A \circ f_{\lambda_k}) \subset S^{2k}$ and

$$i \circ f = A \circ f_{\lambda_k} \qquad (6.13)$$

are valid where $i : S^n \to S^{2k}$ stands for the canonical inclusion map. By (6.13), for each $a \in SO(3)$, we have

$$1 = (f(a(o)),f(a(o))) = (A(f_{\lambda_k}(a(o))),A(f_{\lambda_k}(a(o)))) =$$

$$= (A(\rho(a)v^{\circ}),A(\rho(a)v^{\circ})) = ({}^tAA(\rho(a)v^{\circ}),(\rho(a)v^{\circ})) =$$

$$= ({}^tAA,(\rho(a)v^{\circ})\cdot(\rho(a)v^{\circ})) = ({}^tAA,\rho(a)((v^{\circ})^2)) ,$$

where ρ also stands for the induced representation on the symmetric square $S^2(\mathbf{R}^{2k+1})$. Equivalently,

$$({}^tAA-I_{2k+1},\rho(a)((v^{\circ})^2)) = 0 , \qquad a \in SO(3) ,$$

and hence ${}^tAA-I_{2k+1}$ is orthogonal to the $SO(3)$-submodule

$$W^{o} = \text{span}_{\mathbf{R}}\{\rho(a)((v^{o})^{2}) \mid a\in SO(3)\} \subset S^{2}(\mathbf{R}^{2k+1}) \ .$$

On the other hand, by elementary representation theory, each subrepresentation of $S^2(\mathbf{R}^{2k+1})$ is class 1 for $(SO(3),SO(2))$ (cf. Problem 9 for this chapter). Thus, 6.6 implies $W^{o}= =S^2(\mathbf{R}^{2k+1})$. It follows that ${}^{t}AA-I_{2k+1}=0$, or equivalently, $A\in O(2k+1)$. √

Proof of Theorem 6.5. Let $\{\chi^{j}\}_{j=0}^{n}$ denote the components of $\bar{f}:S_{\kappa}^{2}\to\mathbf{R}^{n+1}$. By Remark 2 after 4.5, we have $\kappa=\kappa(k)$, for some $k\in\mathbf{N}$, and $\{\chi^{j}\}_{j=0}^{n}\subset\tilde{H}_{2}^{k}$. Fullness of f implies that $\{\chi^{j}\}_{j=0}^{n}$ is a linearly independent system. Now, 6.11 applies showing that $n=2k$ and $\{\chi^{j}\}_{j=0}^{2k}\subset\tilde{H}_{2}^{k}$ is an orthonormal base which, via the isomorphism $\tilde{H}_{2}^{k}=\mathbf{R}^{2k+1}$, defines the standard minimal immersion $f_{\lambda_k}=f$. Hence f is standard and rigid. √

To close this section, following the ideas of Do Carmo and Wallach (1969) and (1970), we show that 6.1 and 6.5 are the only rigidity statements valid for minimal immersions between spheres. For this, recall from 4.5 that, for fixed m and k , the equivalence classes of full minimal isometric immersions

$$f:S_{\kappa(k)}^{m}\to S^{n} \ , \qquad n\le n(k) \ ,$$

are parametrized by a compact convex body

$$L=\{C\in W^{\perp} \mid C+I_{n(k)+1} \text{ is positive semidefinite}\}$$

lying in the orthogonal complement of a linear subspace $W\subset S^{2}(\mathbf{R}^{n(k)+1})$ given in (4.6) for $f=f_{\lambda_k}$, $\lambda_k=k(k+m-1)\in\text{Spec}(S^m)$. The homomorphism $\rho:SO(m+1)\to SO(n(k)+1)$ associated with the equivariance of $f_{\lambda_k}:S^{m}\to S^{n(k)}$ induces an (orthogonal) $SO(m+1)$-module structure on $S^{2}(\mathbf{R}^{n(k)+1})$. Further, W (and hence $W^{\perp}$) are $SO(m+1)$-submodules of $S^{2}(\mathbf{R}^{n(k)+1})$. We then have the following nonrigidity theorem of Do Carmo and Wallach.

Theorem 6.14. *For* $m>2$ *and* $k>3$, *we have*

$$\dim \operatorname{int} L = \dim W^{\perp} \geq 18 \ . \tag{6.15}$$

From here on, the rest of this section is devoted to the reformulation of 6.14 in terms of some representations of the special orthogonal group. The dimension estimate (6.15) will then be verified in the next section. First of all, we recall some general facts on representations.

Let G be a compact Lie group with a closed subgroup K and let there be given an orthogonal K-module structure ρ on a finite dimensional vector space V with scalar product $(\,,)$. In what follows, we define an orthogonal G-module structure on the vector space $C_K(G,V)$ of all *continuous* maps $f:G \to V$ which are equivariant with respect to $\rho: K \to O(V)$, where K acts on G by left translations. First, endow $C_K(G,V)$ with the "global" scalar product

$$(f,f') = \int_G (f,f')\,d\lambda \ ,$$

where $d\lambda$ is the normalized Haar measure on G. Setting

$$(a\cdot f)(a') = f(aa') \ , \quad f \in C_K(G,V) \ , \quad a,a' \in G \ ,$$

G acts on $C_K(G,V)$, orthogonally. The orthogonal G-module $C_K(G,V)$ obtained in this way is said to be *induced* by V. The next lemma is known as the *Frobenius reciprocity law.*

Lemma 6.16. *Given a compact Lie group G with a closed subgroup K and finite dimensional orthogonal K-module V, denote by $C_K(G,V)$ the orthogonal G-module induced by V. Then, for any finite dimensional orthogonal G-module $\bar{V}$ the vector space* $\operatorname{Hom}_G(\bar{V}, C_K(G,V))$ *is canonically isomorphic to* $\operatorname{Hom}_K(\bar{V},V)$, *where $\bar{V}$ is considered to be a K-module by restriction.*

Proof. Given a G-module homomorphism $A:\bar{V} \to C_K(G,V)$, define $\phi(A):\bar{V} \to V$ by $\phi(A)(\bar{v}) = A(\bar{v})(1)$, $\bar{v} \in \bar{V}$. As an easy computation shows, $A:\bar{V} \to V$ is a K-module homomorphism and the correspondence $A \to \phi(A)$ establishes the required isomorphism

$\mathrm{Hom}_G(\bar{V}, C_K(G,V)) \cong \mathrm{Hom}_K(\bar{V}, V)$. ✓

For the next lemma, recall that the vector space $\tilde{H}^k_m$ of spherical harmonics of order k on S^m is an irreducible orthogonal $SO(m+1)$-module. Moreover, for fixed orthonormal base $\{f^i_{\lambda_k}\}_{i=0}^{n(k)} \subset \tilde{H}^k_m$, the construction of the standard minimal immersion $f_{\lambda_k}: S^m_{\kappa(k)} \to S^{n(k)}$ in Section 3 shows that the corresponding $SO(m+1)$-module structure ρ on $\tilde{H}^k_m$ is also class 1 for the pair $(SO(m+1), SO(m))$ with $SO(m)$-fixed vector $v^o = f_{\lambda_k}(o) \in \tilde{H}^k_m$. By equivariance,

$$f_{\lambda_k}(a \cdot SO(m)) = \rho(a) v^o \ , \quad a \in SO(m+1) \ ,$$

(which can be thought as a definition of f_{λ_k}) and the proof of 6.6 applies giving the decomposition

$$\tilde{H}^k_m \cong \tilde{H}^o_{m=1} \oplus \tilde{H}^1_{m-1} \oplus \ldots \oplus \tilde{H}^k_{m-1}$$

of $\tilde{H}^k_m$ into irreducible $SO(m)$-submodules, where $\tilde{H}^o_{m-1} = \mathbf{R} \cdot \bar{f}_{\lambda_k}(o)$ and $\tilde{H}^r_{m-1} = \check{O}^r_o$, $r=1,\ldots,k$, and the $SO(m)$-module structure on $\tilde{H}^k_m$ is given by restriction (cf. the argument after (6.8)). The Laplacian $\Delta^{\mathbf{R}^m}$ being of second order, the $SO(m)$-module $\tilde{H}^1_{m-1}$ contains (restrictions) of *all* linear functions and so the symmetric square $S^2(\tilde{H}^1_{m-1})$ is nothing but (or rather, isomorphic to) the $SO(m)$-module of all homogeneous polynomials of degree 2 in m variables, or equivalently, by I.7.6,

$$S^2(\tilde{H}^1_{m-1}) \cong \tilde{H}^o_{m-1} \oplus \tilde{H}^2_{m-1} \tag{6.17}$$

as $SO(m)$-modules.

Lemma 6.18. *Let* $\bar{W}$ *be the sum of those* $SO(m+1)$*-submodules of* $S^2(\tilde{H}^k_m)$ *not containing* $SO(m)$*-submodules isomorphic to* $\tilde{H}^o_{m-1}$ *or* $\tilde{H}^2_{m-1}$ *. Then* $\bar{W} \subset W^\perp$.

Proof. Let Z denote the orthogonal complement of $S^2(\tilde{H}^1_{m-1})$ in $S^2(\tilde{H}^k_m)$ with orthogonal projection $\rho: S^2(\tilde{H}^k_m) \to S^2(\tilde{H}^1_{m-1})$, i.e., as $SO(m)$-modules,

$$S^2(\tilde{H}^k_m) = S^2(\tilde{H}^1_{m-1}) \oplus Z \tag{6.19}$$

and $\ker \pi = Z$. Let $C_{SO(m)}(SO(m+1), S^2(\tilde{H}^1_{m-1}))$ denote the orthogonal $SO(m+1)$-module induced by the $SO(m)$-module $S^2(\tilde{H}^1_{m-1})$. For each $\sigma \in S^2(\tilde{H}^k_m)$, define

$$\psi(\sigma): SO(m+1) \to S^2(\tilde{H}^1_{m-1})$$

by

$$\psi(\sigma)(a) = \pi(\rho(a)\sigma), \quad a \in SO(m+1).$$

Then, $\psi(\sigma) \in C_{SO(m)}(SO(m+1), S^2(\tilde{H}^1_{m-1}))$ since, for each $a \in SO(m)$ and $a' \in SO(m+1)$

$$\psi(\sigma)(aa') = \pi(\rho(aa')\sigma) = \pi(\rho(a)(\rho(a')\sigma)) =$$

$$= \rho(a)(\pi(\rho(a')\sigma)) = (\rho(a) \circ \psi(\sigma))(a').$$

The map ψ which sends $\sigma \in S^2(\tilde{H}^k_m)$ into $\psi(\sigma)$ is actually a homomorphism

$$\psi: S^2(\tilde{H}^k_m) \to C_{SO(m)}(SO(m+1), S^2(\tilde{H}^1_{m-1}))$$

of $SO(m+1)$-modules.

This follows because, for $\sigma \in S^2(\tilde{H}^k_m)$ and $a, a' \in SO(m+1)$, we have

$$\psi(\rho(a)\sigma)(a') = \pi(\rho(a')(\rho(a)\sigma)) = \pi(\rho(a'a)\sigma) =$$

$$= \psi(\sigma)(a'a) = (a \cdot \psi(\sigma))(a').$$

Furthermore, $\ker \psi = W^{\perp}$. Indeed, for $\sigma \in S^2(\tilde{H}^k_m)$, the relation $\psi(\sigma) = 0$ implies

$$\pi(\rho(a)\sigma) = 0, \quad a \in SO(m+1),$$

which, by (6.19), is satisfied if and only if

$$(\rho(a)\sigma, S^2(\tilde{H}^1_{m-1})) = 0 \ , \quad a \in SO(m+1) \ . \tag{6.20}$$

Now, as $\tilde{H}^1_{m-1} \cong \mathfrak{F}^1_o = (f_{\lambda_k})_*(T_o(G/K))^{\vee}$, by equivariance of f_{λ_k}, (6.20) is just a reformulation of the relation $\sigma \in W^{\perp}$. Factorizing out the kernel $W^{\perp}$, we obtain that $W = \mathrm{span}_{\mathbf{R}}\{\rho(a) \cdot (S^2(\tilde{H}^1_{m-1})) \mid a \in SO(m+1)\}$ is contained in $C_{SO(m)}(SO(m+1), S^2(\tilde{H}^1_{m-1}))$ as an $SO(m+1)$-submodule. Using this and the Frobenius reciprocity law 6.16, we have

$$\dim \mathrm{Hom}_{SO(m+1)}(\bar{W}, W) \leq$$

$$\leq \dim \mathrm{Hom}_{SO(m+1)}(\bar{W}, C_{SO(m)}(SO(m+1) \ ,$$

$$S^2(\tilde{H}^1_{m-1}))) = \dim \mathrm{Hom}_{SO(m)}(\bar{W}, S^2(\tilde{H}^1_{m-1})) \ .$$

On the other hand, by hypothesis, $\bar{W}$ does not contain any copy of $\tilde{H}^0_{m-1}$ and $\tilde{H}^2_{m-1}$. This, together with (6.17), implies

$$\mathrm{Hom}_{SO(m)}(\bar{W}, S^2(\tilde{H}^1_{m-1})) = 0 \ .$$

Hence, $\mathrm{Hom}_{SO(m+1)}(\bar{W}, W) = 0$, or equivalently, $\bar{W} \subset W^{\perp}$. ✓

By 6.18, $\dim \bar{W} \leq \dim W^{\perp}$ and (6.15) will then be verified if we prove that, for $m > 2$ and $k > 3$,

$$\dim \bar{W}^k_m \geq 18 \ , \tag{6.21}$$

where we indicate the dependence on m and k by setting $\bar{W} = \bar{W}^k_m$. For our purposes, however, it will be convenient to consider only *complex* $SO(m+1)$-modules. Because we shall deal with absolutely irreducible modules over the special orthogonal group we shall be working in the category of *complex* $SO(m+1)$-modules and the proof of (6.21) will be carried out for the complex dimension of the complexification $\bar{W}^k_m \otimes_{\mathbf{R}} \mathbf{C}$.

7. REPRESENTATIONS OF THE SPECIAL ORTHOGONAL GROUP

The purpose of this section is to prove 6.14 by showing that, for $m>2$ and $k>3$,

$$\dim_{\mathbf{C}}(\bar{W}_m^k \underset{\mathbf{R}}{\otimes} \mathbf{C}) \geq 18 \ . \tag{7.1}$$

To obtain (7.1) basic representation theory of the special orthogonal group is required and, in what follows, we first recall some known facts and results on classification of complex $SO(m+1)$-modules which we use subsequently. (More details can be found e.g. in Boerner (1963), Wallach (1973) and Weyl (1946).)

Let $T \subset SO(m+1)$ denote the (toroidal) subgroup consisting of all matrices

$$\begin{aligned} &\mathrm{diag}(A(\upsilon_1),\ldots,A(\upsilon_\ell),1) \ , \quad \upsilon_1,\ldots,\upsilon_\ell \in \mathbf{R} \ , \\ &\text{if} \quad m=2\ell \end{aligned} \tag{7.2}$$

or

$$\begin{aligned} &\mathrm{diag}(A(\upsilon_1),\ldots,A(\upsilon_\ell)) \ , \quad \upsilon_1,\ldots,\upsilon_\ell \in \mathbf{R} \ , \\ &\text{if} \quad m=2\ell-1 \ , \end{aligned} \tag{7.3}$$

where, for $\upsilon \in \mathbf{R}$, we set

$$A(\upsilon) = \begin{bmatrix} \cos\upsilon & \sin\upsilon \\ -\sin\upsilon & \cos\upsilon \end{bmatrix} \in SO(2) \ .$$

Then, by a standard result of linear algebra, T is a *maximal torus* in $SO(m+1)$ (i.e. each element in $SO(m+1)$ is conjugate to some element in T). Its Lie algebra $\mathcal{T} \subset so(m+1)$ is spanned by the linearly independent system $\{h_j\}_{j=1}^{\ell} \subset so(m+1)$, where

$$h_j = E_{2j-1,2j} - E_{2j,2j-1} \ , \ j=1,\ldots,\ell \ , \tag{7.4}$$

and $E_{jk} \in M(m+1,m+1)$, $j,k=1,\ldots,m+1$, is the matrix with 1 in the (j,k)-position and zeros elsewhere. We order the elements of the (real) dual $\mathcal{T}^*$ lexicographically relative to the dual base $\{\phi_j\}_{j=1}^{\ell} \subset \mathcal{T}^*$ of $\{h_j\}_{j=1}^{\ell}$. If we set $e_j = 2\pi h_j$, $j=1,\ldots,\ell$, the unit lattice $\Gamma_{SO(m+1)}=\{X\in T \mid \exp X=1\}$ is spanned by $\{e_j\}_{j=1}^{\ell}$.

Let V be a (finite dimensional) *complex* $SO(m+1)$-module (i.e. a complex vector space with a fixed complex representation of $SO(m+1)$ on V). On the Lie algebra level, the $SO(m+1)$-module structure of V induces a representation of the Lie algebra $so(m+1)$ on V. A basic notion occurring here is that of *weight vector*, a simultaneous (nonzero) eigenvector in V of all operators contained in the commutative subalgebra T. The corresponding eigenvalue, as a complex scalar on T, is easily seen to have the form $i\phi$, where $\phi\in T^*$ (and $i\in \mathbf{C}$ is the complex unit). The linear functional ϕ on T is said to be a *weight* for V. Equivalently, $(0\neq)\, v\in V$ is a weight vector with weight $\phi\in T^*$ if

$$X\cdot v = i\phi(X)v \tag{7.5}$$

holds for all $X\in T$.

Given a weight $\phi\in T^*$, we denote by V_ϕ the set of all weight vectors corresponding to ϕ; it is, in fact, a linear subspace of V and the action of T on V_ϕ is given by scalar multiplication with $i\phi$. Since T is commutative, it follows from elementary representation theory that V is spanned by weight vectors and that the set $\Phi\subset T^*$ of all weights is finite. For every weight $\phi\in\Phi$, we have $\phi(\Gamma_{SO(m+1)})\subset 2\pi\mathbf{Z}$ and hence ϕ decomposes uniquely as

$$\phi = \sum_{j=1}^{\ell} \rho_j\phi_j , \tag{7.6}$$

where ρ_j is integral, $j=1,\ldots,\ell$. The weight ϕ will then be identified with the ℓ-tuple $\rho=(\rho_1,\ldots,\rho_\ell)\in \mathbf{Z}^{\ell}$.

Remark. Given a (complex) $SO(m+1)$-module V, the universal covering projection $\pi:\mathrm{Spin}(m+1)\to SO(m+1)$ induces a $\mathrm{Spin}(m+1)$-module structure on V. Since $\mathrm{Spin}(m+1)$ is simply connected, it is sometimes more convenient to work with $\mathrm{Spin}(m+1)$-modules (or equivalently, with representations of the Lie algebra $so(m+1)$) for which, in general, the coefficients ρ_j, $j=1,\ldots,\ell$,

in the decomposition above are half-integral as it can be easily seen by comparing the unit lattices $\Gamma_{\mathrm{Spin}(m+1)} \subset \Gamma_{SO(m+1)}$.
The Weyl group of $SO(m+1)$ (i.e. the group of all inner automorphisms of $SO(m+1)$ which leave T invariant) is known to be generated by the symmetric group of ℓ letters (permuting the blocks $A(\upsilon_j)$, $j=1,\dots,\ell$, in (7.2) or (7.3)) and, by the parametrization of T in (7.2) or (7.3), substitutions

$$(\upsilon_1,\dots,\upsilon_\ell) \to (\varepsilon_1\upsilon_1,\dots,\varepsilon_\ell\upsilon_\ell) ,$$

$$\varepsilon_j=\pm 1 , \quad j=1,\dots,\ell , \text{ if } m=2\ell$$

or

$$(\upsilon_1,\dots,\upsilon_\ell) \to (\varepsilon_1\upsilon_1,\dots,\varepsilon_\ell\upsilon_\ell) ,$$

$$\varepsilon_j=\pm 1 , \quad j=1,\dots,\ell , \quad \prod_{j=1}^{\ell} \varepsilon_j=1 , \quad \text{if } m=2\ell-1 .$$

On the other hand, each element of the Weyl group induces, on the Lie algebra level, a linear transformation of T leaving the set of weights Φ invariant. Thus, by the above, if $\phi=(\rho_1,\dots,\rho_\ell)$ is a weight for V then

$$(\rho_{\sigma(1)},\dots,\rho_{\sigma(\ell)}) \in T^* \quad (\sigma \text{ is a permutation of } \ell \text{ letters}),$$

$$(\varepsilon_1\rho_1,\dots,\varepsilon_\ell\rho_\ell) \in T^* , \quad \varepsilon_j=\pm 1 , \quad j=1,\dots,\ell ,$$

$$\text{if } m=2\ell$$

or

$$(\varepsilon_1\rho_1,\dots,\varepsilon_\ell\rho_\ell) \in T^* , \quad \varepsilon_j=\pm 1 , \quad j=1,\dots,\ell ,$$

$$\prod_{j=1}^{\ell} \varepsilon_j=1 , \quad \text{if } m=2\ell-1$$

are also weights for V. From this, it follows that among these weights there is exactly one characterized by the inequalities

$$\rho_1 \geq \rho_2 \geq \dots \geq \rho_{\ell-1} \geq \rho_\ell \geq 0 \quad \text{if } m=2\ell \tag{7.7}$$

or

$$\rho_1 \geq \rho_2 \geq \cdots \geq \rho_{\ell-1} \geq |\rho_\ell| \quad \text{if} \quad m=2\ell-1 \ . \tag{7.8}$$

The following classification theorem is due to É. Cartan.

Theorem 7.9. *For each linear functional* $\phi = \sum_{j=1}^{\ell} \rho_j \phi_j$ *with* $\rho=(\rho_1,\ldots,\rho_\ell) \in Z^\ell$ *satisfying* (7.7) *or* (7.8), *there is an irreducible complex* $SO(m+1)$*-module* V *such that* ϕ *is the highest weight for* V *(relative to the lexicographic order on* T^**). Two irreducible complex* $SO(m+1)$*-modules are equivalent if and only if their highest weights coincide.*

(For the proof, we refer to Boerner (1963) page 265.)
Let V_m^ρ denote the irreducible complex $SO(m+1)$-module with highest weight ρ . In particular, for $k \in Z_+$,

$$V_m^{(k,0,\ldots,0)} \cong \tilde{H}_m^k \otimes_{\mathbf{R}} \mathbf{C} \ . \tag{7.10}$$

The complex dimension of the $SO(m+1)$-module V_m^ρ can be computed in terms of m and the highest weight ρ via the Weyl dimension formula as follows (cf. Boerner (1963) page 266).

Theorem 7.11. *For* $m=2\ell$ *even,*

$$\dim_{\mathbf{C}} V_m^\rho = \frac{\prod_{r<s} (\rho_r-\rho_s-r+s)(\rho_r+\rho_s+2\ell+1-r-s)}{\prod_{r<s} (-r+s)(2\ell+1-r-s)} \cdot \frac{\prod_{t=1}^{\ell} (\rho_t+\ell+\frac{1}{2}-t)}{\prod_{t=1}^{\ell} (\ell+\frac{1}{2}-t)}$$

and, for $m=2\ell-1$ *odd,*

$$\dim_{\mathbf{C}} V_m^\rho = \frac{\prod_{r<s} (\rho_r-\rho_s-r+s)(\rho_r+\rho_s+2\ell-r-s)}{\prod_{r<s} (-r+s)(2\ell-r-s)} \ .$$

Remark. In fact, Theorem 7.11 can be easily deduced from the general Weyl dimension formula

$$\dim_{\mathbf{C}} V = \frac{\prod_{\alpha>0} (\alpha, \rho+\delta)}{\prod_{\alpha>0} (\alpha, \delta)} ,$$

where the products are taken over the positive roots (=weights of the adjoint representation) and $\delta=\frac{1}{2} \sum_{\alpha>0} \alpha$. Indeed, for $m=2\ell$, the positive roots are

$$e_r \pm e_s , \quad 1 \leq r < s \leq \ell , \quad e_t, t=1,\ldots,\ell ,$$

and

$$2\delta = (2\ell-1)e_1 + (2\ell-3)e_2 + \ldots + 3\ell_{e-1} + e_\ell .$$

Thus, we have

$$\dim_{\mathbf{C}} V_m^\rho = \frac{\prod_{\alpha>0} (\alpha,\rho+\delta)}{\prod_{\alpha>0} (\alpha,\delta)} = \frac{\prod_{r<s} (e_r-e_s,\rho+\delta)(e_r+e_s,\rho+\delta)}{\prod_{r<s} (e_r-e_s,\delta)(e_r+e_s,\delta)} \cdot$$

$$\cdot \frac{\prod_{t=1}^{\ell} (e_t,\rho+\delta)}{\prod_{t=1}^{\ell} (e_t,\delta)}$$

and the first dimension formula follows. For $m=2\ell-1$, the positive roots are

$$e_r - e_s , \quad 1 \leq r < s \leq \ell ,$$

$$e_r + e_\ell , \quad 1 \leq r < \ell ,$$

$$e_r + e_s , \quad 1 \leq r < s < \ell ,$$

with

$$2\delta = 2(\ell-1)e_1 + 2(\ell-2)e_2 + \ldots + 2e_{\ell-1} ,$$

and the second dimension formula follows easily.

The following Branching Theorem gives a precise description of how V_m^ρ , when considered as an $SO(m)$-module by restriction, splits into a direct sum of certain irreducible complex $SO(m)$-modules.

Theorem 7.12. *As* $SO(m)$*-modules,*

$$V_m^{\rho} \cong \bigoplus_{\sigma} V_{m-1}^{\sigma} ,$$

where the summation runs over all multi-indices $\sigma \in \mathbf{Z}^{[\frac{m}{2}]}$ *satisfying*

$$\rho_1 \geq \sigma_1 \geq \cdots \geq \rho_\ell \geq |\sigma_\ell| \quad \text{if} \quad m=2\ell$$

or

$$\rho_1 \geq \sigma_1 \geq \cdots \geq \sigma_{\ell-1} \geq |\rho_\ell| \quad \text{if} \quad m=2\ell-1 .$$

(For the proof, see Boerner (1963) pages 266-270.)

Turning to the proof of (7.1) we first state the following.

Theorem 7.13. *For* $m>3$ *and* $k>3$,

$$\bar{W}_m^k \otimes_{\mathbf{R}} \mathbf{C} = \bigoplus_{j=2}^{[\frac{k}{2}]} V_m^{(2k-2j,2j,0,\ldots,0)} \oplus \bar{W}_m^{k-1} \otimes_{\mathbf{R}} \mathbf{C} , \tag{7.14}$$

and, for $m=3$ *and* $k>3$,

$$\bar{W}_3^k \otimes_{\mathbf{R}} \mathbf{C} = \bigoplus_{j=2}^{[\frac{k}{2}]} (V_3^{(2k-2j,2j)} \oplus V_3^{(2k-2j,-2j)} \oplus \tag{7.15}$$

$$\oplus \bar{W}_3^{k-1} \otimes_{\mathbf{R}} \mathbf{C} .$$

Proof. By a result of Do Carmo and Wallach (1971) page 57-62, of representation theory we have the following decomposition of the symmetric square of the $SO(m+1)$-module of *complex* spherical harmonics $\tilde{H}_m^k \otimes_{\mathbf{R}} \mathbf{C}$ of order $k>0$:

$$S^2(\tilde{H}_m^k \otimes_{\mathbf{R}} \mathbf{C}) = \bigoplus_{j=0}^{[\frac{k}{2}]} V_m^{(2k-2j,2j,0,\ldots,0)} \oplus$$

$$\oplus S^2(\tilde{H}_m^{k-1} \otimes_{\mathbf{R}} \mathbf{C}) , \quad m>3 , \tag{7.16}$$

and

$$S^2(\tilde{H}_3^k \otimes_{\mathbf{R}} \mathbf{C}) = \bigoplus_{j=0}^{[\frac{k}{2}]} \{V_3^{(2k-2j,2j)} \oplus V_3^{(2k-2j,-2j)}\} \oplus \tag{7.17}$$

$$\oplus\ S^2(\tilde{H}_3^{k-1} \underset{\mathbf{R}}{\otimes} \mathbf{C}) \ .$$

(Note that the occurrence of the splitting in the bracket {...} is due to the fact that the (complexified) adjoint representation of $SO(4)$ is reducible.)

Setting $m>3$, we first prove (7.14). It follows directly from the Branching Theorem 7.12 that, for $j=0,\ldots,[\frac{k}{2}]$, the $SO(m)$-module $V_m^{(2k-2j,2j,0,\ldots,0)}$ contains $V_{m-1}^{(0,\ldots,0)}$ $(\cong \tilde{H}_{m-1}^{0} \underset{\mathbf{R}}{\otimes} \mathbf{C})$ or $V_{m-1}^{(2,0,\ldots,0)}$ $(\cong \tilde{H}_{m-1}^{2} \underset{\mathbf{R}}{\otimes} \mathbf{C})$, as $SO(m)$-submodules, only if $j=0,1$. Extracting from both sides of (7.16) those $SO(m+1)$-submodules which contain $SO(m)$-submodules isomorphic to $\tilde{H}_{m-1}^{0} \underset{\mathbf{R}}{\otimes} \mathbf{C}$ or $\tilde{H}_{m-1}^{2} \underset{\mathbf{R}}{\otimes} \mathbf{C}$, we arrive at the equality (7.14) directly from the definition of $\bar{W}$ in 6.18.

In case $m=3$, the same argument deduces (7.15) from (7.17).

Proof of Theorem 6.14. Setting $m>2$ and $k>3$, by 7.11, 7.12 and 7.13, we have

$$(\dim \bar{W}_m^k =) \dim_{\mathbf{C}}(\bar{W}_m^k \underset{\mathbf{R}}{\otimes} \mathbf{C}) \geq \dim_{\mathbf{C}}(\bar{W}_m^4 \underset{\mathbf{R}}{\otimes} \mathbf{C}) \geq$$

$$\geq \dim_{\mathbf{C}}(\bar{W}_3^4 \underset{\mathbf{R}}{\otimes} \mathbf{C}) = \dim_{\mathbf{C}}\{V_3^{(4,4)} \oplus V_3^{(4,-4)}\} =$$

$$= 9 + 9 = 18 \ . \ \surd$$

Remark 1. For $k>3$, let $\Delta \subset \mathbf{R}^2$ denote the (closed) convex triangle with vertices $(k,4)$, (k,k), $(2k-4,4) \in \mathbf{R}^2$. Then, by (7.14) and (7.15), we have, for $m>3$ and $k>3$,

$$\bar{W}_m^k \underset{\mathbf{R}}{\otimes} \mathbf{C} = \bigoplus_{\substack{(a,b)\in\Delta \\ a,b \text{ even}}} V_m^{(a,b,0,\ldots,0)}$$

and, for $m=3$ and $k>3$,

$$\bar{W}_3^k \underset{\mathbf{R}}{\otimes} \mathbf{C} = \bigoplus_{\substack{(a,b)\in\Delta \\ a,b \text{ even}}} \{V_3^{(a,b)} \oplus V_3^{(a,-b)}\} ,$$

respectively.

Remark 2. Using the Weyl dimension formula 7.11, $\dim_{\mathbf{C}}(\bar{W}_m^k \underset{\mathbf{R}}{\otimes} \mathbf{C})$ can be computed explicitly.

PROBLEMS FOR CHAPTER IV

1. Derive a first variation formula for the σ-energy E_σ (defined in Remark 3 after the proof of 1.4).
2. Show that the normal nullity of the canonical inclusion map $i:S^m \to S^n$, $m\leq n$, is $(m+1)(n-m)$. (Hint: Use the results of I.7.
3. Given a compact Lie group G acting on the Euclidean n-sphere S^n by isometries, show that there exists a minimal orbit in S^n of highest dimension. (Hint: Use an argument similar to that in the proof of Hsiang's theorem 3.3.)
4. Let G/K be a compact homogeneous manifold with invariant Riemannian metric g such that K acts, via the isotropy representation, transitively on the unit sphere of $T_o(G/K)$. Given a finite dimensional linear subspace $V\subset C^\infty(G/K)$ which is invariant under the induced action of G on $C^\infty(G/K)$ (cf.(3.11)) the elements of the linear subspace $Z(V)=\mathrm{Fix}(K^0,V)\subset V$ are said to be *zonal functions* of V . Show that $Z(V)$ is a non-trivial linear subspace (provided that $V\neq\{0\}$) and if $\dim Z(V)=1$ then the representation of G on V is irreducible.
5. Setting $S^m=SO(m+1)/SO(m)$ prove that, for each $\lambda_k=k(k+m-1)\in$ $\in\mathrm{Spec}(S^m)$, $k\in\mathbf{N}$, the induced representation of $SO(m+1)$ on the invariant linear subspace $V_{\lambda_k}=\tilde{H}_m^k\subset C^\infty(S^m)$ is irreducible. (Hint: The isotropy representation of $SO(m)$, being equivalent to the canonical orthogonal action of $SO(m)$ on $\mathbf{R}^m$, is transitive on the unit sphere of $T_o(S^m)$ and, by Problem 4, it remains to show that $\dim Z(\tilde{H}_m^k)=1$. Now, a zonal function $\mu\in\tilde{H}_m^k$ depends only on the distance function from $o\in S^m$, i.e., for some scalar $\alpha\in C^\infty(\mathbf{R})$, $\mu(x)=\alpha(\mathrm{distance}(o,x))$, $x\in S^m$. By expressing the equation $\Delta^{S^m}\mu=\lambda_k\mu$ in polar coordinates, the

uniqueness of μ follows; cf. also Berger, Gauduchon and Mazet (1971) pages 162-165.)

6. Let G be a compact Lie group endowed with a biinvariant Riemannian metric. Show that, for each closed subgroup $K \subset G$, the left cosets $aK \subset G$, $a \in G$, are totally geodesic submanifolds. (Hint: Use (3.22).)

7. Using Problem 5, construct the standard minimal immersions

$$f_{\lambda_k} : S^m_{\kappa(k)} \to S^{n(k)} , \quad k \in \mathbf{Z}_+ ,$$

via class 1 representations (cf. Theorem 3.17).

8. Show that any (real) odd-dimensional $\mathbf{T}^m$-module V contains a nonzero vector $v^0 \in V$ left fixed by each element of $\mathbf{T}^m$.

9. Prove that each irreducible submodule of the $SO(3)$-module $S^2(\tilde{H}^k_2)$, $k \in \mathbf{N}$, contains a nonzero vector v^0 left fixed by each element of the subgroup $SO(2)(\subset SO(3))$. (Hint: As each irreducible submodule of the $SO(3)$-module $S^2(\tilde{H}^k_2)$ is absolutely irreducible, by Problem 8, it is enough to show that every irreducible *complex* $SO(3)$-module is odd-dimensional (over $\mathbf{C}$). Now, the universal covering group $SU(2)(\cong S^3)$ of $SO(3)$ acts on $\mathbf{C}^2$ by matrix multiplication and this action induces a complex $SU(2)$-module structure on the vector space $P^k_1(\mathbf{C})$ of complex homogeneous polynomials $P: \mathbf{C}^2 \to \mathbf{C}$ of degree $k \in \mathbf{Z}_+$. Then, $\dim_{\mathbf{C}} P^k_1(\mathbf{C}) = k+1$, $k \in \mathbf{Z}_+$, and $\{P^k_1(\mathbf{C}) \mid k \in \mathbf{Z}_+\}$ is a complete system of complex irreducible $SU(2)$-modules. Finally, the $SU(2)$-module structure on $P^k_1(\mathbf{C})$ factors through the canonical projection $SU(2) \to SO(3)$ if and only if k is even; cf. also Hsiang (1975) pages 36-37.)

CHAPTER V

Infinitesimal Deformations of Harmonic Maps into Spheres

In this chapter we return to the study of harmonic maps into spheres from the point of view of deformation theory. Based on Smith's result on infinitesimal deformations of conformal diffeomorphisms of 2-manifolds, presented in Section 1, we define the notion of infinitesimal rigidity of harmonic maps which, being essentially of infinitesimal nature, could, as we shall see later, be considered as a proper generalization of the linear rigidity of minimal isometric immersions into spheres. In Section 2, we also give a variety of examples illustrating this concept. The rest of the chapter is devoted to the study of infinitesimally rigid (equivariant) harmonic maps of a (mostly naturally reductive) homogeneous manifold into spheres. As an application, in Section 4, using Do Carmo-Wallach theory developed in Chapter IV, we give a survey on infinitesimally rigid and flexible (=nonrigid) standard minimal immersions between spheres.

1. CONFORMAL DIFFEOMORPHISMS OF 2-MANIFOLDS

Let $f:M \to N$ be a map of an m-dimensional Riemannian manifold M with metric tensor $g=(,) \in C^\infty(S^2(T^*(M)))$ into an n-dimensional Riemannian manifold N with metric tensor $h=(,) \in$ $\in C^\infty(S^2(T^*(N)))$. We say that f is *conformal* if $f^*(h)=\mu^2 \cdot g$ holds for some scalar $\mu \in C^\infty(M)$. In case μ=const., f is said to be *homothetic*.

Given a vector field v along a map $f:M \to N$, the scalar product

$$\alpha_v = (f_*, dv) \tag{1.1}$$

defines a bilinear form on each of the tangent spaces $T_x(M)$, $x \in M$, or equivalently, $\alpha_v \in T_2^0(M)$. Its transpose ${}^t\alpha_v \in T_2^0(M)$ is given by

$${}^t\alpha_v(Y,Z) = \alpha_v(Z,Y) \ , \quad Y,Z \in V(M) \ .$$

The scalar $\mathrm{div}_f v \in C^\infty(M)$ defined by

$$\mathrm{div}_f v = \mathrm{trace}(f^*, dv) \; (=\mathrm{trace}\ \alpha_v - \mathrm{trace}\ {}^t\alpha_v) \tag{1.2}$$

is said to be the *generalized divergence* of v along f. Finally, $v \in C^\infty(\tau^f)$ is called *infinitesimally conformal* if

$$\frac{2}{n}\,\mathrm{div}_f v \cdot g = \alpha_v + {}^t\alpha_v \ . \tag{1.3}$$

Example 1.4. Setting $M=N$ and $f=id_M$, a vector field X along id_M is nothing but a vector field on M in the ordinary sense and, by definition, the generalized divergence $\mathrm{div}_{id_M} X$ reduces to the classical divergence $\mathrm{div}\ X$ of $X \in V(M)$ (cf. Kobayashi and Nomizu (1963) page 282).
Furthermore, for $Y,Z \in V(M)$, we have

$$\alpha_X(Y,Z) + {}^t\alpha_X(Y,Z) = (Y, \nabla_Z X) + (Z, \nabla_Y X) =$$

$$= (Y,[Z,X]) + (Z,\nabla_X Z) + (Z,[Y,X]) + (Z,\nabla_X Y) =$$

$$= X(Y,Z) - (L_X Y, Z) - (Y, L_X Z) = (L_X g)(Y,Z)$$

and we obtain that $X \in V(M)$ is infinitesimally conformal along id_M if and only if it satisfies the classical condition of infinitesimal conformality

$$\frac{2}{m}\,\mathrm{div}\ X \cdot g = L_X g \tag{1.5}$$

(cf. Kobayashi and Nomizu (1963) page 309).

Proposition 1.6. *Every conformal diffeomorphism* $f: M \to N$ *between 2-dimensional compact oriented Riemannian manifolds is harmonic.*

Proof. Denoting by g and h the respective Riemannian metrics on M and N, we have

$$f^*(h) = \mu^2 \cdot g$$

for some *positive* scalar $\mu \in C^\infty(M)$. Hence $\bar{g}=\mu^2 \cdot g$ can also be considered as a Riemannian metric on M. Endowing M with the Riemannian metric $\bar{g}$, the map f becomes isometric, in particular, harmonic. Now, due to the assumption $\dim M=2$, a change from the Riemannian metric g to the (conformally equivalent) metric $\bar{g}=\mu^2 \cdot g$ on the domain does not have any effect on the energy functional E and hence the first variation formula II: 1.5 implies the result. √

Remark. Conformal diffeomorphisms between 2-dimensional compact oriented Riemannian manifolds M and N are also holomorphic or anti-holomorphic with respect to the Kähler structures on M and N (cf. Eells and Lemaire (1978)).
We can now state R.T. Smith's theorem as follows.

Theorem 1.7. *Every Jacobi field v along a conformal diffeomorphism $f:M \to N$ between 2-dimensional compact oriented Riemannian manifolds is infinitesimally conformal.*

The proof is preceded by two lemmas.

Lemma 1.8. *Given a conformal diffeomorphism $f:M \to N$ between Riemannian manifolds M and N with metric tensors $g=(,)$ and $h(,)$, respectively, an element $v \in C^\infty(\tau^f)$ is infinitesimally conformal along f if and only if $v=X \circ f$, where $X \in$ $\in V(N)$ is a conformal vector field on N.*

Proof. Set $X=v \circ f^{-1} \in V(N)$. By conformality of f there exists a *positive* scalar μ on M with $f^*(h)=\mu^2 \cdot g$. With respect to a local orthonormal moving frame $\{E^i\}_{i=1}^m$, $m=\dim M$, on M, we have

$$\operatorname{div}_f v = \sum_{i=1}^{m} (f_*(E^i), \nabla_{E^i}(X \circ f)) =$$

$$= \sum_{i=1}^{m} (f_*(E^i), \nabla_{f_*(E^i)} X) =$$

$$= \mu^2 \operatorname{trace}((id_N)_*, \nabla X) \circ f = \mu^2 \cdot (\operatorname{div} X) \circ f$$

and so, for $Y, Z \in V(M)$, we obtain

$$\frac{2}{m}(f_*(Y), f_*(Z))(\operatorname{div} X) \circ f =$$

$$= \frac{2}{m} \mu^2 (Y,Z) \cdot (\operatorname{div} X) \circ f = \frac{2}{m}(Y,Z) \operatorname{div}_f v .$$

Moreover,

$$\alpha_v(Y,Z) = \alpha_X(f_*(Y), f_*(Z)) \circ f \quad \text{and}$$

$$^t\alpha_v(Y,Z) = {}^t\alpha_X(f_*(Y), f_*(Z)) \circ f$$

which, by (1.2) and (1.3), completes the proof. √

Remark. By Theorem 1.7, v arises from variation of f through conformal (and hence (anti-) holomorphic) diffeomorphisms.

Now we specialize to the case $\dim M = \dim N = 2$ and prove a formula which is a generalization of one of K. Yano.

Lemma 1.9. *Let $f: M \to N$ be a conformal diffeomorphism of 2-dimensional compact oriented Riemannian manifolds with $f^*(h) = \mu^2 g$, $\mu \in C^\infty(M)$. Then, for any vector field v along f, we have*

$$\int_M \{\frac{1}{\mu^2}(2(\alpha_v, {}^t\alpha_v) - (\operatorname{div}_f v)^2) - \operatorname{trace}(R(f_*, v) f_*, v)\} \cdot \operatorname{vol}(M,g) = 0 , \tag{1.10}$$

where R is the Riemannian curvature tensor of N. Moreover, with respect to an orthonormal base $\{e^1, e^2\} \subset T_x(M)$ at $x \in M$, we have

$$(\alpha_v, {}^t\alpha_v)(x) = \frac{1}{2} \sum_{i,j=1}^{2} \alpha_v(e^i, e^j) \cdot {}^t\alpha_v(e^i, e^j) . \tag{1.11}$$

Proof. Given a vector field v along f, define $X = v \circ f^{-1} \in V(N)$

and set $T=(\nabla_X X)\circ f$. Then, using a local orthonormal moving frame $\{F^1,F^2\}$ on N , we have

$$\mathrm{div}_f T = \mu^2(\mathrm{div}\,\nabla_X X)\circ f = \mu^2 \sum_{j=1}^{2} (F^j,\nabla_{F^j}\nabla_X X)\circ f =$$

$$= \mu^2 \sum_{j=1}^{2} (F^j,\nabla_X\nabla_{F^j}X)\circ f + \mu^2 \sum_{j=1}^{2} (F^j,R(F^j,X)X)\circ f +$$

$$+ \mu^2 \sum_{j=1}^{2} (F^j,\nabla_{[F^j,X]}X)\circ f = \mu^2 \sum_{j=1}^{2} (F^j,(\nabla^2 X)(X,F^j))\circ f -$$

$$- \mu^2 \sum_{j=1}^{2} (X,R(F^j,X)F^j)\circ f + \mu^2 \sum_{j=1}^{2} (F^j,\nabla_{\nabla_{F^j}X}X)\circ f =$$

$$= \mu^2 \mathrm{trace}(\,(id_N)_*,(\nabla^2 X)(X,.)\,)\circ f -$$

$$- \mathrm{trace}(R(f_*,v)f_*,v) + \mu^2 \mathrm{trace}(\,(id_N)_*,\nabla_{\nabla X}X)\circ f \,.$$

To compute the last term, let $(v^i_j)^2_{i,j=1} \in M(2,2)$ be the matrix of the linear map $\nabla v: T_x(M) \to T_{f(x)}(N)$ with respect to fixed orthonormal bases $\{e^1,e^2\} \subset T_x(M)$ and $\{\frac{1}{\mu(x)} f_*(e^1),$ $\frac{1}{\mu(x)} f_*(e^2)\} \subset T_{f(x)}(N)$ at $x \in M$ and $f(x) \in N$, respectively, i.e. set

$$\nabla_{e^i} v = \frac{1}{\mu(x)} \sum_{j=1}^{2} v^i_j f_*(e^j) \;, \; i=1,2 \,.$$

Then,

$$\mu^2(x)\,\mathrm{trace}(\,(id_N)_*,\nabla_{\nabla X}X)_{f(x)} =$$

$$= \sum_{i=1}^{2} (f_*(e^i),\nabla_{\nabla_{f_*(e^i)}X}X) =$$

$$= \sum_{i=1}^{2} (f_*(e^i),\nabla_{\nabla_{e^i}v}X) = \frac{1}{\mu(x)} \sum_{i,j=1}^{2} v^i_j (f_*(e^i),\nabla_{e^j}v) =$$

$$= \sum_{i,j=1}^{2} v^i_j v^j_i = \frac{1}{\mu^2(x)} \sum_{i,j=1}^{2} (f_*(e^i),\nabla_{e^j}v)(f_*(e^j),\nabla_{e^i}v) =$$

$$= \frac{2}{\mu^2(x)} (\alpha_v, {}^t\alpha_v)(x)$$

and thus we obtain

$$\operatorname{div}_f T = \mu^2 \operatorname{trace}(\ (id_N)_*, (\nabla^2 X)(X,.)\ )\circ f + \\ + \frac{2}{\mu^2}(\alpha_v, {}^t\alpha_v) - \operatorname{trace}(R(f_*,v)f_*,v)\ . \tag{1.12}$$

To simplify the first term on the right hand side introduce the vector field

$$S = (\operatorname{div} X)\circ f \cdot v$$

along f and compute its divergence. Using a local orthonormal moving frame $\{E^1,E^2\}$ on M with $e^i = E^i_x$, $i=1,2,$ we have

$$\operatorname{div}_f S = \sum_{i=1}^{2} (f_*(E^i),\ \nabla_{E^i}((\operatorname{div} X)\circ f\cdot v)\ ) =$$

$$= (\operatorname{div} X)\circ f\cdot \sum_{i=1}^{2} (f_*(E^i), \nabla_{E^i} v) +$$

$$+ \sum_{i=1}^{2} (f_*(E^i), E^i((\operatorname{div} X)\circ f)\cdot v) =$$

$$= \frac{(\operatorname{div}_f v)^2}{\mu^2} + \operatorname{trace}(f_*, d((\operatorname{div} X)\circ f)\otimes v)\ .$$

To evaluate the last term at an arbitrary point $x\in M$ choose a local orthonormal moving frame $\{F^1,F^2\}$ (adapted to an orthonormal base $\{F^1_y,F^2_y\}\subset T_y(N)$) over a normal coordinate neighbourhood centered at $y=f(x)\in N$. Then, $\nabla_{f_*(e^i)} F^j = 0$, $i,j=1,2,$ and so

$$e^i((\operatorname{div} X)\circ f) = e^i(\sum_{j=1}^{2} (F^j\circ f, (\nabla_{F^j} X)\circ f)\ ) =$$

$$= \sum_{j=1}^{2} (F^j_y, \nabla_{f_*(e^i)} \nabla_{F^j} X) = \sum_{j=1}^{2} (F^j_y, (\nabla^2 X)(f_*(e^i), F^j_y)\ ) =$$

$$= \operatorname{trace}(\ (id_N)_*, (\nabla^2 X)(f_*(e^i), .)\ )_y$$

and the last term becomes

$$\operatorname{trace}(f_*, d((\operatorname{div} X) \circ f) \otimes v) = \sum_{i=1}^{2} (f_*(E^i), E^i((\operatorname{div} X) \circ f) v) =$$

$$= \sum_{i,j=1}^{2} (f_*(E^i), (F^j \circ f, (\nabla^2 X)(f_*(E^i), F^j \circ f)) v) =$$

$$= \mu^2 \sum_{i,j=1}^{2} (F^i, (F^j, (\nabla^2 X)(F^i, F^j)) X) \circ f =$$

$$= \mu^2 \sum_{j=1}^{2} (F^j, (\nabla^2 X)(\sum_{i=1}^{2} (F^i, X) F^i, F^j)) \circ f =$$

$$= \mu^2 \sum_{j=1}^{2} (F^j, (\nabla^2 X)(X, F^j)) \circ f =$$

$$= \mu^2 \operatorname{trace}(\ (id_N)_*, (\nabla^2 X)(X, .)\ ) \circ f\ .$$

Comparing this with (1.12) we conclude that

$$\operatorname{div}_f T - \operatorname{div}_f S = \frac{1}{\mu^2}(2(\alpha_v, {}^t\alpha_v) - (\operatorname{div}_f v)^2) - \qquad (1.13)$$
$$- \operatorname{trace}(R(f_*, v) f_*, v)\ .$$

Now, by 1.6, f is harmonic and, using Green's theorem, we have

$$\int_M (\operatorname{div}_f T) \operatorname{vol}(M,g) = \int_M \operatorname{trace}(f_*, \nabla T) \operatorname{vol}(M,g) =$$

$$= \int_M \operatorname{trace} \nabla (f_*, T) \operatorname{vol}(M,g) =$$

$$= \int_M \operatorname{div} \gamma^{-1}(f_*, T) \operatorname{vol}(M,g) = 0\ .$$

Similarly, $\int_M (\mathrm{div}_f S)\,\mathrm{vol}(M,g)=0$ and (1.10) follows. √

Proof of Theorem 1.7. We have to show that for a vector field v along $f:M\to N$ the condition

$$\mathrm{trace}\ \nabla^2 v = \mathrm{trace}\ R(f_*,v)f_*$$

implies that v is infinitesimally conformal along f, i.e.

$$\mathrm{div}_f v\cdot g = \alpha_v + {}^t\alpha_v$$

holds, where $g=(,)$ is the Riemannian metric on M. For notational convenience, introduce a scalar product $(,)$ on $T_2^0(M)$ by setting

$$(\sigma,\tau)(x) = \frac{1}{2}\sum_{i,j=1}^{2}\sigma(e^i,e^j)\tau(e^i,e^j),\quad \sigma,\tau\in T_2^0(M),$$

where $\{e^1,e^2\}\subset T_x(M)$, $x\in M$, is an orthonormal base. Then,

$$\|\alpha_v+{}^t\alpha_v-\mathrm{div}_f v\cdot g\|^2 = \|\alpha_v+{}^t\alpha_v\|^2 -$$

$$-\ \mathrm{div}_f v\cdot\mathrm{trace}\{\alpha_v+{}^t\alpha_v\} + (\mathrm{div}_f v)^2,$$

where

$$\|\alpha_v+{}^t\alpha_v\|^2 = 2(\|\alpha_v\|^2 + (\alpha_v,{}^t\alpha_v))$$

and

$$\mathrm{trace}\{\alpha_v+{}^t\alpha_v\} = 2\mathrm{trace}\ \alpha_v = 2\ \mathrm{div}_f v.$$

Furthermore, with respect to a local orthonormal moving frame $\{E^1,E^2\}$ on M, we have

$$\|\alpha_v\|^2 = \frac{1}{2}\sum_{i,j=1}^{2}(f_*(E^i),\nabla_{E^j}v)^2 =$$

$$= \frac{\mu^2}{2}\sum_{i=1}^{2}\|\nabla_{E^i}v\|^2 = \frac{\mu^2}{2}\,\mathrm{trace}\|\nabla v\|^2 .$$

Putting these together, we obtain

$$\| \alpha_v + {}^t\alpha_v - \mathrm{div}_f v \cdot g \|^2 = \mu^2 \mathrm{trace} \| \nabla v \|^2 +$$

$$+ 2(\alpha_v, {}^t\alpha_v) - (\mathrm{div}_f v)^2 .$$

Integrating and using (1.10), we have

$$\int_M \frac{1}{\mu^2} \| \alpha_v - {}^t\alpha_v - \mathrm{div}_f v \cdot g \|^2 \mathrm{vol}(M,g) =$$

$$= \int_M \{ \mathrm{trace} \| \nabla v \|^2 + \frac{1}{\mu^2} (2(\alpha_v, {}^t\alpha_v) -$$

$$- (\mathrm{div}_f v)^2) \} \mathrm{vol}(M,g) = \int_M \{ \mathrm{trace} \| \nabla v \|^2 +$$

$$+ \mathrm{trace}(R(f_*, v) f_*, v) \} \mathrm{vol}(M,g) =$$

$$= \int_M \{ (-\mathrm{trace}\ \nabla^2 v, v) + \mathrm{trace}(R(f_*, v) f_*, v) \} \mathrm{vol}(M,g) . \checkmark$$

2. INFINITESIMAL RIGIDITY

Let $f: M \to S^n$ be a harmonic map of an m-dimensional Riemannian manifold M with metric tensor $(,)$ into the Euclidean n-sphere S^n. According to III.1.19, if $t \to f_t$, $t \in \mathbf{R}$, is a variation of f *through harmonic maps* then $v = \frac{\partial f_t}{\partial t}\Big|_{t=0} \in C^\infty(\tau^f)$ is a Jacobi field along f. Specific examples of such variations are given by $f_t = \phi_t \circ f$, $t \in \mathbf{R}$, where $(\phi_t)_{t \in \mathbf{R}} \subset SO(n+1)$ is a 1-parameter subgroup. A Jacobi field v along f arising in this way enjoys two properties. In the first place the vector field v along f is *projectable* (i.e., for $x, x' \in M$, $f(x) = f(x')$ implies $v_x = v_{x'}$) since $v = \frac{\partial f_t}{\partial t}\Big|_{t=0} = \dot{\phi}_t \circ f = X \circ f$, where $X \in so(n+1)$ is the Killing vector field on S^n induced by $(\phi_t)_{t \in \mathbf{R}} \subset SO(n+1)$. Secondly, the generalized divergence $\mathrm{div}_f v = \mathrm{trace}(f_*, \nabla v) \in C^\infty(M)$ must vanish. To show this, for fixed $x \in M$, choose an orthonormal base $\{e^i\}_{i=1}^m \subset T_x(M)$. Then, by making use of I.8.1 (iii), we have

$$(\mathrm{div}_f v)(x) = \sum_{i=1}^m (f_*(e^i), \nabla_{e^i} v) =$$

$$= \sum_{i=1}^m (f_*(e^i), \nabla_{f_*(e^i)} X) =$$

$$= - \sum_{i=1}^m (f_*(e^i), A_X(f_*(e^i))) = 0 .$$

One of the principal problems to be studied in this chapter is the following:

Does every projectable divergence-free Jacobi field along a harmonic map $f: M \to S^n$ *come from isometric variations (at least infinitesimally)?*

Given a harmonic map $f: M \to S^n$, let $K(f)$ be the vector space of all divergence-free Jacobi fields along f , i.e. set

$$K(f) = \{v \in C^\infty(\tau^f) \mid J_f(v) = 0 \quad \text{and} \quad \mathrm{div}_f v = 0\}$$

and denote by $PK(f) \subset K(f)$ the linear subspace of projectable elements of $K(f)$. (Note that, by III.1.17, $K(f)$ is finite dimensional if M is compact.) The argument above gives the relation

$$so(n+1) \circ f \subset PK(f) \; . \tag{2.1}$$

The harmonic map $f: M \to S^n$ is said to be *infinitesimally rigid* if $so(n+1) \circ f = PK(f)$.

The result of R.T. Smith in 1.7, then implies the following:

Theorem 2.2. *Every conformal diffeomorphism* $f: S^2 \to S^2$ *is infinitesimally rigid.*

Proof. If $v \in K(f) (= PK(f))$ then, by 1.7 and 1.8, we have

$$v = X \circ f \; ,$$

where $X \in V(S^2)$ is an inifinitesimally conformal field on S^2 .

Furthermore, as $\operatorname{div}_f v=0$, the divergence of X vanishes. By virtue of (1.5) this means that X is a Killing vector field on S^2. ✓

Returning to the general situation, our next result gives a simple criterion for infinitesimal rigidity.

Proposition 2.3. *For any harmonic map* $f:M \to S^n$,

$$\dim PK(f) \geq \left(n - \frac{r}{2}\right)(r+1) , \tag{2.4}$$

where $r+1$ *is the dimension of the linear subspace* $\operatorname{span}_{\mathbf{R}}(\operatorname{im} \bar{f}) \subset \mathbf{R}^{n+1}$. *Furthermore, equality holds if and only if* f *is infinitesimally rigid.*

Proof. By (2.1), precomposition with f defines a linear map

$$\Phi : so(n+1) \to PK(f)$$

which is surjective if and only if f is infinitesimally rigid. Thus,

$$\dim PK(f) \geq \dim so(n+1) - \dim \ker \Phi = \frac{n(n+1)}{2} - \dim \ker \Phi$$

with equality if and only if f is infinitesimally rigid. It remains to compute $\dim \ker \Phi$. Now, an element $X \in \ker \Phi$ is represented by a skew-symmetric matrix acting on $\mathbf{R}^{n+1}$ such that $X \circ \bar{f}=0$, i.e., by linearity, $\operatorname{span}_{\mathbf{R}}(\operatorname{im} \bar{f}) \subset \ker X$. Thus, $\ker \Phi$ can be identified with the vector space of skew-symmetric matrices acting on the orthogonal complement of $\operatorname{span}_{\mathbf{R}}(\operatorname{im} \bar{f}) \subset \mathbf{R}^{n+1}$. It follows that $\dim \ker \Phi = \dim so(n-r) = \frac{(n-r)(n-r-1)}{2}$. ✓

For later purposes, it will be convenient to reformulate the condition $v \in K(f)$ in terms of the induced vector function $\check{v} : M \to \mathbf{R}^{n+1}$ via the canonical identification $\check{} : T(\mathbf{R}^{n+1}) \to \mathbf{R}^{n+1}$.

Proposition 2.5. *Let* v *be a vector field along a harmonic map* $f:M \to S^n$. *Then* $v \in K(f)$ *if and only if*

$$\Delta \check{v} = 2e(f)\check{v} , \tag{2.6}$$

where the Laplacian Δ *acts on* $\check{v}$ *component-wise.*

Proof. For each $X \in V(M)$, by II.(2.22), we have

$$(\nabla_X v)\check{} = X(\check{v}) - (X(\check{v}),\bar{f})\bar{f} \qquad (2.7)$$

and hence a straightforward computation yields

$$(\nabla_Y \nabla_X v)\check{} = Y(X(\check{v})) - (Y(X(\check{v})),\bar{f})\bar{f} - (X(\check{v}),\bar{f})Y(\bar{f}) \ ,$$

$X, Y \in V(M)$. Taking traces, we obtain

$$(\text{trace } \nabla^2 v)\check{} = -\Delta\check{v} + (\Delta\check{v},\bar{f})\bar{f} - \text{trace}(d\check{v},\bar{f})\,d\bar{f} \ .$$

Furthermore, as S^n is a space of constant curvature, by Problem 4 for Chapter I, we have

$$(\text{trace } R(f_*,v)f_*)\check{} = (\text{trace}\{(f_*,v)f_*\})\check{} - 2e(f)\check{v} =$$

$$= \text{trace}(d\bar{f},\check{v})\,d\bar{f} - 2e(f)\check{v} = -\text{trace}(\bar{f},d\check{v})\,d\bar{f} - 2e(f)\check{v} \ ,$$

where, in the last equality, we used $(\bar{f},\check{v})=0$. Summarizing, we find that v is Jacobi field along f if and only if

$$\Delta\check{v} - (\Delta\check{v},\bar{f})\bar{f} = 2e(f)\check{v} \ . \qquad (2.8)$$

Furthermore, by (2.7), we have

$$\text{div}_f v = \text{trace}(f_*,\nabla v) = \text{trace}(d\bar{f},(\nabla v)\check{}) =$$

$$= \text{trace}(d\bar{f},d\check{v}) - \text{trace}(d\bar{f},\bar{f})(d\check{v},\bar{f}) \ .$$

The last term vanishes since $\|\bar{f}\|^2=1$. Thus, v is divergence-free if and only if

$$\text{trace}(d\bar{f},d\check{v}) = 0 \ . \qquad (2.9)$$

The harmonicity condition III.(2.2), with $(\bar{f},\check{v})=0$, implies

$$(\Delta\check{v},\bar{f}) = -(\text{trace } \nabla^2\check{v},\bar{f}) = -\text{trace}\{\nabla(d\check{v},\bar{f})\} +$$

$$+ \text{trace}(d\check{v},d\bar{f}) = \text{trace}\{\nabla(\check{v},d\bar{f})\} + \text{trace}(d\check{v},d\bar{f}) =$$

$$= -(\check{v},\Delta\bar{f}) + 2\text{trace}(d\check{v},d\bar{f}) = 2\text{trace}(d\check{v},d\bar{f})$$

since $(\check{v},\Delta\bar{f})=2e(f)(\check{v},\bar{f})=0$. Now, assuming $v\in K(f)$, we obtain $(\Delta\check{v},\bar{f})=0$ and hence equation (2.8) reduces to (2.6). Conversely, multiplying (2.6) with $\bar{f}$, we get $(\Delta\check{v},\bar{f})=0$ and so (2.8) and (2.9) follow, i.e. $v\in K(f)$. √

Proposition 2.5 says that, for a harmonic map $f:M\to S^n$, $K(f)$ is isomorphic with the vector space of all solutions $\check{v}:M\to\mathbf{R}^{n+1}$ of (2.6) satisfying the linear constraint $(\bar{f},\check{v})=0$. As we shall see, this observation will prove to be useful in determining $K(f)$ in specific situations. For example, if the harmonic map $f:M\to S^n$ has constant energy density $e(f)=\frac{\lambda}{2}\in\mathbf{R}$ then, for $v\in K(f)$, each component of $\check{v}$ belongs to the eigenspace V_λ corresponding to the eigenvalue $\lambda\in\text{Spec}(M)$. The next example shows how 2.5 can be used to compute $\dim K(f)$ when the energy density is nonconstant (i.e., when knowledge of the spectrum does not give additional information).

Example 2.10. Recall from II.4.13 that a periodic solution $\alpha\in C^\infty(\mathbf{R})$ (with period $\omega>0$) of the pendulum equation II.(4.14) satisfying the initial data $\alpha(0)=0$ and $\dot{\alpha}(0)=a\ (0<|a|<1)$ gives rise to the harmonic Delaunay map

$$f:\Theta\to S^2\ ,\quad \Theta = \mathbf{R}^2/(\omega\mathbf{Z})\times(2\pi\mathbf{Z})\ ,$$

defined by

$$\bar{f}(\phi,\psi) = (-\cos\psi\cos\alpha(\phi),-\sin\psi\cos\alpha(\phi),\sin\alpha(\phi))\ ,$$

$$\phi,\psi\in\mathbf{R}\ .$$

We claim that f is infinitesimally rigid, i.e., by 2.3, $\dim PK(f)=3$. As $e(f)=\frac{1}{2}(\dot{\alpha}^2+\cos^2\alpha)$, by 2.5, we have to determine the vector space of maps $\check{v}=(v^1,v^2,v^3):\Theta\to\mathbf{R}^3$ satisfying

$$\Delta v^i = (\dot{\alpha}^2+\cos^2\alpha)v^i \ , \quad i=1,2,3, \tag{2.11}$$

and the linear constraint

$$(\bar{f},\check{v})(\phi,\psi) = -\cos\psi\cos\alpha(\phi)v^1(\phi,\psi) -$$

$$-\sin\psi\cos\alpha(\phi)v^2(\phi,\psi) + \sin\alpha(\phi)v^3(\phi,\psi) = 0 \ , \tag{2.12}$$

$$\phi,\psi\in\mathbf{R} \ .$$

For fixed $i=1,2,3,$ the scalar $v^i\in C^\infty(\Theta)$, being considered as a doubly periodic function on $\mathbf{R}^2$, has a Fourier series expansion

$$v^i(\phi,\psi) = p_0(\phi) + \sum_{k=1}^{\infty}\{p_k(\phi)\cos(k\psi)+q_k(\phi)\sin(k\psi)\} \ ,$$

$$\phi,\psi\in\mathbf{R} \ ,$$

which splits (2.11) into the system

$$\ddot{y}_k + (\dot{\alpha}^2+\cos^2\alpha-k^2)y_k = 0 \ , \quad k\in\mathbf{Z}_+ \ , \tag{2.13$_k$}$$

and p_k, q_k are solutions of $(2.13)_k$. On the other hand, $\dot{\alpha}^2+\sin^2\alpha$, being a first integral of II.(4.14), is constant $(=a^2)$ and we have $\dot{\alpha}^2+\cos^2\alpha=a^2-1+2\cos^2\alpha<2$ which implies that $\dot{\alpha}^2+\cos^2\alpha-k^2<0$ for $k\geq 2$.

For periodic solutions y_k of $(2.13)_k$ with period ω, such as p_k and q_k, equation $(2.13)_k$ can be written as

$$\Delta y_k + (\dot{\alpha}^2+\cos^2\alpha-k^2)y_k = 0 \ , \tag{2.14$_k$}$$

where $y_k\in C^\infty(\mathbf{R}/\omega\mathbf{Z})$. As the Laplacian Δ on $\mathbf{R}/\omega\mathbf{Z}$ is positive semidefinite (cf.I.6), we have

$$((\Delta y_k, y_k)) = ((\ (\dot{\alpha}^2+\cos^2\alpha-k^2)y_k, y_k)) \geq 0$$

and hence, for $k\geq 2$, we obtain $y_k=0$. Thus, $p_k=q_k=0$, $k\geq 2$. For $k=1,$ the only ω-periodic solutions of $(2.13)_1$ are the constant multiples of $\cos\alpha$ since the nonperiodic solution

$$\phi \to \cos\alpha(\phi) \int_0^{\phi} \frac{dt}{\cos^2\alpha(t)}, \quad \phi \in \mathbf{R},$$

of $(2.13)_1$ is linearly independent from $\cos\alpha$. Thus,

$$p_1 = a^i \cos\alpha \quad \text{and} \quad q_1 = b^i \cos\alpha,$$

for some $a^i, b^i \in \mathbf{R}$. Finally, $\sin\alpha$ being an ω-periodic solution of $(2.13)_0$, we have

$$p_0 = c^i \sin\alpha + d^i y_0,$$

for some $c^i, d^i \in \mathbf{R}$, where y_0 is the solution of $(2.13)_0$ with initial data $y_0(0)=1$, $\dot{y}_0(0)=0$ and $d^i=0$ if y_0 is not ω-periodic. (Note that y_0 can be expressed in the singular integral form

$$\phi \to \sin\alpha(\phi) \int_0^{\omega/4} \frac{dt}{\sin^2\alpha(t)}, \quad \alpha \in \mathbf{R}.)$$

Summarizing, we obtain that

$$v^i(\phi,\psi) = (a^i \cos\psi + b^i \sin\psi)\cos\alpha(\phi) +$$

$$+ c^i \sin\alpha(\phi) + d^i y_0(\phi), \quad \phi,\psi \in \mathbf{R}, \quad i=1,2,3.$$

Now, it remains to satisfy the orthogonality condition (2.12). Comparing the initial data of the various solutions, a straightforward computation shows that (2.12) holds if and only if $a_1=b_2=c_3=d_1=d_2=d_3=0$, $a_2+b_1=a_3-c_1=b_3-c_2=0$. That is, setting $\rho=b_1$, $\sigma=c_1$ and $\tau=c_2$, we obtain

$$\check{v}(\phi,\psi) = (\rho \sin\psi \cos\alpha(\phi) + \sigma \sin\alpha(\phi),$$

$$-\rho \cos\psi \cos\alpha(\phi) + \tau \sin\alpha(\phi),$$

$$(\sigma\cos\psi + \tau\sin\psi)\cos\alpha(\phi)), \quad \phi,\psi \in \mathbf{R}^2,$$

in particular, $\dim K(f) = \dim PK(f) = 3$. Thus, the Delaunay map is infinitesimally rigid.

Example 2.15. For $t \in \mathbf{R}$, define

$$f_t : \mathbf{T}^2 \to S^3$$

by

$$\bar{f}_t(\phi,\psi) = (\cos t \cos \phi,\ \cos t \sin \phi,\ \sin t \cos \psi,$$

$$\sin t \sin \psi)\ ,\quad \phi,\psi \in \mathbf{R}^2\ .$$

Then, by III.(2.2), the map $f_t : \mathbf{T}^2 \to S^3$ is harmonic with energy density $e(f_t) = \frac{1}{2}$, $t \in \mathbf{R}$. (Note that, for $t = \frac{\pi}{4}$, we recover II. 2.37.) We claim that for $0 < t_o < \frac{\pi}{2}$, the harmonic embedding $f_t : \mathbf{T}^2 \to S^3$ is infinitesimally flexible (=nonrigid). Indeed, differentiating the equation

$$\Delta \bar{f}_t = \bar{f}_t$$

at $t = t_o$, by 2.5, we obtain that $v = \left.\frac{\partial f_t}{\partial t}\right|_{t=t_o} \in K(f_{t_o})\ (=PK(f_{t_o}))$. Assume now, on the contrary, the existence of a Killing vector field X on S^3 with $X \circ f_{t_o} = v$. As X is identified with the corresponding matrix in $so(4)$, this equation becomes

$$\begin{bmatrix} & & \\ & X & \\ & & \end{bmatrix} \begin{bmatrix} \cos t_o \cos \phi \\ \cos t_o \sin \phi \\ \sin t_o \cos \psi \\ \sin t_o \sin \psi \end{bmatrix} = \begin{bmatrix} -\sin t_o \cos \phi \\ -\sin t_o \sin \phi \\ \cos t_o \cos \psi \\ \cos t_o \sin \psi \end{bmatrix}, \quad \phi,\psi \in \mathbf{R}\ .$$

Linear independence of the trigonometric functions implies that $X = \mathrm{diag}(-\tan(t_o), -\tan(t_o), \cot(t_o), \cot(t_o))$ which contradicts the skew-symmetricity of X .

Smith's theorem 2.2 can be specified by saying that the identity $id_{S^2} : S^2 \to S^2$ is an infinitesimally rigid harmonic map, i.e.,

$$K(id_{S^2}) = so(3)\ . \tag{2.16}$$

The next result generalizes this observation to higher dimensions.

Theorem 2.17. *For* $n \in \mathbf{N}$,

$$K(id_{S^n}) = so(n+1) \ , \tag{2.18}$$

or equivalently, $id_{S^n} : S^n \to S^n$ *is infinitesimally rigid.*

Proof. Setting $X \in K(id_{S^n}) \subset V(S^n)$, we first reformulate the Jacobi condition $J_{id_{S^n}}(X)=0$. By I.(4.30),

$$\text{trace}\ R((id_{S^n})_*, X)(id_{S^n})_* = -\text{Ric}(X)\,(=(n+1)X)$$

and hence the Jacobi condition for X takes the form

$$\text{trace}\ \nabla^2 X + \text{Ric}(X) = 0 \ .$$

By duality,

$$\text{trace}\ \nabla^2 \alpha + \text{Ric}(\alpha) = 0 \ , \tag{2.19}$$

where $\alpha=\gamma(X) \in D^1(S^n)$ stands for the 1-form associated to X by the canonical isomorphism $\gamma : T(S^n) \to T^*(S^n)$. Taking into account the Bochner-Weitzenböck formula I.(6.3), this becomes

$$\Delta\alpha - 2\text{Ric}(\alpha) = 0 \ . \tag{2.20}$$

Secondly, we have

$$\text{div}\ X = \text{trace}((id_{S^n})_*, \nabla X) = \text{trace}\{\nabla\alpha\} = -\partial\alpha = 0$$

which, together with (2.20), implies that α satisfies the equation

$$\Delta\alpha - 2\text{Ric}(\alpha) + d(\partial\alpha) = 0 \ .$$

Now, by the Lichnerowicz identity I.(8.4), X is a Killing

vector field on S^n . √

Remark. In fact, as shown by R.T. Smith (1972a) and (1975b), for $n \neq 2$, every Jacobi field along id_{S^n} is automatically divergence-free, i.e. $J(id_{S^{n+1}}) = so(n+1)$. Note, however, that due to the presence of conformal deformations, $\dim J(id_{S^2}) = 6$ (cf. also Smith (1972a) pages 107–109).

Corollary 2.21. *Every harmonic Riemannian submersion $f: M \to S^n$ is infinitesimally rigid.*

Proof. By 2.17, it is enough to show that $v = X \circ f \in PK(f)$, $X \in V(S^n)$, implies $X \in K(id_{S^n})$. Indeed, fix $x \in M$ with $y = f(x) \in S^n$ and choose local orthonormal moving frames $\{E^i\}_{i=1}^m$, $m = \dim M$, and $\{F^j\}_{j=1}^n$ around x and y , respectively, as in the proof of II.2.30. Then, by the argument given there, we have

$$(\text{trace } \nabla^2 v)_x = \sum_{i=1}^m \nabla_{e^i}(\nabla_{E^i}(X \circ f)) -$$

$$- \sum_{i=1}^m \nabla_{\nabla_{e^i} E^i}(X \circ f) = \sum_{i=1}^n \nabla_{e^i}((\nabla_{F^i} X) \circ f) -$$

$$- \sum_{i=n+1}^m \nabla_{f_*(\nabla_{e^i} E^i)} X = \sum_{i=1}^n (\nabla_{F^i}(\nabla_{F^i} X))_y -$$

$$- \nabla_{\sum_{i=n+1}^m f_*(\nabla_{e^i} E^i)} X = \sum_{i=1}^n (\nabla_{F^i} \nabla_{F^i} X)_y = (\text{trace } \nabla^2 X)_y$$

since the fibres of f are minimal and $(\nabla_{F^i} F^i)(y) = 0$, $i = 1, \ldots, n$. Furthermore,

$$\text{trace } R(f_*, v) f_* = \sum_{i=1}^n R(f_*(E^i), v) f_*(E^i) =$$

$$= \sum_{i=1}^n (R(F^i, X) F^i) \circ f = \{\text{trace } R((id_{S^n})_*, X)(id_{S^n})_*\} \circ f ,$$

and so $X \in J(id_{S^n})$. In a similar vein,

$$\operatorname{div}_f v = \sum_{i=1}^{n} (f_*(E^i), \nabla_{E^i}(X \circ f)) =$$

$$= \sum_{i=1}^{n} (F^i, \nabla_{F^i} X) \circ f = (\operatorname{div} X) \circ f = 0$$

and the corollary follows. ✓

Corollary 2.22. *The canonical inclusion map $i: S^m \to S^n$, $m \leq n$, is infinitesimally rigid.*

Proof. First, we determine $K(i)$. Let $\{e^i\}_{i=1}^{n+1} \subset \mathbf{R}^{n+1}$ denote the canonical base. Then the vectors $e^{m+2}, \ldots, e^{n+1} \in \mathbf{R}^{n+1}$ determine $k(=n-m)$ orthonormal parallel sections $W^1, \ldots, W^k$ of the normal bundle ν^i by setting $\check{W}^j = e^{m+j+1}$, $j=1,\ldots,k$. Assuming $v \in K(i)$, the orthogonal decomposition

$$v = J + \sum_{j=1}^{k} v^j \cdot W^j , \quad J \in V(S^m) ,$$

$$v^j \in C^\infty(S^m), \quad j=1,\ldots,k , \tag{2.23}$$

splits the Jacobi condition $\operatorname{trace} \nabla^2 v = \operatorname{trace} R(f_*, v) f_*$ into the system

$$\Delta v^j = m v^j , \quad j=1,\ldots,k , \tag{2.24}$$

$$\operatorname{trace} \nabla^2 J + \operatorname{Ric}(J) = 0 . \tag{2.25}$$

By I.7.5, it follows from (2.24) that, for each $j=1,\ldots,k$, the scalar $v^j \in C^\infty(S^m)$, being an eigenfunction of Δ^{S^m} with eigenvalue m , is the restriction of a homogeneous linear function on $\mathbf{R}^{m+1}$ to S^m , i.e. there exists a unique vector $b^j \in \mathbf{R}^{m+1}$ such that

$$v^j(x) = (b^j, x) , \quad x \in S^m \subset \mathbf{R}^{m+1} .$$

Furthermore, using the decomposition (2.23), we have

$$0 = \operatorname{div}_i v = \operatorname{trace}(i_*, dv) = \operatorname{trace}(i_*, dJ) = \operatorname{div} J$$

which, together with (2.25), means that $J \in K(id_{S^m})$. By (2.18), J is a Killing vector field on S^m, i.e. we obtain $J \in so(n+1)$.

Summarizing, we have shown that $v \in K(i)$ if and only if the tangential part J of v is a Killing vector field on S^m and there exist vectors $b^1, \ldots, b^k \in \mathbf{R}^{m+1}$ such that

$$v_x = J_x + \sum_{j=1}^{k} (b^j, x) W_x^j, \quad x \in S^m \subset \mathbf{R}^{m+1}. \tag{2.26}$$

In other words, the linear map

$$\Phi : K(i) \to so(m+1) \times M(k, m+1)$$

defined by $\Phi(v) = (J, B)$, where J is the tangential part of v and $B \in M(k, m+1)$ consists of row vectors $b^1, \ldots, b^k \in \mathbf{R}^{m+1}$ occurring in (2.26), is a linear isomorphism. In particular, $\dim K(i) = \frac{m(m+1)}{2} + (n-m)(m+1) = (n - \frac{m}{2})(m+1)$ and, by 2.3, $i : S^m \to S^n$ is infinitesimally rigid. ✓

3. HARMONIC MAPS OF A HOMOGENEOUS MANIFOLD INTO SPHERES

Recall from IV.4.1 that a rigid minimal isometric immersion $f: M \to S^n$ of a compact Riemannian manifold M into S^n is equivariant with respect to a homomorphism $\rho : i^0(M) \to SO(n+1)$. For infinitesimally rigid harmonic maps, our analogous result is as follows.

Theorem 3.1. *Let $f: M \to S^n$ be a full infinitesimally rigid harmonic map with constant energy density of a compact Riemannian manifold M into S^n and assume that all elements of $K(f)$ are projectable (e.g. f is an embedding). Then $f: M \to S^n$ is equivariant with respect to a homomorphism $\rho : i^0(M) \to SO(n+1)$.*

The proof of 3.1 is preceded by the following.

Lemma 3.2. *If $f:M \to S^n$ is a harmonic map with constant energy density of a compact Riemannian manifold M into S^n then $f_*(X) \in K(f)$ for all Killing vector field $X \in I(M)$ on M.*

Proof. Let $(\phi_t)_{t\in\mathbf{R}} \subset i^o(M)$ denote the 1-parameter group of isometries induced by X. By II.2.31, the map $f_t = f\circ\phi_t : M \to S^n$, $t \in \mathbf{R}$, is harmonic and hence III.(2.2) yields

$$\Delta \bar{f}_t = 2e(f_t)\bar{f}_t \,, \quad t\in\mathbf{R} \,. \tag{3.3}$$

Furthermore, as ϕ_t is an isometry of f, we have

$$2e(f_t) = \text{trace}\|(f_t)_*\|^2 = \text{trace}\| f_*\circ(\phi_t)_*\|^2 =$$

$$= 2e(f)\circ\phi_t = 2e(f) = \text{const.}$$

and so, differentiating (3.3) at $t=0$,

$$\frac{\partial}{\partial t}(\Delta\bar{f}_t)\Big|_{t=0} = \Delta\Big(\frac{\partial \bar{f}_t}{\partial t}\Big|_{t=0}\Big) = \Delta(f_*(X)^{\vee}) =$$

$$= \frac{\partial}{\partial t}\{2e(f_t)\bar{f}_t\}_{t=0} = 2e(f)\,\frac{\partial \bar{f}_t}{\partial t}\Big|_{t=0} = 2e(f)(f_*(X))^{\vee} \,.$$

Thus, by 2.5, $f_*(X) \in K(f)$. ✓

Proof of Theorem 3.1. Compactness of $i^o(M)$ (cf. I.8) implies that the exponential map $\exp : I(M) \to i^o(M)$ is surjective and hence, for a given element $a \in i^o(M)$ there exists $X \in I(M)$ with $\exp X = a$. In particular, $\phi_1 = a$, where $(\phi_t)_{t\in\mathbf{R}} \subset i^o(M)$ denotes the 1-parameter group of isometries induced by X. By 3.2, $f_*(X) \in K(f) = PK(f)$ and so $f_*(X) = Y\circ f$ holds for some $Y \in so(n+1)$, since, by hypothesis the map f is infinitesimally rigid. The vector fields X and Y being f-related, we have

$$f\circ\phi_t = \psi_t\circ f \,, \quad t\in\mathbf{R} \,,$$

where $(\psi_t)_{t\in\mathbf{R}} \subset SO(n+1)$ is the 1-parameter group of isometries induced by Y. Setting $\rho(a) = \psi_1$, we have

$$f \circ a = \rho(a) \circ f ,$$

where, by fullness of f, $\rho(a)$ is uniquely determined by a. By uniqueness, the correspondence which associates to a the isometry $\rho(a)$ is a homomorphism $\rho : i^{\text{o}}(M) \to SO(n+1)$ of Lie groups. √

Remark 1. The Delaunay map $f : \Theta \to S^2$, in 2.10, is not equivariant though it satisfies $K(f) = PK(f)$; thus the hypothesis $e(f)$=const. is essential in 3.1 (and also in 3.2).

Remark 2. The infinitesimally flexible harmonic embeddings $f_t : \mathrm{T}^2 \to S^3$, $0 < t < \frac{\pi}{2}$, considered in 2.15, are equivariant with respect to the monomorphism

$$\rho : \mathrm{T}^2 \to SO(4)$$

given by

$$\rho(\phi,\psi) = \mathrm{diag}\left(\begin{bmatrix} \cos\phi & -\sin\phi \\ \sin\phi & \cos\phi \end{bmatrix}, \begin{bmatrix} \cos\psi & -\sin\psi \\ \sin\psi & \cos\psi \end{bmatrix} \right), \quad \phi,\psi \in \mathbf{R} .$$

Throughout the rest of this section we specialize our study on infinitesimal rigidity of (full) harmonic embeddings $f : M \to S^n$ to the case when $M = G/K$ is a (compact) naturally reductive (effective) homogeneous manifold with invariant Riemannian metric $g = (\,,\,) \in C^\infty(S^2(T^*(G/K)))$ (cf.IV.3). In view of 3.1, we also assume that f is equivariant with respect to a homomorphism $\rho : G \to SO(n+1)$. Then, as in IV.(5.1), the pull-back G-bundle $\tau^f = f^*(T(S^n))$ splits into the orthogonal direct sum of Riemannian-connected G-bundles

$$\tau^f \cong T(G/K) \oplus \nu^f , \tag{3.4}$$

where ν^f stands for the normal G-bundle of f. Let $\mathrm{inv}_G(\tau^f) \subset C^\infty(\tau^f)$ denote the linear subspace of G-invariant sections of τ^f, i.e. $v \in C^\infty(\tau^f)$ belongs to $\mathrm{inv}_G(\tau^f)$ if

$v\circ a=a\cdot v$, $a\in G$.

Theorem 3.5. *Let* $f:G/K\to S^n$ *be a full equivariant harmonic embedding of a naturally reductive homogeneous manifold* G/K *with invariant Riemannian metric* $g=(,)$ *into* S^n *. Then,*

$$\mathrm{inv}_G(\tau^f)\subset K(f) \quad . \tag{3.6}$$

Proof. Let $\rho:G\to SO(n+1)$ denote the monomorphism associated with the equivariance of f . First we claim that

$$\Delta\rho = 2e(f)\rho \ , \tag{3.7}$$

where ρ is considered as a matrix-valued function on which the Laplacian Δ (with respect to the biinvariant Riemannian metric induced on G) acts component-wise.
Indeed, using the notations of IV.3, we have $a\cdot o=\pi(a)$, $a\in G$, and by equivariance of f

$$(\Delta\rho)(a)\cdot f(o) = (\Delta(f\circ\pi))(a) = (\Delta f)(\pi(a)) =$$

$$= 2e(f)\cdot f(\pi(a)) = 2e(f)\rho(a)\cdot f(o) \ ,$$

where we used the fact that the canonical projection $\pi:G\to G/K$ is a harmonic Riemannian submersion (IV.(3.23)). The same argument works with o replaced by any point of G/K since the group metric is biinvariant. As f is full, (3.7) follows.
Now, putting $v\in\mathrm{inv}_G(\tau^f)$, by 2.5, we have to show that

$$\Delta\check{v} = 2e(f)\check{v} \quad . \tag{3.8}$$

G-invariance of v and (3.7) implies

$$(\Delta\check{v})(\pi(a)) = (\Delta(\check{v}\circ\pi))(a) = (\Delta\rho)(a)\cdot\check{v}(o) =$$

$$= 2e(f)\rho(a)\cdot\check{v}(a) = 2e(f)\check{v}(\pi(a)) \ , \quad a\in G \ ,$$

which completes the proof. √

The crucial property of infinitesimally rigid harmonic maps to be exploited subsequently is contained in the next result.

Theorem 3.9. *Let* $f: G/K \to S^n$ *be a full* ρ*-equivariant infinitesimally rigid harmonic embedding of a naturally reductive homogeneous manifold* G/K *into* S^n *. Then, precomposition with* f *yields a linear isomorphism* $\Phi: Z(\rho(G)) \to \mathrm{inv}_G(\tau^f)$ *, where* $Z(\rho(G))$ *is the Lie algebra of the centralizer of* $\rho(G)$ *in* $SO(n+1)$ *. Furthermore,* Φ *composed with the evaluation map* $\varepsilon(o): C^\infty(\tau^f) \to \tau_o^f$ *maps* $Z(\rho(G))$ *isomorphically to the fixed point set* $\mathrm{Fix}(K, \tau_o^f)$ *of* K *on* τ_o^f *, where* K *acts on* τ_o^f *via* ρ *.*

Proof. Setting $X \in Z(\rho(G))$, consider the 1-parameter group $(\phi_t)_{t\in\mathbb{R}}$ of isometries of S^n induced by X which is contained in the identity component $Z^0(\rho(G))$ of the centralizer $\rho(G)$ in $SO(n+1)$. Then, for $a \in G$ and $x \in G/K$, we have

$$(a\cdot(X\circ f))(x) = \rho(a)_*(X_{f(x)}) = \rho(a)_* \frac{\partial}{\partial t}\{\phi_t(f(x))\}_{t=0} =$$

$$= \frac{\partial}{\partial t}\{(\rho(a)\circ\phi_t)(f(x))\}_{t=0} = \frac{\partial}{\partial t}\{(\phi_t\circ\rho(a))(f(x))\}_{t=0} =$$

$$= \frac{\partial}{\partial t}\{(\phi_t\circ f)(a(x))\}_{t=0} = ((X\circ f)\circ a)(x) ,$$

i.e. the vector field $X\circ f$ along f is G-invariant and setting $\Phi(X) = X\circ f$, $X \in Z(\rho(G))$, we obtain a well-defined linear map $\Phi: Z(\rho(G)) \to \mathrm{inv}_G(\tau^f)$.

If $X \in \ker\Phi$ then the Killing vector field X vanishes on $\mathrm{im}(f) \subset S^n$. As f is full, X vanishes on the whole of S^n and we conclude that Φ is injective.

To prove that Φ is onto, let $v \in \mathrm{inv}_G(\tau^f)$. By (3.5), $v \in K(f)$ $(=PK(f))$ and infinitesimal rigidity of f implies the existence of a Killing vector field X on S^n with $v = X\circ f$. We claim that $X \in Z(\rho(G))$. Indeed, G-invariance of v implies

$$(\rho(a)_* X)\circ f = (X\circ f)\circ a = (X\circ\rho(a))\circ f , \quad a \in G .$$

Again by fullness of f , we obtain

$$\rho(a)_* X = X \circ \rho(a) \quad , \quad a \in G \; . \tag{3.10}$$

Denoting by $(\phi_t)_{t \in \mathbf{R}} \subset SO(n+1)$ the 1-parameter group of isometries induced by X, (3.10) becomes

$$\rho(a) \circ \phi_t = \phi_t \circ \rho(a) \quad , \quad t \in \mathbf{R} \quad , \quad a \in G \; ,$$

or equivalently, $(\phi_t)_{t \in \mathbf{R}} \subset Z^o(\rho(G))$. Thus, $X \in Z(\rho(G))$ and Φ is onto. Finally, the restriction $\varepsilon(o)|\mathrm{inv}_G(\tau^f)$ maps $\mathrm{inv}_G(\tau^f)$ isomorphically onto $\mathrm{Fix}(K, \tau^f_o)$. ✓

In what follows we shall be dealing with compact Lie group actions (mostly on spheres) and, for convenience, we briefly summarize here some basic facts and notions of the theory of compact transformation groups. (For more details, we refer to Bredon (1972).)

Let G be a compact Lie group acting on a Riemannian manifold M *by isometries*. For $x \in M$, the normal bundle ν^f of the inclusion map $f: G(x) \to M$ of the orbit $G(x) = \{a(x) | a \in G\}$ through x into M, inherits a G-bundle structure such that the restriction of the exponential map

$$\exp|\nu^f : \nu^f \to M$$

commutes with the respective actions of G on ν^f and M. The isotropy subgroup $G_x = \{a \in G | a(x) = x\}$ at $x \in M$ acts on the fibre $\nu^f_x \subset T_{f(x)}(M)$ orthogonally and the orbit structure of this action of G_x on ν^f_x is *conical* (i.e. the isotropy subgroups along any ray (less the vertex $0 \in \nu^f_x$) are the same). Then, ν^f_x is said to be a *linear slice* and the action of G_x on ν^f_x is the *slice representation* of G at $x \in M$.

Due to compactness of G, a sufficiently small open invariant neighbourhood B of the zero section of ν^f is mapped diffeomorphically under $\exp|B$ onto an open invariant neighbourhood $U \subset M$ (called a *tubular neighbourhood*) of the orbit $G(x)$. For any point $y \in U$, the isotropy subgroup G_y is conjugate to some subgroup of G_x, or equivalently, G_y is contained in some element of the isotropy type (G_x) (=set of conjugacy classes of G_x in G).

Generalizing this property, for *any* $x,y \in M$, we set

$$(G_x) \leq (G_y)$$

if G_y is contained in some element of (G_x). The relation $\leq$ obtained in this way is a partial ordering on the set of all isotropy types and, according to a result of Montgomery-Samelson-Yang (cf. Bredon (1972) page 179), there exists an absolute maximum (G_z), i.e. each isotropy subgroup G_x, $x \in M$, contains an element of (G_z). Furthermore, the orbits with isotropy type (G_z) (called *principal orbits*) form an open dense subset of M.

An orbit $G(x)$ is principal if and only if the isotropy representation at x is trivial. Nonprincipal orbits on M having the same dimension as that of the principal orbits are said to be *exceptional*. All the remaining orbits are called *singular*, i.e. an orbit $G(x) \subset M$ is singular if $\dim G(x)$ is strictly less than the dimension of the principal orbits.

Using the orbit classification given above, we can now formulate the following:

Theorem 3.11. *Let $f: G/K \to S^n$, $n \geq 2$, be a full ρ-equivariant infinitesimally rigid harmonic embedding of a naturally reductive homogeneous manifold into S^n. Then, either f is onto or $\mathrm{im}(f) \subset S^n$ is a nonprincipal orbit of the action of G on S^n via ρ. In particular, in the latter case, if $K \subset G$ is connected then $\mathrm{im}(f) \subset S^n$ is a singular orbit.*

Proof. Assuming that the orbit $\mathrm{im}(f) \subset S^n$ is principal, we show that $\mathrm{im}(f) = S^n$. As noted above, the isotropy subgroup K acts trivially on ν_o^f (considered as a linear slice at $f(o) \in S^n$), i.e. $\nu_o^f \subset \mathrm{Fix}(K, \tau_o^f)$.

We first claim that the closed subgroup $H = \rho(G) \cdot Z^0(\rho(G))$ acts transitively on S^n. Indeed, the tangent space $T_{f(o)}(H(f(o)))$ clearly contains $T_{f(o)}(\mathrm{im}(f))$ and, by 3.9, it contains $\nu_o^f \subset \varepsilon(o)(Z(\rho(G)))$. It follows that $H(f(o)) \subset S^n$ is open and, H being compact, the claim follows.

Next, we assert that all orbits of G on S^n have the same type. If $y,y'\in S^n$ then choose $h=\rho(a)\cdot z\in H$, $a\in G$, $z\in Z^0(\rho(G))$, such that $y'=h(y)$. Then, $G_{y'}=G_{h(y)}=a\cdot G_{z(y)}\cdot a^{-1}=aG_y\cdot a^{-1}$, or equivalently, $(G_{y'})=(G_y)$. By Borel's classification of actions on S^n with one isotropy type (cf. Bredon (1972) page 196), it follows that G is either transitive on S^n , i.e. we have $\mathrm{im}(f)=S^n$, or $G=S^3$ and G acts freely on S^n . (Note that if $G=S^1$ the harmonic map f is either constant or maps onto a closed geodesic and hence fullness of f implies $n\le 1$ which was excluded.) Assuming $G=S^3$, as G acts freely on S^n , we have $K=\{1\}$; in particular, $\dim Z(\rho(G))=$ $=\dim \mathrm{Fix}(H,\tau_o^f)=\dim \tau_o^f=n$. By the proof of 3.9, $Z^0(\rho(G))$ acts transitively on S^n since $T_{f(o)}(Z^0(\rho(G))(f(o)))=\tau_o^f$. For reasons of dimensions, $Z^0(\rho(G))$ is actually diffeomorphic with S^n and hence $n=3$. (In fact, as is well-known from algebraic topology, S^n , $n\ge 2$, carries a Lie group stucture if and only if $n=3$; cf. Spanier (1966) page 276.) Thus $f:G\to S^3$ is onto. √

For homogeneous hypersurfaces, Theorem 3.11 can be sharpened as follows.

Corollary 3.12. *Let G/K be naturally reductive. Then any full harmonic codimension 1 embedding $f:G/K\to S^n$, $n\ge 2$, with $e(f)=$ $=$const. is infinitesimally flexible.*

Proof. Assuming, on the contrary, that $f:G/K\to S^n$ is infinitesimally rigid, 3.1 and 3.11 imply that the orbit $\mathrm{im}(f)\subset S^n$ is exceptional. (Obviously, $\mathrm{im}(f)\subset S^n$ cannot be singular since otherwise G acts transitively on S^n.) As $\mathrm{im}(f)$ is non-principal, $\mathrm{Fix}(K,\nu_o^f)\ne\nu_o^f$ and hence the isotropy subgroup K acts on the line ν_o^f with trivial fixed point set. According to the terminology of Bredon (1972), $\mathrm{im}(f)\subset S^n$ is special exceptional which contradicts to $n\ge 2$ since $H_1(S^n;Z_2)=0$ (cf. 3.12 Theorem in Bredon (1972) page 185). √

Remark. Fullness in 3.12 is essential since, by 2.22, the canonical inclusion map $i:SO(n)/SO(n-1)\to S^n$ is infinitesimally rigid.

We close this section by showing how the results of IV.4.

extend to harmonic maps with constant energy density (cf. Problem (4.4) in Eells and Lemaire (1983) page 70).
Recall from IV.3 that each eigenvalue $(0\neq)\lambda$ of the Laplacian Δ on an m-dimensional compact oriented irreducible homogeneous manifold G/K with invariant Riemannian metric $g=(,)$ gives rise to a standard minimal immersion

$$f_\lambda : G/K \to S^{n(\lambda)}$$

defined by

$$f_\lambda(aK) = \sum_{i=0}^{n(\lambda)} f_\lambda^i(aK) f_\lambda^i \quad ,$$

where $\{f_\lambda^i\}_{i=0}^{n(\lambda)} \subset V_\lambda$ is an orthonormal base of the corresponding eigenspace $V_\lambda \subset C^\infty(G/K)$. Then, f_λ is a minimal isometric immersion with induced Riemannian metric $\frac{\lambda}{m} g$. Thus, keeping the original Riemannian metric g on G/K, $f_\lambda : G/K \to S^{n(\lambda)}$ is a harmonic (homothetic) immersion with energy density $e(f)=\frac{\lambda}{2}$.

Let $W \subset S^2(\mathbf{R}^{n(\lambda)+1})$ be the linear subspace defined in IV.(4.6) for $f=f_\lambda$ and, as in IV.4.8(ii) set

$$W^0 = \mathrm{span}_{\mathbf{R}}\{(\bar{f}_\lambda(x))^2 \in S^2(\mathbf{R}^{n(\lambda)+1}) \mid x \in G/K\} \subset W \ .$$

Finally, let $(W^0)^\perp$ be the orthogonal complement of W^0 in $S^2(\mathbf{R}^{n(\lambda)+1})$ with respect to the scalar product $(,)$ given in IV.(1.11) and set

$$L^0 = \{C \in (W^0)^\perp \mid C + I_{n(\lambda)+1} \text{ is positive semidefinite}\} \subset (W^0)^\perp \ .$$

Clearly, by IV.(4.7), $L = L_{f_\lambda} \subset L^0$.

For harmonic maps $f : G/K \to S^n$ with constant energy density $\frac{\lambda}{2}$ the classification theorem IV.4.5 takes the form:

Theorem 3.13. *Let* G/K *be a compact oriented irreducible homogeneous manifold with invariant Riemannian metric* $g=(,)$.

Given $\lambda \in \mathrm{Spec}(G/K)$, *the equivalence classes of full harmonic maps* $f: G/K \to S^n$ *with energy density* $e(f)=\frac{\lambda}{2}$ *can be (smoothly) parametrized by the compact convex body* $L^o \subset (W^o)^{\perp}$. *The interior points of* L^o *correspond to full harmonic maps with maximal* $n(=n(\lambda))$. *In particular, any such maps can be deformed into each other by a smooth homotopy of the same type.*

Proof. Given $C \in L^o(\subset S^2(\mathbf{R}^{n(\lambda)+1}))$, define the map

$$\bar{f}_o: G/K \to \mathbf{R}^{n(\lambda)+1}$$

by

$$\bar{f}_o = \sqrt{C+I_{n(\lambda)+1}} \cdot \bar{f}_\lambda \ . \tag{3.14}$$

Then, for $x \in G/K$, we have

$$(\bar{f}_o(x), \bar{f}_o(x)) = ((C+I_{n(\lambda)+1})\bar{f}_\lambda(x), \bar{f}_\lambda(x)) =$$

$$= (C \cdot \bar{f}_\lambda(x), \bar{f}_\lambda(x)) + 1 = (C, (\bar{f}_\lambda(x))^2) + 1 = 1$$

and hence $\bar{f}_o$ gives rise to a map

$$f_o: G/K \to S^{n(\lambda)} \ .$$

Furthermore,

$$\Delta \bar{f}_o = \sqrt{C+I_{n(\lambda)+1}} \cdot \Delta \bar{f}_\lambda = \lambda \sqrt{C+I_{n(\lambda)+1}} \cdot \bar{f}_\lambda = \lambda \bar{f}_o$$

and, by III.(2.2), $f_o: G/K \to S^{n(\lambda)}$ is a harmonic map with energy density $e(f_o)=\frac{\lambda}{2}$. Clearly, f_o defines a *full* harmonic map

$$f: G/K \to S^n$$

with energy density $e(f)=\frac{\lambda}{2}$ such that $i \circ f: G/K \to S^{n(\lambda)}$ is equivalent to f_o, where $i: S^n \to S^{n(\lambda)}$ is the canonical inclusion map, and the equivalence class of f is uniquely determined by C. Associate then to $C \in L^o$ the equivalence class of f obtained in the above way. As for the inverse, let $f: G/K \to S^n$

be a full harmonic map with $e(f)=\frac{\lambda}{2}$. Fullness of f implies the linear independence of the system $\{f^i\}_{i=0}^n$ which, by III.(2.2), is contained in V_λ; in particular, $n \leq n(\lambda)$. Just as in the proof of IV.4.11, we find that there is a positive semi-definite endomorphism $A \in S^2(\mathbf{R}^{n(\lambda)+1})$ such that $\mathrm{im}(A\circ \bar{f}_\lambda) \subset S^{n(\lambda)}$ and $i\circ f$ is equivalent to $A\circ f_\lambda$. Moreover, A is uniquely determined by the equivalence class of f. Associate then to f the matrix

$$C = A^2 - I_{n(\lambda)+1} \,.$$

Then, as $A\circ f_\lambda$ maps into $S^{n(\lambda)}$, for $x \in G/K$, we have

$$(C, (\bar{f}_\lambda(x))^2) = (C\cdot \bar{f}_\lambda(x), \bar{f}_\lambda(x)) =$$

$$= (A\cdot \bar{f}_\lambda(x), A\cdot \bar{f}_\lambda(x)) - 1 = 0 \,,$$

and so $C \in L^0$. Clearly, the correspondence $f \to C$ is the inverse of the one given above. The rest follows by the same reasoning as in the proof of IV.4.5. ✓

A reformulation of infinitesimal rigidity in terms of the parameter space L^0 is given as follows.

Theorem 3.15. *Let G/K and $\lambda \in \mathrm{Spec}(G/K)$ be as in 3.13. Given a full harmonic map $f: G/K \to S^{n(\lambda)}$ with energy density $e(f)=\frac{\lambda}{2}$, we have*

$$K(f) = PK(f) \cong so(n(\lambda)+1) \oplus (W^0)^\perp \,. \tag{3.16}$$

In particular, f is infinitesimally rigid if and only if $W^0 = S^2(\mathbf{R}^{n(\lambda)+1})$, or equivalently, $L^0=\{0\}$.

Proof. By 2.5, $v \in K(f)$ if and only if $\{\check{v}^i\}_{i=0}^{n(\lambda)} \subset V_\lambda$ with $(\bar{f}, \check{v})=0$. As f is full, there exists a *unique* matrix $X \in M(n(\lambda)+1, n(\lambda)+1)$ such that

$$\check{v} = X\circ \bar{f} \,;$$

in particular, $K(f)=PK(f)$. Decomposing X into skew-symmetric and symmetric parts as

$$X = A + B ,$$

where $A\in so(n(\lambda)+1)$ and $B\in S^2(\mathbf{R}^{n(\lambda)+1})$, we have

$$0 = (\bar{f},\check{v}) = (\bar{f},X\bar{f}) = (\bar{f},A\bar{f}) + (\bar{f},B\bar{f}) = -(A\bar{f},\bar{f}) + (\bar{f},B\bar{f})$$

and hence

$$(B\bar{f},\bar{f}) = 0 ,$$

i.e. $B\in(W^0)^{\perp}$. Now, the correspondence which associates to v the pair $(A,B)\in so(n(\lambda)+1)\oplus(W^0)^{\perp}$ is, by 2.5, a linear isomorphism. √

Remark. 3.2 together with the Remark after the proof of IV.4.5 yields that infinitesimal rigidity of the full minimal isometric immersion $f:G/K\to S^{n(\lambda)}$, with induced Riemannian metric $\bar{g}=\frac{\lambda}{m}g$, implies rigidity of f in the sense of IV.4.

Corollary 3.17. *Let G/K and $\lambda\in\mathrm{Spec}(G/K)$ be as in 3.13 and let $f,f':G/K\to S^{n(\lambda)}$ be full harmonic maps with $e(f)=e(f')=$ $=\frac{\lambda}{2}$. Then, if f is infinitesimally rigid there exists an isometry $U\in O(n(\lambda)+1)$ such that $f'=U\circ f$.* √

4. INFINITESIMAL RIGIDITY OF MINIMAL IMMERSIONS BETWEEN SPHERES

Recall from IV.3.30 that each eigenvalue $\lambda_k=k(k+m-1)\in\mathrm{Spec}(S^m)$, $k\in\mathbf{N}$, gives rise to a standard minimal immersion

$$f_{\lambda_k}:S^n_{\kappa(k)}\to S^{n(k)}$$

(for fixed orthonormal base $\{f^i_{\lambda_k}\}_{i=0}^{n(\lambda)}$ of the corresponding eigenspace V_{λ_k}) . The object of this section is to study infinitesimal rigidity and flexiblity of these maps. (As $f_{\lambda_1}:S^m\to S^m$

is an isometry (cf. IV.3.30) and therefore by 2.21 infinitesimally rigid, we may assume that $k>1$.)
Fixing $m>1$ and $k>1$, recall from 3.15 that

$$K(f_{\lambda_k}) = PK(f_{\lambda_k}) = so(n(k)+1) \oplus (W^0)^{\perp} ; \qquad (4.1)$$

in particular, f_{λ_k} is infinitesimally rigid if and only if $W^0 = S^2(\tilde{H}_m^k)$.
In the course of verifying IV.6.11, we showed that, for $m=2$, $W^0 = S^2(\tilde{H}_2^k)$, i.e. we have the following:

Theorem 4.2. *Each standard minimal immersion*

$$f_{\lambda_k} : S^2_{\kappa(k)} \to S^{2k} , \quad \kappa(k) = \frac{2}{k(k+1)} , \quad k>1 ,$$

is infinitesimally rigid. √

As shown below the remaining cases $m>2$ and $k>1$ belong to the infinitesimally flexible range. For the following theorem, recall that the homomorphism $\rho : SO(m+1) \to SO(n(k)+1)$ associated with the equivariance of f_{λ_k} induces an (orthogonal) $SO(m+1)$-module structure on $S^2(\mathbf{R}^{n(k)+1})$ such that W^0 (and hence $(W^0)^{\perp}$) are $SO(m+1)$-submodules in $S^2(\mathbf{R}^{n(k)+1})$.

Theorem 4.3. *For* $m>2$ *and* $k>1$ *, the standard minimal immersion*

$$f_{\lambda_k} : S^m_{\kappa(k)} \to S^{n(k)}$$

is infinitesimally flexible. Furthermore, denoting by V_m^{ρ} *the irreducible complex* $SO(m+1)$*-module with highest weight* $\rho = (\rho_1, \ldots, \rho_{\ell}) \in \mathbf{Z}^{\ell}$, $\ell = [\frac{m+1}{2}]$ *, we have, for* $m>3$ *,*

$$(W^0)^{\perp} \otimes_{\mathbf{R}} \mathbf{C} = \bigoplus_{\substack{(a,b)\in\Delta_0 \\ a,b \text{ even}}} V_m^{(a,b,0,\ldots,0)} \qquad (4.4)$$

and, for $m=3$ *,*

$$(W^{\circ})^{\perp} \underset{\mathbf{R}}{\otimes} \mathbf{C} = \bigoplus_{\substack{(a,b)\in\Delta_{o} \\ a,b \text{ even}}} \{V_3^{(a,b)} \oplus V_3^{(a,-b)}\} , \tag{4.5}$$

where $\Delta_{o} \subset \mathbf{R}^2$ *is the closed triangle with vertices* $(2,2)$, (k,k) *and* $(2k-2,2)$.

Proof. We first claim that every class 1 representation V of $(SO(m+1),SO(m))$ is equivalent to some $\tilde{H}^j_m$, $j\in\mathbf{Z}_+$ (considered as an irreducible $SO(m+1)$-module; cf. Problem 5 for Chapter IV).

Indeed, by the argument given in the proof of IV.3.17, V gives rise to a minimal immersion

$$f:S^m_{\kappa} \to S \ (\kappa>0 \ ; \ \ S=\text{unit sphere in } V)$$

defined by

$$f(a\cdot SO(m)) = \rho(a)v^{\circ} , \quad a\in SO(m+1) ,$$

where $\rho:SO(m+1) \to SO(V)$ is the homomorphism corresponding to the $SO(m+1)$-module structure on V and $v^{\circ}\in V$ is an $SO(m)$-fixed unit vector. Furthermore, fixing a base $\{e^i\}_{i=0}^n \subset V$, $\dim V=n+1$, $e^{\circ}=v^{\circ}$, induces a monomorphism $\phi:V \to C^{\infty}(S^m)$ of $SO(m+1)$-modules. Now, $\bigoplus_{j\in\mathbf{Z}_+} \tilde{H}^j_m \subset C^{\infty}(S^m)$ is a dense linear subspace (cf. the proof of I.7.5) and the claim follows.
As every irreducible $SO(m+1)$-submodule of W° is class 1 for $(SO(m+1),SO(m))$, by IV.6.6, the direct sum of all irreducible subrepresentations of $SO(m+1)$ in $S^2(\tilde{H}^k_m)$ which are class 1 for $(SO(m+1),SO(m))$ is $W^{\circ}\subset S^2(\tilde{H}^k_m)$. Thus, by IV.(7.10), the complexification $W^{\circ} \underset{\mathbf{R}}{\otimes} \mathbf{C}$ is the sum of all complex irreducible $SO(m+1)$-modules in $S^2(\tilde{H}^k_m \underset{\mathbf{R}}{\otimes} \mathbf{C})$ which have the form $V_m^{(j,0,\dots,0)}$, $j\in\mathbf{Z}_+$. Now, by a simple discussion, IV.(7.16) and IV.(7.17) imply (4.4) and (4.5), respectively. √

Remark. For the *linearly flexible* range $m>2$ and $k>3$, IV. 6.14 automatically implies the first statement of 4.3 since

$L \subset L^{0}$ holds for the corresponding parameter spaces.

Once $\dim_{\mathbf{C}} V_m^{(a,b,0,\ldots,0)}$ is known for each $(a,b)\in\Delta_0, a,b$ even, we can compute $\dim(W^0)^{\perp}=\dim L^0$ via (4.4) and (4.5) (and hence $\dim K(f_{\lambda_k})=\dim PK(f_{\lambda_k})$, cf. (4.1) with IV.(3.35)). In what follows, using the Weyl dimension formula, we determine $\dim_{\mathbf{C}} V_m^{(a,b,0,\ldots,0)}$.

(i) $m=2\ell$ *even*, $\ell>2$. By IV.7.11,

$$\dim_{\mathbf{C}} V_m^{(a,b,0,\ldots,0)} = \frac{(a-b+1)(a+b+2\ell-2)}{2\ell-2} \cdot$$

$$\cdot \prod_{r=3}^{\ell} \frac{(a+r-1)(a+2\ell-r)}{(r-1)(2\ell-r)} \cdot \prod_{r=3}^{\ell} \frac{(b+r-2)(b+2\ell-r-1)}{(r-2)(2\ell-r-1)} \cdot$$

$$\cdot \frac{(2a+2\ell-1)(2b+2\ell-3)}{(2\ell-1)(2\ell-3)} = \frac{(a-b+1)(a+b+2\ell-2)}{2\ell-2} \cdot$$

$$\cdot \frac{1}{a+1}\binom{a+2\ell-3}{a}\binom{b+2\ell-4}{b}\frac{(2a+2\ell-1)(2b+2\ell-3)}{(2\ell-1)(2\ell-3)} =$$

$$= \frac{(a-b+1)(a+b+m-2)(2a+m-1)(2b+m-3)}{(a+1)(m-1)(m-2)(m-3)}\binom{a+m-3}{a}\binom{b+m-4}{b} .$$

(ii) $m=2\ell-1$ *odd*, $\ell>2$. By similar computation we obtain the same formula for $\dim_{\mathbf{C}} V_m^{(a,b,0,\ldots,0)}$ as above.

(iii) $m=3$ *and* $m=4$. Again by IV.7.11,

$$\dim_{\mathbf{C}} V_3^{(a,b)} = \dim_{\mathbf{C}} V_3^{(a,-b)} = (a-b+1)(a+b+1)$$

and

$$\dim_{\mathbf{C}} V_4^{(a,b)} = \frac{1}{6}(a-b+1)(a+b+2)(2a+3)(2b+1) .$$

Example 4.6. We claim that for the standard minimal immersion

$$f_{\lambda_2}: S^m_{\kappa(2)} \to S^{n(2)} , \quad m>1 ,$$

we have

$$\dim(W^{0})^{\perp} = \frac{1}{12}(m^4+4m^3-m^2-16m) - 1 \qquad (4.7)$$

and hence, as $n(2)=\frac{m(m+3)}{2}$ (cf. IV.3.35)

$$\dim K(f_{\lambda_2}) = \dim PK(f_{\lambda_2}) = \frac{1}{24}(5m^4+26m^3+19m^2-50m) - 1 .$$

Indeed, in this case the triangle $\Delta_0 \subset \mathbf{R}^2$ reduces to the point (2,2) and so, for $m>3$,

$$\dim(W^{0})^{\perp} = \dim_{\mathbf{C}} V_m^{(2,2,0,\ldots,0)}$$

and, for $m=3$,

$$\dim(W^{0})^{\perp} = \dim_{\mathbf{C}}\{V_3^{(2,2)} \oplus V_3^{(2,-2)}\} .$$

For $m=2\ell$ and $m=2\ell-1$, $\ell>2$, we have

$$\dim_{\mathbf{C}} V_m^{(2,2,0,\ldots,0)} = \frac{(m+2)(m+3)(m+1)}{3(m-1)(m-2)(m-3)}\binom{m-1}{2}\binom{m-2}{2} =$$

$$= \frac{1}{12}(m+2)(m+3)(m+1)(m-2) = \frac{1}{12}(m^4+4m^3-m^2-16m) - 1 .$$

Furthermore,

$$\dim_{\mathbf{C}} V_4^{(2,2)} = 35$$

and

$$\dim_{\mathbf{C}} V_3^{(2,2)} = \dim_{\mathbf{C}} V_3^{(2,-2)} = 5$$

and, in each case, (4.7) follows.

PROBLEMS FOR CHAPTER V

1. In the use of III.1.19, give an other proof of Lemma 3.2. (Hint: Let $\{E^i\}_{i=1}^m$ be a local orthonormal moving frame on $M\backslash\mathrm{Zero}(X)$ such that $[X,E^i]=0$, $i=1,\ldots,m$. Use I.8.1 and symmetry of the second fundamental form $\beta(f)$ to derive, by local computation, $\mathrm{div}_f\, f_*(X)=0$ on $M\backslash\mathrm{Zero}(X)$. Now, for $X\neq 0$,

$\text{Zero}(X) \subset M$ is nowhere dense.)

2. Given a harmonic map $f: M \to S^n$ of a compact Riemannian manifold M into S^n and a vector field v along f show that $f_t = \exp \circ (tv): M \to S^n$ is harmonic for all $t \in \mathbf{R}$ if and only if $v \in K(f)$ and $\|v\| = \text{const}$. (Hint: By the formula III.(2.1) of the Riemannian curvature tensor R of S^n, the initial value problem II.(3.10)-(3.11) can be solved explicitly for $P_v(t) = = \tau^t (f_t)_* \in \mathcal{D}^1(\tau^f)$, $t \in \mathbf{R}$. Substituting the solution into II.(3.12) we obtain a Fourier polynomial whose coefficients have to vanish; cf. Tóth (1981).)

3. Let $f: M \to S^n$ be a full infinitesimally rigid harmonic embedding with constant energy density of a compact *symmetric* Riemannian manifold M into S^n. Prove that $\beta(f)$ takes its values in the normal bundle ν^f. (Hint: By 3.1, f is ρ-equivariant and hence the pull-back $\omega \in C^\infty(S^2(T^*(M)))$ of the Riemannian metric of S^n via f is $i^0(M)$-invariant. As M is symmetric, we have $\nabla\omega = 0$ (cf. Lichnerowicz (1958) page 174), or equivalently

$$(\beta(f)(X,Y), f_*(Z)) = -(\beta(f)(X,Z), f_*(Y)), \quad X,Y,Z \in \mathcal{V}(M).$$

Using symmetry of $\beta(f)$ (II.2.7) both sides reduce to zero.)

4. In the use of Problem 3, verify that a diffeomorphism $f: S^m \to \to S^m$ is isometric if and only if f is an infinitesimally rigid harmonic map with constant energy density.

5. For G/K symmetric, derive Theorem 3.11 from Problem 3. (Hint: It is enough to show that $(\beta(f)(X', X''), v)$ holds for $X', X'' \in \mathcal{V}(G/K)$ and $v \in \text{inv}_G(\nu^f)$. By infinitesimal rigidity, $v = Y \circ f$, for some $Y \in so(n+1)$. Choose $Y', Y'' \in \mathcal{V}(S^n)$ with $f_*(X') = Y' \circ f$ and $f_*(X'') = Y'' \circ f$ and compute directly $(\beta(f)(X', X''), Y \circ f)$.)

6. Prove Corollary 3.12 without the use of 3.11. (Hint: As in the end of the proof of 3.12, $\text{im}(f) \subset S^n$ is principal. Assuming infinitesimal rigidity of f, we have $X \circ f = v$ for some $X \in \in so(n+1)$, where $v \in \text{inv}_G(\nu^f) \subset K(f)$ is a unit section. Then, $\|X\| = 1$ and hence $\nabla_X X = 0$ on S^n. Each integral curve of X is a closed geodesic of S^n and the projection $\pi: S^n \to B$ onto the orbit space B is a (smooth) principal bundle with structure

group S^1. The composition $\tilde{f}=f\circ\pi:G/K\to B$ is a covering and, as B is simply connected, a diffeomorphism. Then, $f\circ\tilde{f}^{-1}$ is a global section with contradicts simply connectedness of S^n.)

7. Let G/K and $\lambda\in\mathrm{Spec}(G/K)$ be as in Theorem 3.13; denote by $\rho:G\to SO(n(\lambda)+1)$ the homomorphism associated with the equivariance of a standard minimal immersion $f_\lambda:G/K\to S^{n(\lambda)}$. Show that W^0 and L^0 are invariant subspaces in $S^2(\mathbf{R}^{n(\lambda)+1})$ with respect to the linear action of G on $S^2(\mathbf{R}^{n(\lambda)+1})$ defined by ρ. Using the proof of Theorem 3.13, verify that the action of G on the parameter space L^0 is induced by the action of G on the set of full harmonic maps $f:G/K\to S^n$ with $e(f)=\frac{\lambda}{2}$ given by $f\to f\circ a$, $a\in G$.

8. Retaining the notation of Problem 7, for a given full harmonic map $f:G/K\to S^n$ with $e(f)=\frac{\lambda}{2}$, define

$$G_f=\{a\in G\,|\,\text{there exists } A\in SO(n+1) \text{ such that } f\circ a=A\circ f\}.$$

Show that $G_f\subset G$ is a closed subgroup and there exists a homomorphism $\rho_f:G_f\to SO(n+1)$ such that

$$f\circ a=\rho_f(a)\circ f,\qquad a\in G_f.$$

9. Given a closed subgroup $H\subset G$, call a full harmonic map $f:G/K\to S^n$ (with $e(f)=\frac{\lambda}{2}$) H-equivariant if $H\subset G_f$ (cf. Problem 8). Denoting by $SO(m)\subset SO(m+1)$ the subgroup induced by the inclusion $\mathbf{R}^m\subset\mathbf{R}^{m+1}$, show that two $SO(m)$-equivariant harmonic maps $f,f':S^m\to S^n$ with $e(f)=e(f')=\text{const.}$ are equivalent. (Hint: By the definition of equivalence and by Problem 7, $\mathrm{Fix}(SO(m),L^0)\subset L^0$ is the parameter space of equivalence classes of $SO(m)$-equivariant full harmonic maps (with energy density $e(f)=\frac{\lambda}{2}$). As $\mathrm{Fix}(SO(m),L^0)\subset\mathrm{Fix}(SO(m),(W^0)^\perp)$ the statement is equivalent to the triviality of the linear subspace $\mathrm{Fix}(SO(m),(W^0)^\perp)\subset(W^0)^\perp$. Now, as the $SO(m+1)$-module structure of $(W^0)^\perp\otimes_{\mathbf{R}}\mathbf{C}$ is known from Theorems 4.2-4.3, verify that $\mathrm{Fix}(SO(m),(W^0)^\perp\otimes_{\mathbf{R}}\mathbf{C})=\{0\}$ via the Branching Theorem applied to the inclusion $SO(m)\subset SO(m+1)$.)

10. Give examples of non-equivalent $SO(2)$-equivariant full harmonic maps $f, f' : S^3 \to S^n$ with $e(f) = e(f')$. (Hint: The Hopf map $f : S^3 \to S^2$ is T^2-equivariant with respect to the standard maximal torus $T^2 \subset SO(4)$.)

11. Check the validity of (4.7) by elementary computation using the explicit formula in IV.3.30 for the standard minimal immersion $f_{\lambda_2} : S^m_{\kappa(2)} \to S^{n(2)}$. (Hint: Using the notation of IV.3.30, $v \in K(f_{\lambda_2})$ if and only if

$$\check{v} = \sum_{k=1}^{m+1} a_k \phi_k + 2 \sum_{i<j} b_{ij} \phi_{ij} \ , \quad \sum_{k=1}^{m+1} a_k = 0 \ ,$$

where the scalars $a_k, b_{ij} \in C^\infty(S^m)$ have the form

$$a_r = \sum_{k=1}^{m+1} a_k^r \phi_k + 2 \sum_{i<j} b_{ij}^r \phi_{ij} \ , \quad \sum_{k=1}^{m+1} a_k^r = 0 \ , \quad r = 1, \ldots, m+1 \ ,$$

and

$$b_{pq} = \sum_{k=1}^{m+1} a_k^{pq} \phi_k + 2 \sum_{i<j} b_{ij}^{pq} \phi_{ij} \ , \quad \sum_{k=1}^{m+1} a_k^{pq} = 0 \ , \quad 1 \le p < q \le m+1 \ ,$$

with $a_k^r, b_{ij}^r, a_k^{pq}, b_{ij}^{pq} \in \mathbf{R}$. Now, $\dim K(f_{\lambda_2})$ is determined by taking into account the orthogonality relation $(\bar{f}_{\lambda_2}, \check{v}) = 0$ and computing the maximal number of linearly independent coefficients.)

CHAPTER VI

Applications in Theoretical Physics

The differential calculus for twisted tensor bundles developed in Section 1, Chapter II, can also be used to give *global formulation* for various physical theories such as σ-models and general relativity. Among these a peculiarly important part is played by the theory of Yang-Mills fields. In the following two sections we discuss the beginning of this topic by developing the elements of the theory in a rather similar way as was done for the energy and volume functionals. In Section 1, we derive the first and second variation formulas for the Yang-Mills functional acting on the (affine) space of covariant differentiations of a fixed vector bundle. These formulas are then exploited in Section 2 to prove Simons' instability theorem for Yang-Mills fields over S^n , $n \geq 5$. Though we make use of some results of the earlier chapters, the presentation follows closely the work of Bourguignon and Lawson (1981)† (see also Bourguignon, Lawson and Simons (1979)). We assume that the reader has a prior knowledge of the theory of connections on principal fibre bundles, covariant differentiations on (associated) vector bundles, etc.

Note that harmonic maps can also serve as models of physical theories but, as a detailed account on this would far exceed the scope of this book, for the interested reader we only refer to the paper of Misner (1978) and (without completeness) the works of D'Adda, Lüschner and Di Vecchia (1978), Din and Zakrzewski (1980a-b) and (1982), Eichenherr and Forger (1979), Golo and Peremolov (1978), Pohlmeyer (1976), Zakharov and Mikhailov (1978) and Zakrzewski (1982).

† References to theoretical physics are contained in the Specific Bibliography for Chapter VI.

1. FIRST AND SECOND VARIATION FORMULAS FOR THE YANG-MILLS FUNCTIONAL

Let M be an m-dimensional compact oriented Riemannian manifold with metric tensor $g=(,)\in C^{\infty}(S^2(T^*(M)))$ and G a compact Lie group with Lie algebra G. We also fix a principal fibre bundle $\pi:P\to M$ over M with structure group G. An *inner automorphism* of the principal fibre bundle $\pi:P\to M$ is a diffeomorphism $\pi:P\to P$ which commutes with the (right) action of G on P (i.e., for $a\in G$ and $u\in P$, we have $\tau(u\cdot a)=\tau(u)\cdot a$) and projects down (via π) to the identity of M. The group of all inner automorphisms $\mathbf{G}_P$ is said to be the *gauge group* of $\pi:P\to M$. Denoting by G_P the fibre bundle (in fact, the bundle of groups) over M associated with $\pi:P\to M$ by the adjoint representation Ad of G on G (i.e. the total space of G_P is the twisted product $P\times_{\mathrm{Ad}}G$; cf. Kobayashi and Nomizu (1963) pages 54-55), the gauge group $\mathbf{G}_P$ can be identified with the group $C^{\infty}(G_P)$ of all sections of G_P. (Indeed, given an inner automorphism $\tau\in\mathbf{G}$, as G acts freely on P, for $u\in P$, $\pi(u)=x$, there is a unique $b\in G$ with $\tau(u)=u\cdot b$. Now, at $x\in M$, the section of G_P associated to τ takes the value $\{(u\cdot a,a^{-1}ba)\,|\,a\in G\}\in(G_P)_x$.)
Let G_P denote the vector bundle (in fact, the bundle of Lie algebras) over M associated with $\pi:P\to M$ by the adjoint representation ad of G on G. The total space of G_P is then precisely the twisted product $P\times_{\mathrm{ad}}G$ and the identification $\mathbf{G}_P=C^{\infty}(G_P)$ allows us to define the *infinitesimal gauge group* as the Lie algebra of all sections of G_P. The exponential map $\exp:G\to G$ is easily seen to give rise to maps

$$\exp_P:G_P\to G_P \tag{1.1}$$

and

$$\exp_P:C^{\infty}(G_P)\to C^{\infty}(G_P)=\mathbf{G}_P\,. \tag{1.2}$$

For our purposes, however, it will be more convenient to work on vector bundles rather than on principal fibre bundles. Therefore, we fix a monomorphism $\rho:G\to \mathrm{O}(n)$ and form the vector bundle ξ over M associated with $\pi:P\to M$ by ρ. Then, ξ inherits a Euclidean structure (,) from that of $\mathbf{R}^n$. Let $O(\xi)$ and $so(\xi)$ denote the fibre bundles over M such that, for $x\in M$,

$$O(\xi)_x = \{\text{group of orthogonal transformations of } \xi_x\}$$

and

$$so(\xi)_x = \{\text{Lie algebra of skew-symmetric linear endomorphisms of } \xi_x\},$$

respectively. The monomorphism $\rho:G\to O(n)$ induces a bundle map

$$G_P \to O(\xi)$$

which embeds each fibre $(G_P)_x$, $x\in M$, into $O(\xi)_x$. (In fact, given $\tau_x=\{(u\cdot a,a^{-1}ba)\,|\,a\in G\}\in(G_P)_x$, $u\in P$, $\pi(u)=x$, and $b\in G$, the element u defines a linear isomorphism $u:\mathbf{R}^n\to\xi_x$; cf. Kobayashi and Nomizu (1963) page 55. Associate then to τ_x the orthogonal transformation $u\circ\rho(b)\cdot u^{-1}\in O(\xi)_x$.) The image of G_P under this embedding is denoted by $G_\xi\subset O(\xi)$. The gauge group $\mathbf{G}_P=C^\infty(G_P)$ can then be identified with the group $\mathbf{G}_\xi$ of all sections of G_ξ.
Similarly, on the Lie algebra level, G_P embeds onto a sub-bundle G_ξ of $so(\xi)$ and the infinitesimal gauge group $C^\infty(G_P)$ gets identified with the Lie algebra $C^\infty(G_\xi)$.

We shall be primarily interested in the space C_P of connections on the principal fibre bundle $\pi:P\to M$. Each connection $\Gamma\in C_P$ is well-known to induce an *orthogonal* covariant differentiation $\nabla\in\mathrm{Diff}_1(\xi,\mathrm{Hom}(T(M),\xi))$ on the associated vector bundle ξ (cf. Kobayashi and Nomizu (1963) pages 113-117). The totality of covariant differentiations ∇ on ξ obtained from the elements of C_P is denoted by C_ξ.

Remark. The physical motivation for studying C_P can be given as follows. We regard the base manifold M as the possible locations of a particle which carries an internal structure, its set of states, represented at a location $x\in M$ by the fibre

$G_x=\pi^{-1}(x)\subset P$ over x. Moving along a curve $c:[0,1]\to M$, the particle is assumed to carry its internal states. The internal state u at $c(0)$, gives rise to a lift $\tilde{c}:[0,1]\to P$ of c (i.e. $\pi\circ\tilde{c}=c$ and $\tilde{c}(0)=u$) which represents the change of its states. The correspondence $c\to\tilde{c}$ can then be thought of as the horizontal lift with respect to a connection $\Gamma\in C_P$. Physically, Γ is given by the presence of an external field.

By the results of I.4, any two covariant differentiations $\nabla,\nabla'\in C_\xi$ differ by a zero order operator $A(=\nabla'-\nabla)\in\mathcal{D}^1(\mathrm{Hom}(\xi,\xi))$. For $X\in V(M)$ and $v,v'\in C^\infty(\xi)$, orthogonality of ∇ and ∇' (cf. I.(4.13)) imply

$$X(v,v') = (\nabla_X v,v') + (v,\nabla_X v')$$

and

$$X(v,v') = (\nabla'_X v,v') + (v,\nabla'_X v') .$$

Substracting these two equations we obtain that $A(X)=\nabla'_X-\nabla_X$ is a skew-symmetric linear endomorphism on each of the fibres of ξ, i.e., actually, $A\in\mathcal{D}^1(so(\xi))$. Furthermore, using the fact that ∇ and ∇' are induced by connections Γ and Γ' in C_P, respectively, A is easily seen to belong to $\mathcal{D}^1(G_\xi)\subset$ $\subset\mathcal{D}^1(so(\xi))$. It follows that C_ξ is an affine space with $\mathcal{D}^1(G_\xi)$ as the group of translations.
In a similar vein, for each $\nabla\in C_\xi$, the corresponding curvature tensor $R\in\mathcal{D}^2(so(\xi))$ (I.4.18) of ξ is actually, a 2-form on M with values in G_ξ.
As noted above, each element $\tau\in\mathbf{G}_\xi$ can be considered as an orthogonal transformation on each of the fibres of ξ. Setting

$$\nabla^\tau = \tau\circ\nabla\circ\tau^{-1} , \quad \nabla\in C_\xi , \tag{1.3}$$

it follows easily that $\nabla^\tau\in C_\xi$ and hence the gauge group $\mathbf{G}_\xi(=\mathbf{G}_P)$ acts on the space of covariant differentiations C_ξ. Furthermore, for fixed $\nabla\in C_\xi$ with curvature tensor $R\in\mathcal{D}^2(G_\xi)$, letting R^τ, $\tau\in\mathbf{G}_\xi$, denote the curvature tensor of ∇^τ, we have

$$R^\tau = \tau\circ R\circ\tau^{-1} . \tag{1.4}$$

The Euclidean structure on ξ induces a fibre metric (,) on $so(\xi)$ given by

$$(\sigma,\sigma') = \text{trace}\ {}^{t}\sigma'\circ\sigma\ , \qquad \sigma,\sigma'\in C^{\infty}(so(\xi))\ , \tag{1.5}$$

where the trace is taken with respect to the fibre metric on ξ. By restriction, this fibre metric defines a Euclidean structure (,) on $G_{\xi}\subset so(\xi)$.

Each covariant differentiation $\nabla\in C_{\xi}$ induces canonically a covariant differentiation ∇ on $so(\xi)\subset \text{Hom}(\xi,\xi)\,(\tilde{=}\xi^{*}\otimes\xi)$ which preserves the subbundle $G_{\xi}\subset so(\xi)$ (i.e., for $X\in V(M)$, we have $\nabla_{X}(C^{\infty}(G_{\xi}))\subset C^{\infty}(G_{\xi})$). Orthogonality of $\nabla\in C_{\xi}$ implies the orthogonality of the induced covariant differentiation ∇^{G} on G_{ξ}. Thus, for each $\nabla\in C_{\xi}$, the bundle of Lie algebras G_{ξ} is Riemannian-connected: in particular, the differential operators d,∂ and Δ are defined acting on forms of M with values in G_{ξ} (cf. II.1). Note also that the curvature tensor R^{G} of G_{ξ} with respect to the induced covariant differentiation ∇^{G} on G_{ξ} is given as

$$\begin{aligned} R^{G}(X,Y)\sigma &= R(X,Y)\circ\sigma - \sigma\circ R(X,Y) = \\ &= [R(X,Y),\sigma], \quad X,Y\in V(M), \quad \sigma\in C^{\infty}(G_{\xi})\ , \end{aligned} \tag{1.6}$$

where, at $x\in M$, $[R(X_{x},Y_{x}),\sigma_{x}]\in(G_{\xi})_{x}\subset so(\xi)_{x}$ is the Lie bracket of the skew-symmetric linear endomorphisms $R(X_{x},Y_{x}),\sigma_{x}\in(G_{\xi})_{x}$.

Given a covariant differentiation $\nabla\in C_{\xi}$, by (1.2), each section $\sigma\in C^{\infty}(G_{\xi})\,(=C^{\infty}(G_{P}))$ defines a curve $t\to\tau_{t}=\exp_{P}(t\cdot\sigma)$, $t\in\mathbf{R}$, in $G_{\xi}\,(=G_{P})$. Then, by (1.3), for $X\in V(M)$, we have

$$\begin{aligned} \frac{\partial}{\partial t}\{\nabla_{X}^{\tau_{t}}\}_{t=0} &= \frac{\partial}{\partial}\{\tau_{t}\circ\nabla_{X}\circ\tau_{t}^{-1}\}_{t=0} = \\ &= \frac{\partial}{\partial t}\{\exp_{P}(t\cdot\sigma)\circ\nabla_{X}\circ\exp_{P}(-t\cdot\sigma)\}_{t=0} = \\ &= \sigma\circ\nabla_{X}-\nabla_{X}\circ\sigma = \nabla_{X}\sigma = (d\sigma)(X)\ , \end{aligned}$$

where d is the exterior differentiation of the Riemannian-connected vector bundle $\Lambda(T^{*}(M))\otimes G_{\xi}$ with respect to $\nabla\in C_{\xi}$.

It follows that the tangent space to the orbit $\mathbf{G}_\xi(\nabla) \subset C_\xi$ of the gauge group $\mathbf{G}_\xi$ at ∇ is nothing but $d(C^\infty(G_\xi)) \subset \mathcal{D}^1(G_\xi)$. By the definition of the adjoint ∂ of d, a transversal subspace to $d(C^\infty(G_\xi))$ is given by $\ker \partial \subset \mathcal{D}^1(G_\xi)$ which is said to be the *space of infinitesimal deformations* of ∇. Given a covariant differentiation $\nabla \in C_\xi$, recall that the corresponding curvature tensor R of ξ is a 2-form on M with values in G_ξ. We define the *Yang Mills action* on $\nabla \in C_\xi$ as the integral (with respect to the volume form $\mathrm{vol}(M,g)$ on M) of the scalar $\frac{1}{2}\|R\|^2$, i.e., we set

$$YM(\nabla) = \frac{1}{2}\int_M \|R\|^2 \,\mathrm{vol}(M,g) \ . \tag{1.7}$$

Thus, (1.7) defines a functional $YM: C_\xi \to \mathbf{R}$ (called the *Yang-Mills functional*) on the space of covariant differentiations C_ξ. As in the cases of energy and volume functionals treated in Chapters III-IV, we are interested in the critical points of YM which are then called *Yang-Mills covariant differentiations*. The corresponding curvatures are said to be *Yang-Mills fields*. Note that, by (1.3), the Yang-Mills functional is invariant under the action of $\mathbf{G}_\xi$ on C_ξ.
The first variation formula which follows gives a characterization of Yang-Mills fields in terms of the Euler-Lagrange equation associated to the Yang-Mills functional YM.

Theorem 1.8. *Given a (smooth) variation*

$$t \to \nabla^t \ , \qquad t \in \mathbf{R} \ , \tag{1.9}$$

of a covariant differentiation ∇ *in* C_ξ *(i.e.* $\nabla^0 = \nabla$*), we have*

$$\left.\frac{\partial YM(\nabla^t)}{\partial t}\right|_{t=0} = \int_M (\partial R, B)\,\mathrm{vol}(M,g) \ , \tag{1.10}$$

where $R \in \mathcal{D}^2(G_\xi)$ *is the curvature tensor of* ξ *with respect to* ∇ *and* $B = \left.\frac{\partial \nabla^t}{\partial t}\right|_{t=0} \in \mathcal{D}^1(G_\xi)$. *In particular,* ∇ *is a Yang-Mills covariant differentiation if and only if* $\partial R = 0$.

Proof. The space of covariant differentiations C_ξ being affine, the variation (1.9) gives rise to a 1-parameter family

$$t \to A^t \in \mathcal{D}^1(G_\xi) \ , \quad t \in \mathbf{R} \ ,$$

such that

$$\nabla^t = \nabla + A^t \ , \quad t \in \mathbf{R} \ , \tag{1.11}$$

with $A^0 = 0$. Denoting by $R^t \in \mathcal{D}^2(G_\xi)$ the curvature tensor of ∇^t , by II.(1.1), for $X, Y \in V(M)$, we have

$$\begin{aligned} R^t(X,Y) &= \nabla^t_X \circ \nabla^t_Y - \nabla^t_Y \circ \nabla^t_X - \nabla^t_{[X,Y]} = R(X,Y) + \\ &+ \nabla_X \circ A^t(Y) - A^t(Y) \circ \nabla_X + A^t(X) \circ \nabla_Y - \nabla_Y \circ A^t(X) - \\ &- A^t([X,Y]) + A^t(X) \circ A^t(Y) - A^t(Y) \circ A^t(X) = \\ &= R(X,Y) + \nabla_X A^t(Y) - \nabla_Y A^t(X) - A^t([X,Y]) + \\ &+ [A^t(X), A^t(Y)] = R(X,Y) + (dA^t)(X,Y) + \\ &+ [A^t(X), A^t(Y)] \ , \end{aligned} \tag{1.12}$$

where $[\ ,\ ]$ stands for the Lie bracket on $C^\infty(G_\xi)$. Differentiating (1.12) at $t=0$, by (1.11) and $A^0=0$, we obtain

$$\frac{\partial}{\partial t}\{R^t\}_{t=0} = d\{\frac{\partial A^t}{\partial t}\Big|_{t=0}\} = d\{\frac{\partial \nabla^t}{\partial t}\Big|_{t=0}\} = dB \ .$$

Thus, we have

$$\begin{aligned} \frac{\partial YM(\nabla^t)}{\partial t}\Big|_{t=0} &= \frac{1}{2}\int_M \frac{\partial}{\partial t}\{\| R^t \|^2\}_{t=0} \operatorname{vol}(M,g) = \\ &= \int_M (\frac{\partial}{\partial t}\{R^t\}_{t=0}, R) \operatorname{vol}(M,g) = \\ &= \int_M (dB, R) \operatorname{vol}(M,g) = \int_M (\partial R, B) \operatorname{vol}(M,g) \ . \ \checkmark \end{aligned}$$

Given a covariant differentiation $\nabla \in C_\xi$, by I.(4.9) and II.(1.1), the corresponding curvature tensor $R \in \mathcal{D}^2(G_\xi)$ is easily seen to satisfy the identity

$$dR = 0 \ . \tag{1.13}$$

In particular, by 1.7, $\nabla \in C_\xi$ is a Yang-Mills covariant differentiation if and only if its curvature tensor R is a harmonic 2-form on M with values in G_ξ (i.e., $\Delta R=0$).

Remark. Of great physical interest is the particular case when the base manifold M is 4-dimensional. A curvature tensor $R \in \mathcal{D}^2(G_\xi)$ corresponding to a covariant differentiation $\nabla \in C_\xi$ is then said to be *self-dual* (resp., *anti-self-dual*) if $\star R=R$ (=resp., $\star R=-R$), where $\star : \Lambda^r(T^*(M)) \otimes G_\xi \to \Lambda^{4-r}(T^*(M)) \otimes G_\xi$, $r=1,2,3,4$, is the canonical extension of the Hodge star operator (cf. Problem 5 for Chapter I). As the formula

$$\partial = d^* = \star^{-1} \circ d \circ \star$$

holds on the twisted bundle $\Lambda^2(T^*(M)) \otimes G_\xi$ as well (cf. Problem 5(iii) for Chapter I and II.1), by making use of (1.13), we find that self-dual (or anti-self-dual) curvature tensors are automatically Yang-Mills fields. Note that, on $S^4=M$, self-dual and anti-self-dual Yang-Mills fields are known to exist and, for any base M, the corresponding Yang-Mills covariant differentiations are absolute minima of the Yang-Mills functional YM. (For further details, see Bourguignon and Lawson (1981).)

In physical applications it is important to find all Yang-Mills fields as they are thought to be relevant in understanding the structure of the quantum Yang-Mills field theory. (This was initiated by Belavin, Polyakov, Schwartz and Tyupkin (1975); see also Atiyah, Drinfeld, Hitchin and Manin (1978), Forgács, Horváth and Palla (1981a-f) and (1982a-b) and Stora (1980). For detailed accounts, see Atiyah (1979), Coleman (1979) and Rajaraman (1982).) For certain base manifolds, the solutions of the self-dual (or anti-self-dual) equations describe magnetic monopoles (or anti-monopoles). (For details, see Bogomolny (1976), Corrigan and Goddard (1981), Forgács, Horváth and Palla (1981a-f) and (1982a), Hitchin (1982), Prasad and Rossi (1981), Prasad and Sommerfield (1975), Ward (1981a-c) and (1982).)

In what follows, analogous to the energy and volume functionals, we study the qualitative behaviour of the Yang-Mills functional near a critical point $\nabla \in C_\xi$ by means of the following second variation formula.

Theorem 1.14. *Let $\nabla \in C_\xi$ be a Yang-Mills covariant differentiation. Given a 2-parameter variation*

$$(s,t) \to \nabla^{(s,t)}, \quad (s,t) \in \mathbf{R}^2, \tag{1.15}$$

of ∇ in C_ξ (i.e. $\nabla^{(0,0)} = \nabla$) such that

$$B = \left.\frac{\partial \nabla^{(s,0)}}{\partial s}\right|_{s=0} \in \ker \partial \quad \textit{and}$$

$$C = \left.\frac{\partial \nabla^{(0,t)}}{\partial t}\right|_{t=0} \in \ker \partial , \tag{1.16}$$

we have

$$\left.\frac{\partial^2 YM(\nabla^{(s,t)})}{\partial s \partial t}\right|_{s,t=0} = \int_M (-\mathrm{trace}(\nabla^G)^2 B + B \circ \mathrm{Ric}^M + 2S(B), C)\, \mathrm{vol}(M,g), \tag{1.17}$$

where, with respect to an orthonormal base $\{e^i\}_{i=1}^m \subset T_x(M)$, $x \in M$,

$$S(B)(X_x) = \sum_{i=1}^m [R(e^i, X_x), B(e^i)] = \sum_{i=1}^m R^G(e^i, X_x)(B(e^i)), \quad X_x \in T_x(M) \tag{1.18}$$

(cf. (1.6)).

Proof. The variation (1.15) gives rise to a 2-parameter family

$$(s,t) \to A^{(s,t)} \in \mathcal{D}^1(G_\xi) \ , \quad (s,t) \in \mathrm{R}^2 \ ;$$

such that

$$\nabla^{(s,t)} = \nabla + A^{(s,t)} \ , \quad (s,t) \in \mathrm{R}^2 \ , \tag{1.19}$$

with $A^{(0,0)}=0$. Let $R^{(s,t)} \in \mathcal{D}^2(G_\xi)$, $(s,t) \in \mathrm{R}^2$, denote the curvature tensor of $\nabla^{(s,t)}$ and, for $t \in \mathrm{R}$, set $\nabla^t = \nabla^{(0,t)}$, $A^t = A^{(0,t)}$, $R^t = R^{(0,t)}$ and $B^t = \left.\frac{\partial A^{(s,t)}}{\partial s}\right|_{s=0} \in \mathcal{D}^1(G_\xi)$. Keeping $t \in \mathrm{R}$ fixed, by the first variation formula (1.10), we obtain

$$\left.\frac{\partial YM(\nabla^{(s,t)})}{\partial s}\right|_{t=0} = \int_M (\partial^t R^t, B^t) \operatorname{vol}(M,g) \ , \tag{1.20}$$

where ∂^t stands for the exterior codifferentiation induced by $\nabla^t \in C_\xi$. Differentiating (1.20) at $t=0$ and using $\partial R=0$, with obvious notations, we have

$$\left.\frac{\partial^2 YM(\nabla^{(s,t)})}{\partial s\, \partial t}\right|_{s=t=0} = \int_M \left.\frac{\partial}{\partial t}(\partial^t R^t, B^t)\right|_{t=0} \operatorname{vol}(M,g) =$$

$$= \int_M \left(\frac{\partial}{\partial t}\{\partial^t R^t\}_{t=0}, B\right) \operatorname{vol}(M,g) =$$

$$= \int_M \left.\frac{\partial}{\partial t}(\partial^t R^t, B)\right|_{t=0} \operatorname{vol}(M,g) =$$

$$= \int_M \left.\frac{\partial}{\partial t}(R^t, d^t B)\right|_{t=0} \operatorname{vol}(M,g) =$$

$$= \int_M \left(\left.\frac{\partial R^t}{\partial t}\right|_{t=0}, dB\right) \operatorname{vol}(M,g) \ +$$

$$+ \int_M \left(R, \frac{\partial}{\partial t}\{d^t B\}_{t=0}\right) \operatorname{vol}(M,g) =$$

$$= \int_M (\partial dB, C) \operatorname{vol}(M,g) + \int_M \left(R, \frac{\partial}{\partial t}\{d^t B\}_{t=0}\right) \operatorname{vol}(M,g) \ ,$$

where, in the last equality, we used the fact that

$$\frac{\partial R^t}{\partial t}\bigg|_{t=0} = d\{\frac{\partial A^t}{\partial t}\bigg|_{t=0}\} = dC$$

(cf. the proof of 1.8). Furthermore, for $X,Y\in V(M)$, we compute

$$\frac{\partial}{\partial t}\{d^t B\}_{t=0}(X,Y) = \frac{\partial}{\partial t}\{(d^t B)(X,Y)\}_{t=0} =$$

$$= \frac{\partial}{\partial t}\{\nabla_X^t(B(Y)) - \nabla_Y^t(B(X)) - B([X,Y])\}_{t=0} =$$

$$= \frac{\partial}{\partial t}\{A^t(X)\circ B(Y) - B(Y)\circ A^t(X) - A^t(Y)\circ B(X) + B(X)\circ A^t(Y)\}_{t=0} =$$

$$= C(X)\circ B(Y) - B(Y)\circ C(X) - C(Y)\circ B(X) + B(X)\circ C(Y) =$$

$$= [C(X),B(Y)] - [C(Y),B(X)] .$$

Hence, with respect to an orthonormal base $\{e^i\}_{i=1}^m \subset T_x(M)$, $x\in M$, we have

$$(R,\frac{\partial}{\partial t}\{d^t B\}_{t=0})_x = \sum_{i<j} (R(e^i,e^j),[C(e^i),B(e^j)] -$$

$$- [C(e^j),B(e^i)]) = 2\sum_{i<j} (R(e^i,e^j),[C(e^i),B(e^j)]) =$$

$$= \sum_{i,j=1}^m (R(e^i,e^j),[C(e^i),B(e^j)]) =$$

$$= \sum_{i,j=1}^m ([B(e^j),R(e^i,e^j)],C(e^i)) =$$

$$= \sum_{i,j=1}^m ([R(e^j,e^i),B(e^j)],C(e^i)) =$$

$$= \sum_{i=1}^m (S(B)(e^i),C(e^i)) = (S(B),C) .$$

Summarizing, we obtain

$$\left.\frac{\partial^2 YM(\nabla^{(s,t)})}{\partial s \partial t}\right|_{s=t=o} = \int_M (\partial dB+S(B),C)\,\mathrm{vol}(M,g). \tag{1.21}$$

Assume now that $B\in\mathcal{D}^1(G_\xi)$ is contained in the kernel of ∂. Then, by the Bochner-Weitzenböck formula II.(1.13) (applied to the vector bundle G_ξ), we have

$$\partial dB = \partial dB + d\partial B = \Delta B = -\mathrm{trace}(\nabla^G)^2 + B\circ\mathrm{Ric}^M + S(B)$$

and so (1.21) implies (1.17). ✓

Remark. The second variation formula (1.17) defines a symmetric bilinear pairing H_∇ on the space of infinitesimal deformations $\ker\partial \subset \mathcal{D}^1(G_\xi)$ of ∇ (called the *Hessian* of the Yang-Mills covariant differentiation $\nabla\in C_\xi$) by setting

$$H_\nabla(B,C) = \int_M (-\mathrm{trace}(\nabla^G)^2B+B\circ\mathrm{Ric}^M+S(B),C)\mathrm{vol}(M,g), B,C \in$$

$$\in \ker\partial .$$

Given a Yang-Mills covariant differentiation $\nabla\in C_\xi$, the differential operator $J_\nabla\in\mathrm{Diff}_2(T^*(M)\otimes G_\xi,\ T^*(M)\otimes G_\xi)$ defined by

$$J_\nabla(B) = -\mathrm{trace}(\nabla^G)^2B + B\circ\mathrm{Ric}^M + S(B),\quad B\in\mathcal{D}^1(G_\xi), \tag{1.22}$$

(occurring in (1.17)) is elliptic (cf. II.1.10) and self-adjoint. As in the case of energy and volume functionals, the *nullity* and the *Morse index* of $\nabla\in C_\xi$ given by

$$\dim\ker\{J_\nabla|\ker\partial\} \quad \text{and}$$

$$\mathrm{card}\{\text{negative eigenvalues of } J_\nabla|\ker\partial\}$$

are finite. Yang-Mills covariant differentiations $\nabla\in C_\xi$ with zero Morse index are said to be *weakly stable*. Note that weak stability of a Yang-Mills covariant differentiation $\nabla\in C_\xi$ is equivalent to saying that the corresponding Hessian H_∇ is positive semidefinite (on the space of infinitesimal deformations $\ker \partial \subset \mathcal{D}^1(G_\xi)$ of ∇).

2. INSTABILITY

The object of this section is to prove the following instability theorem of J. Simons.

Theorem 2.1. *For $m \geq 5$, any Yang-Mills covariant differentiation $\nabla \in C_\xi$ over S^m has strictly positive Morse index.*

For the proof of 2.1 we need some preliminary results. In what follows, we fix a Yang-Mills covariant differentiation $\nabla \in C_\xi$ over $M = S^m$, $m \in \mathbf{N}$, with harmonic curvature tensor $R \in \mathcal{D}^2(G_\xi)$. We shall use the notations introduced in III.2, namely, for a given vector $\check{Z} \in \mathbf{R}^{m+1}$, the tangential part of the uniform extension $Z \in \mathcal{V}(\mathbf{R}^{m+1})$ of $\check{Z}$ is denoted by $Z^\top \in \mathcal{V}(S^m)$.

Lemma 2.2. *Given a Yang-Mills field $R \in \mathcal{D}^2(G_\xi)$ over S^m, $m \in \mathbf{N}$, for each $\check{Z} \in \mathbf{R}^{m+1}$, the 1-form $B = \iota_{Z^\top} R \in \mathcal{D}^1(G_\xi)$ with values in G_ξ belongs to the space of infinitesimal deformations* $\ker \partial$, *i.e.*

$$\partial B = \partial(\iota_{Z^\top} R) = 0, \tag{2.3}$$

where $\iota_{Z^\top}$ stands for the canonical extension of the interior product with respect to $Z^\top \in \mathcal{V}(S^m)$ to $\mathcal{D}(G_\xi)$.

Proof. For fixed $x \in M$, let $\{E^i\}_{i=1}^m$ be a local orthonormal moving frame (adapted to an orthonormal base $\{e^i\}_{i=1}^m \subset T_x(S^m)$) over a normal coordinate neighbourhood centered at x. Then, by III.(2.6), we have

$$(\partial B)_x = -\sum_{i=1}^m (\nabla_{e^i} B)(e^i) = -\sum_{i=1}^m \nabla_{e^i}(B(E^i)) =$$

$$= -\sum_{i=1}^m \nabla_{e^i}(R(Z^\top, E^i)) = -\sum_{i=1}^m (\nabla_{e^i} R)(Z^\top_x, e^i) -$$

$$- \sum_{i=1}^{m} R(\nabla^{\top}_{e^i} Z, e^i) = -(\partial R)(Z^{\top}_x) + (\check{Z}, x) \cdot \sum_{i=1}^{m} R(e^i, e^i) = 0. \checkmark$$

Lemma 2.4. *For each* $\check{Z} \in \mathbf{R}^{m+1}$,

$$\text{trace } \nabla^2 Z^{\top} = -Z^{\top} . \tag{2.5}$$

Proof. Choosing $\{E^i\}_{i=1}^m$ as above and making use of III.(2.6), we compute

$$\text{trace}\{\nabla^2 Z^{\top}\}_x = \sum_{i=1}^{m} \nabla_{e^i}(\nabla_{E^i} Z^{\top}) = - \sum_{i=1}^{m} \nabla_{e^i}((\check{Z},.)E^i) =$$

$$= - \sum_{i=1}^{m} e^i(\check{Z},.)e^i = - \sum_{i=1}^{m} (\check{Z}, \check{e}^i)e^i = -Z^{\top}_x ,$$

where the last but one equality holds because

$$e^i(\check{Z},.) = \frac{\partial}{\partial t}(\check{Z}, \cos t \cdot x + \sin t \cdot \check{e}^i)\Big|_{t=0} = (\check{Z}, \check{e}^i) . \checkmark$$

Proof of Theorem 2.1. For $x \in S^m$, setting

$$q_x(\check{Z}) = (-\text{trace}(\nabla^G)^2 B + B \circ \text{Ric}^{S^m} + S(B), B)_x , \quad \check{Z} \in \mathbf{R}^{m+1} ,$$

with $B = \iota_{Z^{\top}} R \in \ker \partial$ (cf.(2.3)), we obtain a quadratic form q_x on $\mathbf{R}^{m+1}$ and, by the definition of the Hessian H_{∇} of $\nabla \in C_{\xi}$,

$$H_{\nabla}(B,B) = \int_{S^m} q(\check{Z}) \text{vol}(S^m, g) . \tag{2.6}$$

In what follows, we simplify the expression of $q_x(\check{Z})$. First, we have

$$\text{trace}(\nabla^G)^2 B = \iota_{Z^{\top}} \text{ trace}(\nabla^G)^2 R - B . \tag{2.7}$$

Indeed, for $x \in S^m$, let $\{E^i\}_{i=1}^m$ be a local orthonormal moving frame (adapted to an orthonormal base $\{e^i\}_{i=1}^m \subset T_x(S^m)$) over a normal coordinate neighbourhood centered at x . Then, by III. (2.6) and (2.5) above, for $i=1,\ldots,m$, we have

$$\text{trace}\{(\nabla^G)^2 B\}(e^i) = \sum_{j=1}^m \{\nabla^G_{e^j}(\nabla^G_{E^j}B)\}(e^i) =$$

$$= \sum_{j=1}^m \nabla^G_{e^j}\{(\nabla^G_{E^j}B)(E^i)\} = \sum_{j=1}^m \nabla^G_{e^j}\{\nabla^G_{E^j}(R(Z^\top,E^i)) -$$

$$- R(Z^\top,\nabla_{E^j}E^i)\} = \sum_{j=1}^m \nabla^G_{e^j}\{(\nabla^G_{E^j}R)(Z^\top,E^i)+R(\nabla_{E^j}Z^\top,E^i)\} =$$

$$= \text{trace}\{(\nabla^G)^2 R\}(Z^\top_x,e^i) + 2\sum_{j=1}^m (\nabla^G_{e^j}R)(\nabla_{e^j}Z^\top,e^i) +$$

$$+ R(\text{trace}\{\nabla^2 Z^\top\}_x,e^i) = \text{trace}\{(\nabla^G)^2 R\}(Z^\top_x,e^i) +$$

$$+ 2(\check{Z},x)(\partial R)(e^i) - R(Z^\top_x,e^i)$$

and (2.7) follows since $\partial R=0$.

Next, we apply the Bochner-Weitzenböck formula given in the Remark after the proof of II.1.12 to the *harmonic* 2-*form* R with values in G_ξ to get

$$0 = \Delta R = -\text{trace}\{(\nabla^G)^2 R\} + R(R) \quad . \tag{2.8}$$

Here, the second term on the right hand side is computed, making use of III.(2.1) and (1.6) and (1.18) above, as follows:

$$R(R)(Z^\top_x,e^i) = \sum_{j=1}^m R^G(e^j,Z^\top_x)(R(e^j,e^i)) -$$

$$- \sum_{j=1}^m R(R^{S^m}(e^j,Z^\top_x)e^j,e^i) - \sum_{j=1}^m R(e^j,R^{S^m}(e^j,Z^\top_x)e^i) -$$

$$- \sum_{j=1}^m R^G(e^j,e^i)(R(e^j,Z^\top_x)) + \sum_{j=1}^m R(R^{S^m}(e^j,e^i)e^j,Z^\top_x) +$$

$$+ \sum_{j=1}^m R(e^j,R^{S^m}(e^j,e^i)Z^\top_x) = \sum_{j=1}^m \{[R(e^j,Z^\top_x),R(e^j,e^i)] -$$

$$- [R(e^j,e^i),R(e^j,Z_x^\top)]\} + 2(m-2)R(Z_x^\top,e^i) =$$

$$= 2\sum_{j=1}^{m} [R(e^j,e^i),R(Z_x^\top,e^j)] + 2(m-2)R(Z_x^\top,e^i) =$$

$$= 2S(B)(e^i) + 2(m-2)B(e^i) .$$

Using this, by (2.7) and (2.8), we obtain

$$q_x(\check{Z}) = -2(m-2)\|B\|_x^2 + \|B\|_x^2 + (m-1)\|B\|_x^2 = \\ = (4-m)\|B\|_x^2 . \tag{2.9}$$

Now, let $\{\check{Z}^j\}_{j=0}^m \subset \mathbf{R}^{m+1}$ denote the orthonormal base with $\check{Z}^0=x$ and $\check{e}^j=\check{Z}^j$, $j=1,\ldots,m$. Then, by (2.9), we have

$$\text{trace } q_x = \sum_{j=0}^{m} q_x(\check{Z}^j) = (4-m)\sum_{j=0}^{m} \|\iota_{(Z^j)^\top}R\|_x^2 =$$

$$= (4-m)\sum_{i,j=1}^{m} \|R(e^j,e^i)\|^2 = 2(4-m)\sum_{i<j} \|R(e^j,e^i)\|^2 =$$

$$= 2(4-m)\|R\|_x^2 .$$

Assuming that $\nabla\in C_\xi$ is weakly stable, by (2.6),

$$\int_{S^m} \text{trace } q \text{ vol}(S^m,g) = 2(4-m)\int_{S^m} \|R\|^2 \text{vol}(S^m,g) \geq 0$$

and we obtain that $m\leq 4$. √

Remark. As noted in the previous section, for $m=4$, self-dual and anti-self-dual Yang-Mills fields are absolute minima for the Yang-Mills functional. In Bourguignon and Lawson (1981) it is also shown that, for $G=SU(2)$, $SU(3)$ or $U(2)$, a weakly stable Yang-Mills covariant differentiation $\nabla\in C_\xi$ over S^4 has necessarily self-dual or anti-self-dual curvature tensor

$R \in \mathcal{D}^2(G_\xi)$.

Returning again to physical applications, it would be of crucial importance to know whether other (non weakly stable) critical points exist. The analogous problem for a 2-dimensional physical model has been solved and it was shown that all solutions of the corresponding Euler-Lagrange equations necessarily satisfy the respective first order equations (cf. Jaffe and Taubes (1980)). In 3-dimensions, for the static monopole case, the existence of a critical point has been shown which does not satisfy the first order equations (cf. Taubes (1982a)). Therefore, it is a very intriguing open question to settle the problem for the 4-dimensional Yang-Mills case.

SPECIFIC BIBLIOGRAPHY FOR CHAPTER VI

Atiyah, M.F. (1979). *Geometry of Yang-Mills fields*, Academia Nazionale dei Lincei Scuola Normale Superiore, Lezioni Fermiane, Pisa.

Atiyah, M.F., Drinfeld, V.G., Hitchin, N.J. and Manin, Yu.J. (1978). Construction of instantons, *Phys. Lett.* 65A, 185-187.

Belavin, A.A., Polyakov, A.M., Schwartz, A.S. and Typkin, Yu.S. (1975). Pseudoparticle solutions of the Yang-Mills equations, *Phys. Lett.* 59B, 85-87.

Bogomolny, E.B. (1976). The stability of classical solutions, *Sov. J. Nucl. Phys.* 24, 449-454.

Bourguignon, J.-P. and Lawson, H.B. (1981). Stability and isolation phenomena for Yang-Mills fields, *Comm. Math. Phys.* 79, no.2, 189-230.

Bourguignon, J.-P., Lawson, H.B. and Simons, J. (1979). Stability and gap phenomena for Yang-Mills fields, *Proc. Natl. Acad. Sci. USA*, Vol.76, No.4, 1550-1553.

Coleman, S. (1979). Uses of instantons, in: "*The whys of subnuclear physics*", Ed. A. Zichichi, Plenum Press, New York.

Corrigan, E. and Goddard, P. (1981). An n monopole solution with $4n-1$ degrees of freedom, *Comm. Math. Phys.* 80, 575-587.

D'Adda, A., Lüschner, M. and Di Vecchia, P. (1978). A $1/n$ expandable series of non linear σ models with instantons,

Nucl. Phys. B146, 63-67.

Din, A.M. and Zakrzewski, W.J. (1980a). Stability properties of classical solutions to nonlinear σ models, *Nuclear Physics* B168, 173-180.

Din, A.M. and Zakrzewski, W.J. (1980b). Embeddings of classical solutions of O_{2n+1} nonlinear σ-models in CP^{n-1} models, *Lettere al Nuovo Cimento*, Vol.28, N.4, 121-127.

Din, A.M. and Zakrzewski, W.J. (1981). Classical solutions in grassmannian σ-models, *Letters in Math. Phys.* 5, 553-561.

Din, A.M. and Zakrzewski, W.J. (1982). Quantum fluctuations around a non-instanton solution, *Letters in Math. Phys.* 6, 373-378.

Eichenherr, H. and Forger, M. (1979). On the dual symmetry of the nonlinear sigma models, *Nucl. Phys.* B155, 381-393.

Forgács, P., Horváth, Z. and Palla, L. (1980). Generating the Bogomolny-Prasad-Sommerfield one monopole solution by a Bäcklund transformation, *Phys. Rev. Lett.* 45, 505-508.

Forgács, P., Horváth, Z. and Palla, L. (1981a). Exact multi-monopole solutions in the Bogomolny-Prasad-Sommerfield limit, *Phys. Lett.* 99B, 232-236.

Forgács, P., Horváth, Z. and Palla, L. (1981b). Generating monopoles of arbitrary charge by Bäcklund transformation, *Phys. Lett.* 102B, 131-135.

Forgács, P., Horváth, Z. and Palla, L. (1981c). Towards complete integrability of the self-duality equations, *Phys. Rev.* D23, 1876-1879.

Forgács, P., Horváth, Z. and Palla, L. (1981d). Nonlinear superposition of monopoles, *Nucl. Phys.* B192, 141-158.

Forgács, P., Horváth, Z. and Palla, L. (1981e). Soliton theoretic framework for generating multimonopoles, *Ann. Phys.* 136, 371-397.

Forgács, P., Horváth, Z. and Palla, L. (1981f). Exact, fractionally charged self-dual solution, *Phys. Rev. Lett.* 46, 392-394.

Forgács, P., Horváth, Z. and Palla, L. (1982a). Finitely separated monopoles generated as solitons, *Phys. Lett.* 109B, 200-204.

Forgács, P., Horváth, Z. and Palla, L. (1982b). Physicists'

techniques for multimonopole solutions, in: *Monopoles in quantum field theory*, ed. N. Craigie et al, World Scientific, Singapore, 21-57.

Forgács, P., Horváth, Z. and Palla, L. (1982c). One can have noninteger topological charge, *Zeitschrift für Phys.* C12, 359-360.

Golo, V.L. and Peremolov, A.M. (1978). Solution of the duality equations for the two dimensional *SU*(*N*) invariant chiral model, *Phys. Lett.* 79B, 112-113.

Hitchin, N.J. (1982). Monopoles and geodesics, *Comm. Math. Phys.* 83, 579-602.

Jaffe, A. and Taubes, C. (1980). Vortices and monopoles, *Progress in Physics* 2, Birkhäuser.

Misner, Ch.W. (1978). Harmonic maps as models for physical theories, *Preprint*, Maryland.

Pohlmeyer, K. (1976). Integrable Hamiltonian systems and interactions through quadratic constraints, *Comm. Math. Phys.* 46, 207-221.

Prasad, M.K. and Rossi, P. (1981). Construction of exact Yang-Mills-Higgs multimonopoles of arbitrary charge, *Phys. Rev. Lett.* 46, 806-809.

Prasad, M.K. and Sommerfield, C.M. (1975). Exact classical solution for the 't Hooft monopole and the Julia-Zee dyon, *Phys. Rev. Lett.* 35, 760-762.

Rajaraman, R. (1982). *Solitons and instantons*, North Holland.

Stora, R. (1980). Yang-Mills fields: Semi classical aspects, in: *Bifurcation phenomena in mathematical physics and related topics*, Eds.: C. Bardos, D. Bessis, D. Reidel Publ. Co., Doldrecht, Holland.

Taubes, C.H. (1982a). The existence of a non-minimal solution to the *SU*(2) Yang-Mills-Higgs equations on R^3, I. *Comm. Math. Phys.* 86, 257-298.

Taubes, C.H. (1982b). The existence of a non-minimal solution to the *SU*(2) Yang-Mills-Higgs equations on R^3, II, *Comm. Math. Phys.* 86, 299-320.

Ward, R.S. (1981a). A Yang-Mills-Higgs monopole of charge 2, *Comm. Math. Phys.* 79, 317-325.

Ward, R.S. (1982b). The Yang-Mills-Higgs monopoles close together,

Phys. Lett. 102B, 136-138.

Ward, R.S. (1981c). Magnetic monopoles with gauge group $SU(3)$ broken to $U(2)$, *Phys. Lett.* 107B, 281-284.

Ward, R.S. (1982). Deformation of the embedding of the $SU(2)$ monopole solution in $SU(3)$, *Comm. Math. Phys.* 86, 437-448.

Zakharov, V.E. and Mikhailov, A.V. (1978). Relativistically invariant two dimensional models of field theory which are integrable by means of the inverse scattering problem method, *Sov. Phys. JETP* 47, 1017-1027.

Zakrzewski, W.J. (1982). Classical solutions of CP^{n-1} models and their generalizations, *Czech. J. Phys.* B32, 589-596.

Summary of Basic Notations

1. $\mathbf{N}$: set of positive integers

   $\mathbf{Z}_+$: set of nonnegative integers

   $\mathbf{Z}$: ring of integers

   $\mathbf{Z}_p$: ring of integers modulo $p \in \mathbf{Z}$

   $\mathbf{Q}$: rational number field

   $\mathbf{R}$: real number field

   $\mathbf{R}_+$: set of nonnegative real numbers

   $\mathbf{C}$: complex number field

   $\mathbf{R}^m$: vector space of m-tuples of real numbers $x=(x^1,\ldots,x^m)$ with standard inner product $(x,x') = \sum_{i=1}^{m} x^i x'^i$, $x,x' \in \mathbf{R}^m$

   $\mathbf{C}^m$: vector space of m-tuples of complex numbers $z=(z^1,\ldots,z^m)$

   S^m: Euclidean m-sphere

   $\mathbf{R}P^m$: real projective m-space

   $\mathbf{T}^m$: m-dimensional torus

   $M(p,q)$: vector space of $(p\times q)$-matrices with real entries

   ${}^tA$: transpose of a matrix $A \in M(p,q)$

   I_p: identity matrix in $M(p,p)$

   V : real vector space of dimension p

   $T^s_r(V)$: tensor space of type (s,r) over V

   $\Lambda^r(V)$: r-th exterior power of V

   $\Lambda(V) = \bigoplus_{r=o}^{p} \Lambda^r(V)$: exterior algebra of V with exterior multiplication $\wedge$

   $S^r(V)$: r-th symmetric power of V

   $S(V) = \bigoplus_{r=1}^{\infty} S^r(V)$: symmetric algebra of V with multiplication $\cdot$

   $(\,,)$: scalar product on V

   V^*: dual of V

   $\gamma: V \to V^*$: canonical isomorphism for a vector space V with inner product $(\,,)$

2. *M denotes an m-dimensional manifold*

   (U,ϕ): local coordinate neighbourhood of M

   $\{{}^M\partial_i\}_{i=1}^m$: base fields associated to (U,ϕ)

   $C^\infty(M)$: algebra of scalars on M

ξ : real vector bundle over M with fibre ξ_x at $x \in M$

$C^\infty(\xi)$: $C^\infty(M)$-module of sections of ξ

ξ^*: dual vector bundle of ξ

ε_M^V: trivial vector bundle over M with fibre V

$T(M)$: tangent bundle of M with fibre $T_x(M)$ (tangent space) at $x \in M$

$c: I \to M$: curve defined on an open interval $I \subset \mathbf{R}$ with tangent vector $\dot{c}(t) \in T_{c(t)}(M)$ at $c(t)$, $t \in I$

$V(M) = C^\infty(T(M))$: Lie algebra of vector fields on M

$T^*(M)$: cotangent bundle of M with fibre $T_x^*(M)$ (cotangent space) at $x \in M$

$\Lambda^r(T^*(M))$: vector bundle of r-covectors on M

$\Lambda(T^*(M)) = \bigoplus_{r=0}^{m} \Lambda^r(T^*(M))$: vector bundle of covectors on M

$\mathcal{D}^r(M) = C^\infty(\Lambda^r(T^*(M)))$: $C^\infty(M)$-module of r-forms on M

$\mathcal{D}(M) = \bigoplus_{r=0}^{m} \mathcal{D}^r(M)$: exterior algebra of M with exterior multiplication $\wedge$

ι_X: interior product with respect to a vector field $X \in V(M)$

d : exterior differentiation

$T_r^s(M)$: tensor bundle of type (s,r) on M

$\mathcal{T}_r^s(M) = C^\infty(T_r^s(M))$: $C^\infty(M)$-module of tensor fields of type (s,r) on M

L_X: Lie differentiation with respect to $X \in V(M)$

3. ξ *denotes a vector bundle over* M †

$(\,,\,)_\xi$: fibre metric on ξ

∇^ξ: covariant differentiation on ξ

R^ξ: curvature of ξ with respect to a covariant differentiation ∇^ξ on ξ

4. M *denotes an* m*-dimensional Riemannian manifold with metric tensor* $g(\,,\,) \in C^\infty(S^2(T^*(M)))$

$\gamma: T(M) \to T^*(M)$: canonical isomorphism induced by the metric tensor g

$\mathrm{vol}(M,g)$: volume form on M

† In what follows, if there is no danger of confusion the indices M and ξ will be omitted.

∇^M: Levi-Civita covariant differentiation with respect to a vector field $X \in V(M)$

${}^M\Gamma^i_{jk}$: Christoffel symbols with respect to a local coordinate neighbourhood of M

$\{e^i\}_{i=1}^m$: orthonormal base of $T_x(M)$ at $x \in M$

$\{E^i\}_{i=1}^m$: local orthonormal moving frame ((adapted to the base $\{e^i\}_{i=1}^m$) on a normal coordinate neighbourhood U centered at $x \in M$)

R^M: Riemannian curvature tensor of M

Ric^M: Ricci tensor field on M

∂ : adjoint of d

Δ^M: Laplacian on M

Spec(M): spectrum of Δ acting on $C^\infty(M)$

V_λ: eigenspace of Δ corresponding to the eigenvalue $\lambda \in \mathrm{Spec}(M)$

$H^r(M)$: vector space of harmonic r-forms on M

$H(M) = \bigoplus_{r=0}^m H^r(M)$: vector space of harmonic forms on M

$P^r(M)$: vector space of parallel r-forms on M

$P(M) = \bigoplus_{r=0}^m P^r(M)$: vector space of parallel forms on M

$\tilde{H}^r_p$: vector space of spherical harmonics of order r on S^p

5. ξ *is a Riemannian-connected vector bundle over a Riemannian manifold* M *of dimension* m

$\Lambda^r(T^*(M)) \otimes \xi$: twisted bundle of r-covectors on M

$D^r(\xi) = C^\infty(\Lambda^r(T^*(M)) \otimes \xi)$: $C^\infty(M)$-module of r-forms on M with values in ξ

$D(\xi) = \bigoplus_{r=0}^m D^r(\xi)$: $C^\infty(M)$-module of forms on M with values in ξ

d, ∂, Δ: exterior differential, its adjoint and Laplacian acting on $D(\xi)$, resp.

$H^r(\xi)$: vector space of harmonic r-forms on M with values in ξ

$H(\xi) = \bigoplus_{r=0}^m H^r(\xi)$: vector space of harmonic forms on M with values in ξ

$P^r(\xi)$: vector space of parallel r-forms on M with values in ξ

$P(\xi) = \bigoplus_{r=0}^{m} P^r(\xi)$: vector space of parallel forms on M with values in ξ

6. *M and N denote manifolds of dimensions m and n, resp.*
   - $f: M \to N$: map
   - f_*: differential of f
   - $f^*\alpha$: pull-back of an r-form α on N via f
   - ξ^f: pull-back of a vector bundle ξ over N via f
   - $\tau^f = T(N)^f$: pull-back of the tangent bundle $T(N)$ via f

7. *M and N denote Riemannian manifolds with metric tensors g and h, resp.*
   - $e(f)$: energy density of f
   - $E(f)$: energy of f
   - $\tau(f)$: tension field of f
   - $\beta(f)$: second fundamental form of f
   - J_f: Jacobi operator associated with f
   - $J(f)$: vector space of Jacobi fields along f
   - $K(f)$: vector space of divergence-free Jacobi fields along f
   - H_f: Hessian of f

8. *$f: M \to N$ denotes an isometric immersion*
   - $\mathrm{Vol}(f)$: volume of f
   - $\perp, \top$: normal and tangential projections
   - ν^f: normal bundle of f
   - $\mu(f)$: mean curvature of f
   - $\nabla^\perp$: induced connection on ν^f
   - $A(f)$: shape operator
   - $J_f^\perp$: normal Jacobi operator associated with f
   - $J^\perp(f)$: vector space of normal Jacobi fields along f
   - $H_f^\perp$: normal Hessian of f

9. *G denotes a Lie group*
   - G^o: identity component of G
   - G : Lie algebra of G
   - L_a: left translation with $a \in G$
   - R_a: right translation with $a \in G$

$\mathrm{Ad}_G(a)$: inner automorphism by $a \in G$

$\mathrm{ad}_G(a)$: adjoint representation of G on G

$K \subset G$: subgroup with Lie subalgebra $K \subset G$

$\pi : G \to G/K$: canonical projection

$i(M)$: full isometry group of a Riemannian manifold M

10. *M denotes an m-dimensional manifold with base point o*

$\tilde{M}$: universal covering manifold of M with base point $\tilde{o}$

$\pi_M : \tilde{M} \to M$: universal covering projection

$H_r(M; A)$: r-th smooth singular homology group with coefficients in an Abelian group A

$H^r_{de\,R}(M; \mathbf{R})$: r-th de Rham cohomology group of M

$b_r(M)$: r-th Betti number of M

$\pi_r(M, o)$: r-th homotopy group of M (with base point o)

$\deg(f)$: degree of a map $f : M \to N$

$*$: join of maps

Bibliography

Agmon, S. (1965). *Lectures on elliptic boundary value problems*, Van Nostrand.

Baird, P. (1983). *Harmonic maps with symmetry, harmonic morphisms and deformations of metrics*, Research Notes in Math. Pitman Press.

Baird, P. and Eells, J. (1981). A conservation law for harmonic maps, in *Springer Notes* 894.

Berger, M., Gauduchon, P. and Mazet, E. (1971). *Le spectre d'une variété Riemannienne*, Springer Notes 194.

Boothby, W.M. (1975). *An introduction to differentiable manifolds and Riemannian geometry*, Academic Press, New York.

Boerner, H. (1963). *Representations of groups*, North Holland, Amsterdam.

Boutot, J.F. (1974). Théorème de Hodge, *Astérisque* 17, 3-27.

Bredon, G. (1972). *Introduction to compact transformation groups*, Academic Press, New York.

Calabi, E. (1967). Minimal immersions of surfaces in Euclidean spheres, *J. Diff. Geom.*, 1, 111-125.

Cheeger, J. and Ebin, D.G. (1975). *Comparison theorems in Riemannian geometry*, North Holland, Amsterdam.

Cheeger, J. and Gromoll, D. (1972). On the structure of complete manifolds of nonnegative curvature, *Ann. of Math.*, 96, 413-443.

Chevalley, C. (1946). *Theory of Lie groups*, Vol.I. Princeton Univ. Press, Princeton, New Jersey.

Delaunay, C. (1841). Sur la surface de révolution dont la courbure moyenne est constante, *J. Math. Pures et Appl.* Ser 1(6), 309-320.

Dieudonné, J. (1972). *Treatise on analysis*, Vol.II. Academic Press, New York.

Dieudonné, J. (1974). *Treatise on analysis*, Vol.IV. Academic Press, New York.

Do Carmo, M.P. and Wallach, N.R. (1969). Minimal immersions of spheres into spheres, *Proc. Nat. Acad. Sci. USA.* 63, 640-642.

Do Carmo, M.P. and Wallach, N.R. (1970). Representations of compact groups and minimal immersions into spheres, *J. Diff.*

Geom., 4, 91-104.

Do Carmo, M.P. and Wallach, N.R. (1971). Minimal immersions of spheres into spheres, *Ann. of Math.*, 93, 43-62.

Eells, J. (1978). On the surfaces of Delaunay and their Gauss maps, *Proc. IV. Int. Colloq. Diff. Geometry*, Santiago de Compestela, 97-116.

Eells, J. (1979). Minimal graphs, *Manuscripta Math.* 28, 101-108.

Eells, J. and Lemaire, L. (1978). A report on harmonic maps, *Bull. London Math. Soc.* 10, 1-68.

Eells, J. and Lemaire, L. (1980). On the construction of harmonic and holomorphic maps between surfaces, *Math. Ann.* 252, 27-52.

Eells, J. and Lemaire, L. (1983). *Selected topics in harmonic maps*, CBMS Reg. Conf. Series, No.50.

Eells, J. and Lemaire, L. (1984). Examples of harmonic maps from disks to hemispheres, *Math. Z.* (to appear).

Eells, J. and Sampson, J.H. (1964). Harmonic mappings of Riemannian manifolds, *Amer. J. Math.*, 86, 109-160.

Eells, J. and Wood, J.C. (1976). Restrictions on harmonic maps of surfaces, *Topology* 15, 263-266.

Eells, J. and Wood, J.C. (1981). Maps of minimum energy, *J. London Math. Soc.*, 23, 303-310.

Ehresmann, Ch. (1947). Sur les espaces fibrés différentiables, *C.R. Paris*, 1611.

Fischer, A.E. and Wolf, J.A. (1975). The structure of compact Ricci-flat Riemannian manifolds, *J. Diff. Geom.*, 10, 277-288.

Fuglede, B. (1978). Harmonic morphisms between Riemannian manifolds, *Ann. Inst. Fourier*, Grenoble 28, 2, 107-144.

Fuglede, B. (1979). Harnack sets and openness of harmonic morphisms, *Math. Ann.*, 241, 181-186.

Fuller, F.B. (1954). Harmonic mappings, *Proc. Nat. Acad. Sci.*, 40, 987-991.

Giaquinta, M. and Giusti, E. (1982). On the regularity of the minima of variational integrals, *Acta Math.* 148, 31-46.

Giaquinta, M. and Giusti, E. The singular set of the minima of certain quadratic functionals, *Analysis* (to appear).

Hartman, P. (1964). *Ordinary differential equations*, Wiley, New York.

Hartman, P. (1967). On homotopic harmonic maps, *Canad. J. Math.*, 19, 673-687.

Helgason, S. (1978). *Differential geometry, Lie groups, and symmetric spaces*, Acad. Press, New York.

Hermann, R. (1969). A sufficient condition that a map of Riemannian manifolds be a fibre bundle, *Proc. AMS*, 11, 236-242.

Hochschild, G. (1965). *The structure of Lie groups*, Holden Day, San Francisco.

Hopf, E. (1931). Über der Funktionalen, insbesondere den analytischen Character der Lösungen elliptischer Differentialgleichungen zweiter Ordnung, *Math. Z.*, 34, 194-233.

Hsiang, W.Y. (1966). On the compact, homogeneous minimal submanifolds, *Proc. Nat. Acad. Sci. USA*. 56, 5-6.

Hsiang, W.Y. (1967). Remarks on closed minimal submanifolds in the standard Riemannian m-sphere, *J. Diff. Geom.*, 1, 257-267.

Hsiang, W.Y. (1975). *Cohomology theory of topological transformation groups*, Springer, Berlin.

Hsiang, W.Y. and Lawson, H.B. (1971). Minimal submanifolds of low cohomogeneity, *J. Diff. Geom.*, 5, 1-38.

Hu, S.T. (1959). *Homotopy theory*, Academic Press, New York.

Husemoller, D. (1966). *Fibre bundles*, McGraw-Hill, New York.

Karcher, H. and Wood, J.C. (1983). Non existence results and growth properties for harmonic maps and forms, *Preprint* (Bonn).

Kobayashi, S. (1972). *Transformation groups in differential geometry*, Ergebnisse der Math. Band 70, Springer.

Kobayashi, S. and Nomizu, K. (1963). *Foundations of differential geometry*, Vol.I. Wiley (Interscience), New York.

Kobayashi, S. and Nomizu, K. (1969). *Foundations of differential geometry*, Vol.II. Wiley (Interscience), New York.

Lawson, H.B. (1970). *Lectures on minimal submanifolds*, IMPA.

Lemaire, L. (1975). Applications harmoniques de surfaces, *C.R. Paris* A 280, 897-899.

Lemaire, L. (1977). Applications harmoniques de variétés produits, *Comm. Math. Helv.*, 52, 11-24.

Lemaire, L. (1978). Applications harmoniques de surfaces riemanniennes, *J. Diff. Geom.*, 13, 51-78.

Leung, P.F. (1981). On the stability of harmonic maps, *Univ.*

Notre Dame Thesis.

Li, P. (1981). Minimal immersions of compact irreducible homogeneous Riemannian manifolds, *J. Diff. Geom.*, 16, 105-115.

Lichnerowicz, A. (1955). *Théorie globale des connexions et des groupes d'holonomie*, Cremonese, Roma.

Lichnerowicz, A. (1958). *Géométrie des groupes de transformations*, Dunod, Paris.

Lichnerowicz, A. (1971). Variétés Kähleriennes à première classe de Chern non negative et variétés Riemanniennes à courbure de Ricci généralisée non negative, *J. Diff. Geom.*, 6, 47-94.

Mazet, E. (1973). La formule de la variation seconde de l'énergie au voisinage d'une application harmonique, *J. Diff. Geom.*, 8, 279-296.

Milnor, J.W. (1963). *Morse theory*, Annals of Math. Studies, 51.

Milnor, J.W. and Stasheff, J. (1974). *Characteristic classes*, Princeton Univ. Press, Princeton, New Jersey.

Myers, S.B. and Steenrod, N. (1939). The group of isometries of a Riemannian manifold, *Ann. of Math.*, 40, 400-416.

Nagano, T. and Smyth, B. (1975). Minimal varieties and harmonic maps into tori, *Comm. Math. Helv.*, 50, 249-265.

Nagura, T. (1981). On the Jacobi differential operators associated to minimal isometric immersions of symmetric spaces into spheres, I, *Osaka J. Math.*, 18, 115-145.

Nagura, T. (1982). On the Jacobi differential operators associated to minimal isometric immersions of symmetric spaces into spheres, II, *Osaka J. Math.*, 79-124.

Nagura, T. On the Jacobi differential operators associated to minimalisometric immersions of symmetric spaces into spheres, III, *Osaka J. Math.*, (to appear).

Narasimhan, R. (1968). *Analysis on real and complex manifolds*, Masson & Cie, Paris, North-Holland, Amsterdam.

O'Neil, B. (1966). The fundamental equations of a submersion, *Michigan Math. J.*, 13, 459-469.

Palais, R.S. (1965). *Seminar on the Atiyah-Singer index theorem*, Princeton Univ. Press, Princeton, New Jersey.

Petrovski, J. (1939). On the analytic nature of solutions of systems of differential equations, *Mat. Sbornik*, 47, 3-70.

de Rham, G. (1955). *Variétés différentiables*, Herman & Cie, Paris

Sacks, J. and Uhlenbeck, K. (1977). The existence of minimal immersions of two-spheres, *Bull. Amer. Math. Soc.*, 83, 1033-1036.

Sampson, J.H. (1978). Some properties and applications of harmonic mappings, *Ann. Ec. Norm. Sup.* XI, 211-228.

Sampson, J.H. (1982). On harmonic mappings, *Symposia Mathematica*, Vol.XXVI, Bologna, 197-210.

Schoen, R. and Uhlenbeck, K. (1982). A regularity theory for harmonic maps, *J. Differential Geom.* 17, 307-335.

Schoen, R. and Yau, S.T. (1979). Compact group actions and the topology of manifolds with non-positive curvature, *Topology*, 18, 361-380.

Simons, J. (1968). Minimal varieties in Riemannian manifolds, *Ann. of Math.*, 88, 62-105.

Siu, Y.T. (1979). Some remarks on the complex analyticity of harmonic maps, *Southeast Asian Bull. Math.*, 3, 240-253.

Siu, Y.T. and Yau, S.T. (1979). Compact Kähler manifolds of positive bisectional curvature, *Preprint*, Stanford Univ.

Smith, R.T. (1972a). *Harmonic mappings of spheres*, Thesis, Warwick Univ.

Smith, R.T. (1972b). Harmonic mappings of spheres, *Bull. Amer. Math. Soc.*, 78, 593-596.

Smith, R.T. (1975a). Harmonic mappings of spheres, *Amer. J. Math.*, 97, 364-385.

Smith, R.T. (1975b). The second variation formula for harmonic mappings, *Proc. Amer. Math. Soc.*, 47, 229-236.

Spanier, E. (1966). *Algebraic topology*, McGraw Hill, New York.

Steenrod, N. (1951). *Topology of fibre bundles*, Princeton Univ. Press, Princeton, New Jersey.

Sunada, T. (1979). Rigidity of certain harmonic mappings, *Inventiones Math.*, 51, 297-307.

Takahashi, T. (1966). Minimal immersions of Riemannian manifolds, *J. Math. Soc. Japan*, 18, 380-385.

Tischler, D. (1970). On fibering certain foliated manifolds over S^1, *Topology*, 9, 153.

Toda, H. (1962). *Composition methods in homotopy groups of spheres*, Annals of Math. Studies, 49.

Tóth, G. (1981). On variations of harmonic maps into spaces of constant curvature, *Ann. Mat. Pura Appl.* (IV), Vol.CXXVIII, 389-399.

Tóth, G. (1982a). On harmonic maps into locally symmetric Riemannian manifolds, *Symposia Mathematica*, Vol.XXVI, Bologna 69-94.

Tóth, G. (1982b). On rigidity of harmonic mappings into spheres, *J. London Math. Soc.* (2), 26, 475-486.

Uhlenbeck, K. (1970). Harmonic maps: A direct method in the calculus of variations, *Bull. Amer. Math. Soc.*, 76, 1082-1087.

Vilms, J. (1970). Totally geodesic maps, *J. Diff. Geom.*, 4, 73-79.

Wallach, N.R. (1970). Extension of locally defined minimal immersions of spheres into spheres, *Arch. Math.*, 21, 210-213.

Wallach, N.R. (1972). Minimal immersions of symmetric spaces into spheres, in *Symmetric spaces*, Dekker, New York, 1-40.

Wallach, N.R. (1973). *Harmonic analysis on homogeneous spaces*, Dekker, New York.

Warner, F. (1970). *Foundations of differentiable manifolds and Lie groups*, Scott Foresman, Glenview, Illinois.

Watson, B. (1975). δ-commuting mappings and Betti numbers, *Tôhoku Math. J.*, 27, 135-152.

Wells, R.O. (1973). *Differential analysis on complex manifolds*, Prentice-Hall, Inc. Englewood Cliffs, N.J.

Weyl, H. (1946). *The classical groups*, Princeton Univ. Press, Princeton, New Jersey.

Wood, J.C. (1976). Harmonic maps and complex analysis, *Proc. Summer Course in Complex Analysis*, Trieste, Vol.III. 289-308.

Wood, R. (1968). Polynomial maps from spheres to spheres, *Inventiones Math.*, 5, 163-168.

Xin, Y.L. (1980). Some results on stable harmonic maps, *Duke Math. J.* 47, 609-613.

Index